The Philosophy of Mathematics Education

Paul Ernest

The Falmer Press
(A member of the Taylor & Francis Group)
London • New York • Philadelphia

UK The Falmer Press, Rankine Road, Basingstoke, Hampshire RG24 0PR

USA The Falmer Press, Taylor & Francis Inc., 1900 Frost Road, Suite 101, Bristol, PA 19007

First published 1991

British Library Cataloguing in Publication Data
Ernest, Paul
The philosophy of mathematics education.
1. Education. Curriculum subjects: Mathematics. Teaching.
I. Title
510.7

ISBN 1-85000-666-0
ISBN 1-85000-667-9 pbk

Library of Congress Cataloging-in-Publication Data is available on request

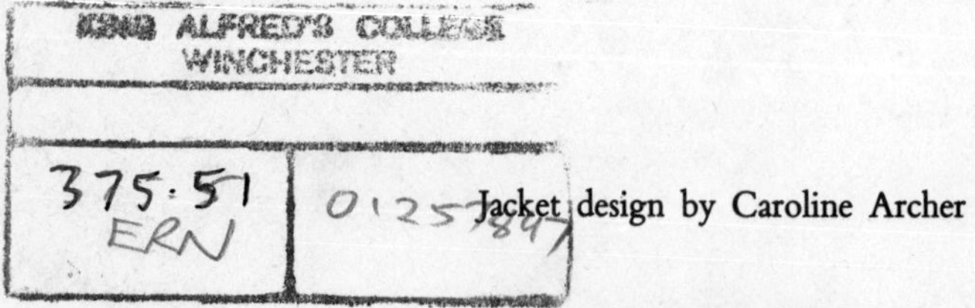

Jacket design by Caroline Archer

Typeset in 11/12 Bembo by
Chapterhouse, The Cloisters, Formby L37 3PX
Printed in Great Britain by
Burgess Science Press, Basingstoke

Contents

List of Tables and Figures

Acknowledgments

I wish to acknowledge some of the support, help and encouragement I have received in writing this book, from a large number of friends, colleagues and scholars. A number of colleagues at Exeter have been very helpful. Charles Desforges was instrumental in helping me to conceive of the book, and has made useful criticisms since. Bob Burn has been my fiercest and hence most useful critic. Neil Bibby, Mike Golby, Phil Hodkinson, Jack Priestley, Andy Sparkes and Rex Stoessiger have all made helpful suggestions and criticisms. Neil first suggested the term 'social constructivism' to me, which I have adopted.

Colleagues at the Research into Social Perspectives of Mathematics Education seminar, convened by Steve Lerman and Marilyn Nickson, have heard early versions of some of the theses of the book, and contributed, through their criticism. I have also learned a great deal from the papers given by others at this seminar, since 1986. Members of the group, especially Leone Burton, Steve Lerman and Stuart Plunkett, have made very useful criticisms of the text, helping me to reconceptualize the book and its purpose (but not always in the way suggested). I also owe the title of the book to Steve. Several others, such as Barry Cooper, deserve thanks for spending some time looking at draft chapters.

I am grateful to David Bloor, Reuben Hersh, Moshe Machover and Sal Restivo who have each seen portions of the book and responded positively. This was most encouraging, especially since I hold their work in very high regard. Sal Restivo is also encouraging me to expand the first part of the book into a freestanding work for his series Science, Technology and Society, with SUNY Press.

The tolerance and indulgence of Malcolm Clarkson and Christine Cox at the Falmer Press has been a great help, as I overshot self-imposed deadlines and word-counts.

It would not have been possible for me to write this book without the support and encouragement of my family. Indeed, my intellectual progress from the absolutist position I held twenty years ago is in no small part due to my incessant arguments and discussions with Jill about life, the universe and everything. In addition, Jill has read much of the book critically, and helped me to improve it. Jane and Nuala have also been encouraging and supportive, in their own inimitable ways.

Paul Ernest
University of Exeter,
School of Education
May 1990

Introduction

1. Rationale

The philosophy of mathematics is in the midst of a Kuhnian revolution. For over two thousand years, mathematics has been dominated by an absolutist paradigm, which views it as a body of infallible and objective truth, far removed from the affairs and values of humanity. Currently this is being challenged by a growing number of philosophers and mathematicians, including Lakatos (1976), Davis and Hersh (1980) and Tymoczko (1986). Instead, they are affirming that mathematics is fallible, changing, and like any other body of knowledge, the product of human inventiveness.

This philosophical shift has a significance that goes far beyond mathematics. For mathematics is understood to be the most certain part of human knowledge, its cornerstone. If its certainty is questioned, the outcome may be that human beings have no certain knowledge at all. This would leave the human race spinning on their planet, in an obscure corner of the universe, with nothing but a few local myths for consolation. This vision of human insignificance may be too much, or rather too little, for some to bear. Does the last bastion of certainty have to be relinquished? In the modern age uncertainty has been sweeping through the humanities, ethics, the empirical sciences: is it now to overwhelm all our knowledge?

However, in relinquishing the certainty of mathematics it may be that we are giving up the false security of the womb. It may be time to give up this protective myth. Perhaps human beings, like all creatures, are born into a world of wonders, an inexhaustible source of delight, which we will never fathom completely. These include the crystal worlds and rich and ornate webs which the human imagination weaves in mathematical thought. In these are infinite worlds beyond the infinite, and wondrous long and tight chains of reasoning. But it could be that such imaginings are part of what it means to be human, and not the certain truths we took them to be. Perhaps facing up to uncertainty is the next stage of maturity for the human race. Relinquishing myths of certainty may be the next act of decentration that human development requires.

Mathematics and Education

How mathematics is viewed is significant on many levels, but nowhere more so than in education and society. For if mathematics is a body of infallible, objective knowledge, then it can bear no social responsibility. Thus, the underparticipation of sectors of the population, such as women; the sense of cultural alienation from mathematics felt by many groups of students; the relationship of mathematics to human affairs such as the transmission of social and political values; its role in the distribution of wealth and power; none of these issues are relevant to mathematics.

On the other hand, if it is acknowledged that mathematics is a fallible social construct, then it is a process of inquiry and coming to know, a continually expanding field of human creation and invention, not a finished product. Such a dynamic view of mathematics has powerful educational consequences. The aims of teaching mathematics need to include the empowerment of learners to create their own mathematical knowledge; mathematics can be reshaped, at least in school, to give all groups more access to its concepts, and to the wealth and power its knowledge brings; the social contexts of the uses and practices of mathematics can no longer be legitimately pushed aside, the implicit values of mathematics need to be squarely faced. When mathematics is seen in this way it needs to be studied in living contexts which are meaningful and relevant to the learners, including their languages, cultures and everyday lives, as well as their school based experiences. This view of mathematics provides a rationale, as well as a foundation for multicultural and girl-friendly approaches to mathematics. Overall, mathematics becomes responsible for its uses and consequences, in education and society. Those of us in education have a special reason for wanting this more human view of mathematics. Anything else alienates and disempowers learners.

2. The Philosophy of Mathematics Education

This book is called *The philosophy of mathematics education*, and one task for this introduction is to explain the title. Higginson (1980) has identified a number of foundation disciplines for mathematics education including philosophy. A philosophical perspective on mathematics education, he argues, draws together a different set of problems from that seen from any other vantage point.

At least four sets of problems and issues for the philosophy of mathematics education can be distinguished.

1. The Philosophy of Mathematics

What is mathematics, and how can we account for its nature? What philosophies of mathematics have been developed? Whose?

2. The Nature of Learning

What philosophical assumptions, possibly implicit, underpin the learning of mathematics? Are these assumptions valid? Which epistemologies and learning theories are assumed?

3. The Aims of Education

What are the aims of mathematics education? Are these aims valid? Whose aims are they? For whom? Based on which values? Who gains and who loses?

4. The Nature of Teaching

What philosophical assumptions, possibly implicit, does mathematics teaching rest on? Are these assumptions valid? What means are adopted to achieve the aims of mathematics education? Are the ends and means consistent? These provide a starting point for the philosophy of mathematics education. Of particular importance is the philosophy of mathematics[1].

> In fact, whether one wishes it or not, all mathematical pedagogy, even if scarcely coherent, rests on a philosophy of mathematics.
>
> Thom (1971, p. 204).

> The issue, then, is not, What is the best way to teach? but, What is mathematics really all about? . . . Controversies about . . . teaching cannot be resolved without confronting problems about the nature of mathematics.
>
> Hersh (1979, p. 34)

A number of authors further develop these views (Steiner, 1987). For example, curriculum developments depend greatly on the underlying philosophy of mathematics (Confrey, 1981; Robitaille and Dirks, 1982; Lerman, 1983; Ernest, 1985a), as does the view of mathematics they communicate to learners (Erlwanger, 1973). One large scale investigation of 16 years olds found two distinct personal philosophies of mathematics, dependent on the text: (A) algorithmic, mechanical and somewhat stereotyped; and (B) open-ended, intuitive and heuristic.

> [For] the students taking the S.M.P. course . . . The level of factor B score indicates that these students see mathematics in a wider context of application, that they have a more strongly developed sense of intuition and their approach to problems allows greater flexibility.
>
> Preston (1975, pages 4–5)[2]

In addition to curriculum philosophies, teachers' personal philosophies of mathematics also have a powerful impact on the way mathematics is taught (Davis, 1967; Cooney, 1988; Ernest, 1988b, 1989c). One influential study concluded:

> The observed consistency between the teachers' professed conceptions of mathematics and the way they typically presented the content strongly suggests that the teachers' views, beliefs and preferences about mathematics do influence their instructional practice.
>
> Thompson (1984, page 125)

Such issues are central to the philosophy of mathematics education, and have important practical outcomes for the teaching and learning of mathematics.

3. This book

The first part of the book treats the philosophy of mathematics. It contains both a critique of existing approaches, and a new philosophy of mathematics. For although the traditional paradigm is under attack, the novel and promising ideas in the *Zeitgeist* have not yet been synthesized. Social constructivism is offered to fill this vacuum.

The second part explores the philosophy of mathematics education. It shows that many aspects of mathematics education rest on underlying philosophical assumptions. By uncovering some of them, the aim is to put a critical tool into the hands of teachers and researchers.

Notes

1 A systematic ambiguity should be signalled. *The* philosophy of mathematics is the overall field of philosophical inquiry into the nature of mathematics. In contrast, *a* philosophy of mathematics is a particular account or view of the nature of mathematics. In general, these meanings are signalled by the use of the definite or indefinite article (or the plural form), respectively.

2 It should be mentioned that a more negative attitude to mathematics was associated with view (B) of the SMP students.

PART 1

The Philosophy of Mathematics

1

A Critique of Absolutist Philosophies of Mathematics

1. Introduction

The main purpose of this chapter is to expound and criticize the dominant epistemological perspective of mathematics. This is the *absolutist* view that mathematical truth is absolutely certain, that mathematics is the one and perhaps the only realm of certain, unquestionable and objective knowledge. This is to be contrasted with the opposing *fallibilist* view that mathematical truth is corrigible, and can never be regarded as being above revision and correction.

Much is made of the absolutist-fallibilist distinction because, as is shown subsequently, the choice of which of these two philosophical perspectives is adopted is perhaps the most important epistemological factor underlying the teaching of mathematics.

2. The Philosophy of Mathematics

The philosophy of mathematics is the branch of philosophy whose task is to reflect on, and account for the nature of mathematics. This is a special case of the task of epistemology which is to account for human knowledge in general. The philosophy of mathematics addresses such questions as: What is the basis for mathematical knowledge? What is the nature of mathematical truth? What characterises the truths of mathematics? What is the justification for their assertion? Why are the truths of mathematics necessary truths?

A widely adopted approach to epistemology, is to assume that knowledge in any field is represented by a set of propositions, together with a set of procedures for verifying them, or providing a warrant for their assertion. On this basis, mathematical knowledge consists of a set of propositions together with their proofs. Since mathematical proofs are based on reason alone, without recourse to empirical data, mathematical knowledge is understood to be the most certain of all knowledge. Traditionally the philosophy of mathematics has seen its task as providing a foundation for the certainty of mathematical knowledge. That is, providing a system into which mathematical knowledge can be cast to systematically establish its truth. This depends on an assumption, which is widely adopted, implicitly if not explicitly.

Assumption

The role of the philosophy of mathematics is to provide a systematic and absolutely secure foundation for mathematical knowledge, that is for mathematical truth.[1]

This assumption is the basis of foundationism, the doctrine that the function of the philosophy of mathematics is to provide certain foundations for mathematical knowledge. Foundationism is bound up with the absolutist view of mathematical knowledge, for it regards the task of justifying this view to be central to the philosophy of mathematics.

3. The Nature of Mathematical Knowledge

Traditionally, mathematics has been viewed as the paradigm of certain knowledge. Euclid erected a magnificent logical structure nearly 2,500 years ago in his *Elements*, which until the end of the nineteenth century was taken as the paradigm for establishing truth and certainty. Newton used the form of the *Elements* in his *Principia*, and Spinoza in his *Ethics*, to strengthen their claims to systematically expound truth. Thus mathematics has long been taken as the source of the most certain knowledge known to humankind.

Before inquiring into the nature of mathematical knowledge, it is first necessary to consider the nature of knowledge in general. Thus we begin by asking, what is knowledge? The question of what constitutes knowledge lies at the heart of philosophy, and mathematical knowledge plays a special part. The standard philosophical answer to this question is that knowledge is justified belief. More precisely, that propositional knowledge consists of propositions which are accepted (i.e., believed), provided there are adequate grounds available for asserting them (Sheffler, 1965; Chisholm, 1966; Woozley, 1949).

Knowledge is classified on the basis of the grounds for its assertion. *A priori* knowledge consists of propositions which are asserted on the basis of reason alone, without recourse to observations of the world. Here reason consists of the use of deductive logic and the meanings of terms, typically to be found in definitions. In contrast, empirical or *a posteriori* knowledge consists of propositions asserted on the basis of experience, that is, based on the observations of the world (Woozley, 1949).

Mathematical knowledge is classified as *a priori* knowledge, since it consists of propositions asserted on the basis of reason alone. Reason includes deductive logic and definitions which are used, in conjunction with an assumed set of mathematical axioms or postulates, as a basis from which to infer mathematical knowledge. Thus the foundation of mathematical knowledge, that is the grounds for asserting the truth of mathematical propositions, consists of deductive proof.

The proof of a mathematical proposition is a finite sequence of statements ending in the proposition, which satisfies the following property. Each statement is an axiom drawn from a previously stipulated set of axioms, or is derived by a rule of inference from one or more statements occurring earlier in the sequence. The term 'set of axioms' is conceived broadly, to include whatever statements are admitted into a proof without demonstration, including axioms, postulates and definitions.

An example is provided by the following proof of the statement '$1+1=2$' in the axiomatic system of Peano Arithmetic. For this proof we need the definitions and axioms $s0=1$, $s1=2$, $x+0=x$, $x+sy=s(x+y)$ from Peano Arithmetic, and the logical rules of inference $P(r), r=t \Rightarrow P(t)$; $P(v) \Rightarrow P(c)$ (where r, t; v; c; and P(t) range over terms; variables; constants; and propositions in the term t, respectively, and '$\Rightarrow$' signifies logical implication).[2] The following is a proof of $1+1=2$: $x+sy=s(x+y)$, $1+sy=s(1+y)$, $1+s0=s(1+0)$, $x+0=x$, $1+0=1$, $1+s0=s1$, $s0=1$, $1+1=s1$, $s1=2$, $1+1=2$.

An explanation of this proof is as follows. $s0=1$ [D1] and $s1=2$ [D2] are definitions of the constants 1 and 2, respectively, in Peano Arithmetic. $x+0=x$ [A1] and $x+sy=s(x+y)$ [A2] are axioms of Peano Arithmetic. $P(r), r=t \Rightarrow P(t)$ [R1] and $P(v) \Rightarrow P(c)$ [R2], with the symbols as explained above, are logical rules of inference. The justification of the proof, statement by statement as shown in Table 1.1.

Table 1.1: Proof of 1 + 1 = 2 with justification

Step	Statement	Justification of Statement
S1	$x+sy=s(x+y)$	A2
S2	$1+sy=s(1+y)$	R2 applied to S1, using $v=x$, $c=1$
S3	$1+s0=s(1+0)$	R2 applied to S2, using $v=y$, $c=0$
S4	$x+0=x$	A1
S5	$1+0=1$	R2 applied to S4, using $v=x$, $c=1$
S6	$1+s0=s1$	R1 applied to S3 and S5, using $r=1+0$, $t=1$
S7	$s0=1$	D1
S8	$1+1=s1$	R1 applied to S6 & S7, using $r=s0$, $t=1$
S9	$s1=2$	D2
S10	$1+1=2$	R1 applied to S8 & S9, using $r=s1$, $t=2$

This proof establishes '$1+1=2$' as an item of mathematical knowledge or truth, according to the previous analysis, since the deductive proof provides a legitimate warrant for asserting the statement. Furthermore it is *a priori* knowledge, since it is asserted on the basis of reason alone.

However, what has not been made clear are the grounds for the assumptions made in the proof. The assumptions made are of two types: mathematical and logical assumptions. The mathematical assumptions used are the definitions (D1 and D2) and the axioms (A1 and A2). The logical assumptions are the rules of inference used (R1 and R2), which are part of the underlying proof theory, and the underlying syntax of the formal language.

We consider first the mathematical assumptions. The definitions, being explicit definitions, are unproblematic, since they are eliminable in principle. Every occurrence of the defined terms 1 and 2 can be replaced by what it abbreviates (s0 and ss0, respectively). The result of eliminating these definitions is the abbreviated proof: $x + sy = s(x + y)$, $s0 + sy = s(s0 + y)$, $s0 + s0 = s(s0 + 0)$, $x + 0 = x$, $s0 + 0 = s0$, $s0 + s0 = ss0$; proving '$s0 + s0 = ss0$', which represents '$1 + 1 = 2$'. Although explicit definitions are eliminable in principle, it remains an undoubted convenience, not to mention an aid to thought, to retain them. However, in the present context we are concerned to reduce assumptions to their minimum, to reveal the irreducible assumptions on which mathematical knowledge and its justification rests.

If the definitions had not been explicit, such as in Peano's original inductive definition of addition (Heijenoort, 1967), which is assumed above as an axiom, and not as a definition, then the definitions would not be eliminable in principle. In this case the problem of the basis of a definition, that is the assumption on which it rests, is analogous to that of an axiom.

The axioms in the proof are not eliminable. They must be assumed either as self-evident axiomatic truths, or simply retain the status of unjustified, tentative assumptions, adopted to permit the development of the mathematical theory under consideration. We will return to this point.

The logical assumptions, that is the rules of inference (part of the overall proof theory) and the logical syntax, are assumed as part of the underlying logic, and are part of the mechanism needed for the application of reason. Thus logic is assumed as an unproblematic foundation for the justification of knowledge.

In summary, the elementary mathematical truth '$1 + 1 = 2$', depends for its justification on a mathematical proof. This in turn depends on assuming a number of basic mathematical statements (axioms), as well as on the underlying logic. In general, mathematical knowledge consists of statements justified by proofs, which depend on mathematical axioms (and an underlying logic).

This account of mathematical knowledge is essentially that which has been accepted for almost 2,500 years. Early presentations of mathematical knowledge, such as Euclid's Elements, differ from the above account only by degree. In Euclid, as above, mathematical knowledge is established by the logical deduction of theorems from axioms and postulates (which we include among the axioms). The underlying logic is left unspecified (other than the statement of some axioms concerning the equality relation). The axioms are not regarded as temporarily adopted assumptions, held only for the construction of the theory under consideration. The axioms are considered to be basic truths which needed no justification, beyond their own self evidence (Blanche, 1966).[3] Because of this, the account claims to provide certain grounds for mathematical knowledge. For since logical proof preserves truth and the assumed axioms are self-evident truths, then any theorems derived from them must also be truths (this reasoning is implicit, not explicit in Euclid). However, this claim is no longer accepted because Euclid's axioms and postulates are not considered to be basic and incontrovertible truths, none of which can be negated or denied without resulting in contradiction. In fact, the denial of some of them, most notably the Parallel Postulate, merely leads to other bodies of geometric knowledge (non-euclidean geometry).

Beyond Euclid, modern mathematical knowledge includes many branches which depend on the assumption of sets of axioms which cannot be claimed to be basic universal truths, for example, the axioms of group theory, or of set theory (Maddy, 1984).

4. The Absolutist View of Mathematical Knowledge

The absolutist view of mathematical knowledge is that it consists of certain and unchallengeable truths. According to this view, mathematical knowledge is made up of absolute truths, and represents the unique realm of certain knowledge, apart from logic and statements true by virtue of the meanings of terms, such as 'All bachelors are unmarried'.

Many philosophers, both modern and traditional, hold absolutist views of mathematical knowledge. Thus according to Hempel:

> the validity of mathematics derives from the stipulations which determine the meaning of the mathematical concepts, and that the propositions of mathematics are therefore essentially 'true by definition'.
>
> (Feigl and Sellars, 1949, page 225)

Another proponent of the certainty of mathematics is A.J. Ayer who claims the following.

> Whereas a scientific generalisation is readily admitted to be fallible, the truths of mathematics and logic appear to everyone to be necessary and certain.
>
> The truths of logic and mathematics are analytic propositions or tautologies.
>
> The certainty of *a priori* propositions depends on the fact that they are tautologies. A proposition [is] a tautology if it is analytic. A proposition is analytic if it is true solely in the virtue of the meaning of its consistituent symbols, and cannot therefore be either confirmed or refuted by any fact of experience.
>
> (Ayer, 1946, pages 72, 77 and 16, respectively).

The deductive method provides the warrant for the assertion of mathematical knowledge. The grounds for claiming that mathematics (and logic) provide absolutely certain knowledge, that is truth, are therefore as follows. First of all, the basic statements used in proofs are taken to be true. Mathematical axioms are assumed to be true, for the purposes of developing that system under consideration, mathematical definitions are true by *fiat*, and logical axioms are accepted as true. Secondly, the logical rules of inference preserve truth, that is they allow nothing but truths to be deduced from truths. On the basis of these two facts, every statement in a deductive proof, including its conclusion, is true. Thus, since mathematical theorems are all established by means of deductive proofs, they are all certain truths. This constitutes the basis of the claim of many philosophers that mathematical truths are certain truths.

This absolutist view of mathematical knowledge is based on two types of assumptions: those of mathematics, concerning the assumption of axioms and definitions, and those of logic concerning the assumption of axioms, rules of inference and the formal language and its syntax. These are local or micro-assumptions. There is also the possibility of global or macro-assumptions, such as whether logical deduction suffices to establish all mathematical truths. I shall subsequently argue that each of these assumptions weakens the claim of certainty for mathematical knowledge.

The absolutist view of mathematical knowledge encountered problems at the beginning of the twentieth century when a number of antinomies and contradictions were derived in mathematics (Kline, 1980; Kneebone, 1963; Wilder, 1965). In a series of publications Gottlob Frege (1879, 1893) established by far the most rigorous formulation of mathematical logic known to that time, as a foundation for mathematical knowledge. Russell (1902), however, was able to show that Frege's system was inconsistent. The problem lay in Frege's Fifth Basic Law, which allows a set to be created from the extension of any concept, and for concepts or properties to be applied to this set (Furth, 1964). Russell produced his well-known paradox by defining the property of 'not being an element of itself'. Frege's law allows the extension of this property to be regarded as a set. But then this set is an element of itself if, and only if, it is not; a contradiction. Frege's Law could not be dropped without seriously weakening his system, and yet it could not be retained.

Other contradictions also emerged in the theory of sets and the theory of functions. Such findings have, of course, grave implications for the absolutist view of mathematical knowledge. For if mathematics is certain, and all its theorems are certain, how can contradictions (i.e., falsehoods) be among its theorems? Since there was no mistake about the appearance of these contradictions, something must be wrong in the foundations of mathematics. The outcome of this crisis was the development of a number of schools in the philosophy of mathematics whose aims were to account for the nature of mathematical knowledge and to re-establish its certainty. The three major schools are known as logicism, formalism and constructivism (incorporating intuitionism). The tenets of these schools of thought were not fully developed until the twentieth century, but Korner (1960) shows that their philosophical roots can be traced back at least as far as Leibniz and Kant.

A. Logicism

Logicism is the school of thought that regards pure mathematics as a part of logic. The major proponents of this view are G. Leibniz, G. Frege (1893), B. Russell (1919),

A.N. Whitehead and R. Carnap (1931). At the hands of Bertrand Russell the claims of logicism received the clearest and most explicit formulation. There are two claims:

1 All the concepts of mathematics can ultimately be reduced to logical concepts, provided that these are taken to include the concepts of set theory or some system of similar power, such as Russell's Theory of Types.
2 All mathematical truths can be proved from the axioms and rules of inference of logic alone.

The purpose of these claims is clear. If all of mathematics can be expressed in purely logical terms and proved from logical principles alone, then the certainty of mathematical knowledge can be reduced to that of logic. Logic was considered to provide a certain foundation for truth, apart from over-ambitious attempts to extend logic, such as Frege's Fifth Law. Thus if carried through, the logicist programme would provide certain logical foundations for mathematical knowledge, re-establishing absolute certainty in mathematics.

Whitehead and Russell (1910–13) were able to establish the first of the two claims by means of chains of definitions. However logicism foundered on the second claim. Mathematics requires non-logical axioms such as the Axiom of Infinity (the set of all natural numbers is infinite) and the Axiom of Choice (the Cartesian product of a family of non-empty sets is itself non-empty). Russell expressed it himself as follows.

> But although all logical (or mathematical) propositions can be expressed wholly in terms of logical constants together with variables, it is not the case that, conversely, all propositions that can be expressed in this way are logical. We have found so far a necessary but not a sufficient criterion of mathematical propositions. We have sufficiently defined the character of the primitive *ideas* in terms of which all the ideas of mathematics can be *defined*, but not of the primitive *propositions* from which all the propositions of mathematics can be *deduced*. This is a more difficult matter, as to which it is not yet known what the full answer is.
>
> We may take the axiom of infinity as an example of a proposition which, though it can be enunciated in logical terms, cannot be asserted by logic to be true.
>
> (Russell, 1919, pages 202–3, original emphasis)

Thus not all mathematical theorems and hence not all the truths of mathematics can be derived from the axioms of logic alone. This means that the axioms of mathematics are not eliminable in favour of those of logic. Mathematical theorems depend on an irreducible set of mathematical assumptions. Indeed, a number of important mathematical axioms are independent, and either they or their negation can be adopted, without inconsistency (Cohen, 1966). Thus the second claim of logicism is refuted

To overcome this problem Russell retreated to a weaker version of logicism called 'if-thenism', which claims that pure mathematics consists of implication statements of the form '$A \rightarrow T$'. According to this view, as before, mathematical truths are established as theorems by logical proofs. Each of these theorems (T) becomes the consequent in an implication statement. The conjunction of mathematical axioms (A) used in the proof are incorporated into the implication statement as its antecedent (see Carnap, 1931). Thus all the mathematical assumptions (A) on which the theorem (T) depends are now incorporated into the new form of the theorem (A→T), obviating the need for mathematical axioms.

This artifice amounts to an admission that mathematics is an hypothetico-deductive system, in which the consequences of assumed axiom sets are explored, without asserting their necessary truth. Unfortunately, this device also leads to failure, because not all mathematical truths, such as 'Peano arithmetic is consistent,' can be expressed in this way as implication statements, as Machover (1983) argues.

A second objection, which holds irrespective of the validity of the two logicist claims, constitutes the major grounds for the rejection of formalism. This is Godel's Incompleteness Theorem, which establishes that deductive proof is insufficient for demonstrating all mathematical truths. Hence the successful reduction of mathematical axioms to those of logic would still not suffice for the derivation of all mathematical truths.

A third possible objection concerns the certainty and reliability of the underlying logic. This depends on unexamined and, as will be argued, unjustified assumptions.

Thus the logicist programme of reducing the certainty of mathematical knowledge to that of logic failed in principle. Logic does not provide a certain foundation for mathematical knowledge.

B. Formalism

In popular terms, formalism is the view that mathematics is a meaningless formal game played with marks on paper, following rules. Traces of a formalist philosophy of mathematics can be found in the writings of Bishop Berkeley, but the major proponents of formalism are David Hilbert (1925), early J. von Neumann (1931) and H. Curry (1951). Hilbert's formalist programme aimed to translate mathematics into uninterpreted formal systems. By means of a restricted but meaningful meta-mathematics the formal systems were to be shown to be adequate for mathematics, by deriving formal counterparts of all mathematical truths, and to be safe for mathematics, through consistency proofs.

The formalist thesis comprises two claims.

1 Pure mathematics can be expressed as uninterpreted formal systems, in which the truths of mathematics are represented by formal theorems.
2 The safety of these formal systems can be demonstrated in terms of their freedom from inconsistency, by means of meta-mathematics.

Kurt Godel's Incompleteness Theorems (Godel, 1931) showed that the programme could not be fulfilled. His first theorem showed that not even all the truths of arithmetic can be derived from Peano's Axioms (or any larger recursive axiom set).

This proof-theoretic result has since been exemplified in mathematics by Paris and Harrington, whose version of Ramsey's Theorem is true but not provable in Peano Arithmetic (Barwise, 1977). The second Incompleteness Theorem showed that in the desired cases consistency proofs require a meta-mathematics more powerful than the system to be safeguarded, which is thus no safeguard at all. For example, to prove the consistency of Peano Arithmetic requires all the axioms of that system and further assumptions, such as the principle of transfinite induction over countable ordinals (Gentzen, 1936).

The formalist programme, had it been successful, would have provided support for an absolutist view of mathematical truth. For formal proof, based in consistent formal mathematical systems, would have provided a touchstone for mathematical truth. However, it can be seen that both the claims of formalism have been refuted. Not all the truths of mathematics can be represented as theorems in formal systems, and furthermore, the systems themselves cannot be guaranteed safe.

C. Constructivism

The constructivist strand in the philosophy of mathematics can be traced back at least as far as Kant and Kronecker (Korner, 1960). The constructivist programme is one of reconstructing mathematical knowledge (and reforming mathematical practice) in order to safeguard it from loss of meaning, and from contradiction. To this end, constructivists reject non-constructive arguments such as Cantor's proof that the Real numbers are uncountable, and the logical Law of the Excluded Middle.

The best known constructivists are the intuitionists L.E.J. Brouwer (1913) and A. Heyting (1931, 1956). More recently the mathematician E. Bishop (1967) has carried the constructivist programme a long way, by reconstructing a substantial portion of Analysis, by constructive means. Various forms of constructivism still flourish today, such as in the work of the philosophical intuitionist M. Dummett (1973, 1977). Constructivism includes a whole range of different views, from the ultra-intuitionists (A. Yessenin-Volpin), via what may be termed strict philosophical intuitionists (L. E. J. Brouwer), middle-of-the-road intuitionists (A. Heyting and early H. Weyl), modern logical intuitionists (A. Troelstra) to a range of more or less liberal constructivists including P. Lorenzen, E. Bishop, G. Kreisel and P. Martin-Lof.

These mathematicians share the view that classical mathematics may be unsafe, and that it needs to be rebuilt by 'constructive' methods and reasoning. Constructivists claim that both mathematical truths and the existence of mathematical objects must be established by constructive methods. This means that mathematical constructions are needed to establish truth or existence, as opposed to methods relying on proof by contradiction. For constructivists knowledge must be established through constructive proofs, based on restricted constructivist logic, and the meaning of mathematical terms/objects consists of the formal procedures by which they are constructed.

Although some constructivists maintain that mathematics is the study of constructive processes performed with pencil and paper, the stricter view of the intuitionists, led by Brouwer, is that mathematics takes place primarily in the mind, and that written mathematics is secondary. One consequence of that is that Brouwer regards all axiomatizations of intuitionistic logic to be incomplete. Reflection can always uncover further intuitively true axioms of intuitionistic logic, and so it can never be regarded as being in final form.

Intuitionism represents the most fully formulated constructivist philosophy of mathematics. Two separable claims of intuitionism can be distinguished, which Dummett terms the positive and the negative theses.

> The positive one is to the effect that the intuitionistic way of construing mathematical notions and logical operations is a coherent and legitimate one, that intuitionistic mathematics forms an intelligible body of theory. The negative thesis is to the effect that the classical way of construing mathematical notions and logical operations is incoherent and illegitimate, that classical mathematics, while containing, in distorted form, much of value, is, nevertheless, as it stands unintelligible.
>
> (Dummett, 1977, page 360).

In restricted areas where there are both classical and constructivist proofs of a result, the latter is often preferable as more informative. Whereas a classical existence proof may merely demonstrate the logical necessity of existence, a constructive existence proof shows how to construct the mathematical object whose existence is asserted. This lends strength to the positive thesis, from a mathematical point of view. However, the negative thesis is much more problematic, since it not only fails to account for the substantial body of non-constructive classical mathematics, but also denies its validity. The constructivists have *not* demonstrated that there are inescapable problems facing classical mathematics nor that it is incoherent and invalid. Indeed both pure and applied classical mathematics have gone from strength to strength since the constructivist programme was proposed. Therefore, the negative thesis of intuitionism is rejected.

Another problem for the constructivist view, is that some of its results are inconsistent with classical mathematics. Thus, for example, the real number continuum, as defined by the intuitionists, is countable. This contradicts the classical result not because there is an inherent contradiction, but because the definition of real numbers is different. Constructivist notions often have a different meaning from the corresponding classical notions.

From an epistemological perspective, both the positive and negative theses of intuitionism are flawed. The intuitionists claim to provide a certain foundation for their version of mathematical truth by deriving it (mentally) from intuitively certain axioms, using intuitively safe methods of proof. This view bases mathematical knowledge exclusively on subjective belief. But absolute truth (which the intuitionists claim to provide) cannot be based on subjective belief alone. Nor is there any guarantee that different intuitionists' intuitions of basic truth will coincide, as indeed they do not.

> Intuitionism sacrificed large parts of mathematics in exchange for the soothing reassurance that what remained was justified by our 'primordial intuition' *(Urintuition)*. But intuition is subjective, and not intersubjective enough to prevent intuitionists from differing about what their 'primordial intuitions' should enshirine as the basis of mathematics.
>
> (Kalmar, 1967, page 190).

Thus the positive thesis of intuitionism does not provide a certain foundation for even a subset of mathematical knowledge. This criticism extends to other forms of constructivism which also claim to base constructive mathematical truth on a foundation of self-evident constructivist assumptions.

The negative thesis of intuitionism (and of constructivism, when it is embraced), leads to the unwarranted rejection of accepted mathematical knowledge, on the grounds that it is unintelligible. But classical mathematics *is* intelligible. It differs from constructivist mathematics largely in the assumptions on which it is based.[4] Thus constructivism is guilty of what is analogous to a Type I Error in statistics, namely the rejection of valid knowledge.

5. The Fallacy of Absolutism

We have seen that a number of absolutist philosophies of mathematics have failed to establish the logical necessity of mathematical knowledge. Each of the three schools of thought logicism, formalism and intuitionism (the most clearly enunciated form of constructivism) attempts to provide a firm foundation for mathematical truth, by deriving it by mathematical proof from a restricted but secure realm of truth. In each case there is the laying down of a secure base of would-be absolute truth. For the logicists, formalists and intuitionists this consists of the axioms of logic, the intuitively certain principles of meta-mathematics, and the self-evident axioms of 'primordial intuition', respectively. Each of these sets of axioms or principles is assumed without demonstration. Therefore each remains open to challenge, and thus to doubt. Subsequently each of the schools employs deductive logic to demonstrate the truth of the theorems of mathematics from their assumed bases. Consequently these three schools of thought fail to establish the absolute certainty of mathematical truth. For deductive logic only transmits truth, it does not inject it, and the conclusion of a logical proof is at best as certain as the weakest premise.

It can be remarked that all three schools' attempts also fail to provide a foundation for the full range of would-be mathematical truths by these means. For as Godel's first Incompleteness theorem shows, proof is not adequate to demonstrate all truths. Thus there are truths of mathematics not captured by the systems of these schools.

The fact that three schools of thought in the philosophy of mathematics have failed to establish the certainty of mathematical knowledge does not settle the general issue. It is still possible for other grounds to be found for asserting the certainty of mathematical truth. Absolute truth in mathematics still remains a possibility.

However this possibility is denied by powerful general arguments for refusing to accord the status of certainty to the truths of mathematics. These resemble the common argument used above to criticize the three schools, since they all rely on deductive systems.

Lakatos (1962) shows that the quest for certainty in mathematics leads inevitably to a vicious circle. Any mathematical system depends on a set of assumptions, and trying to establish their certainty by proving them, leads to an infinite regression. There is no way of discharging the assumptions. Without proof, the assumptions remain fallible beliefs, and not certain knowledge. All we can do is to minimize them, to get a reduced set of axioms, which we have to either accept without proof, thus breaking the vicious circle at the cost of forfeiting certainty, or which we can replace. Replacement sends us on a further circuit of the vicious circle. The reduced set of axioms can only be dispensed with by replacing it with assumptions of at least the same strength. Thus we cannot establish the certainty of mathematics without making assumptions, which thereby fails to be absolute certainty.

It should be understood that this argument is directed at the whole of mathematical knowledge, and it is not framed for a single formal system or language. Many attempts to provide a foundation for mathematics in such a language manage to reduce assumptions *in that formal system or language.* What has been done in such a case is to push some or all of the basic assumptions into the meta-language, as was the formalists' explicit strategy. Wherever it remains, and somewhere it must, there is a kernel of assumptions which introduces truth into the system, and which deduction transmits to all the theorems of the system (provided the system is safe, i.e., consistent).

As Lakatos says: 'we have to admit that *meta-mathematics does not stop the infinite regress in proofs which now reappears in the infinite hierarchy of ever richer meta-theories'* (Lakatos, 1978, page 22).

Mathematical truth ultimately depends on an irreducible set of assumptions, which are adopted without demonstration. But to qualify as true knowledge, the assumptions require a warrant for their assertion. There is no valid warrant for mathematical knowledge other than demonstration or proof.[5] Therefore the assumptions are beliefs, not knowledge, and remain open to challenge, and thus to doubt.

This is the central argument against the possibility of certain knowledge in mathematics. It directly contradicts the absolutist claims of the foundationist schools of thought. Beyond the foundationist schools, it is considered to be an unanswerable refutation of absolutism by a number of authors.

> [T]he absolute truth point of view must be discarded. The 'facts' of any branch of pure mathematics must be recognised as being assumptions (postulates or axioms), or definitions or theorems . . . The most that can be claimed is that *if* the postulates are true *and* the definitions are accepted, and *if* the methods of reasoning are sound, *then* the theorems are true. In other words we arrive at a concept of *relative truth* (of theorems in relation to

postulates, definitions, and logical reasoning) to replace the *absolute truth* point of view.

(Stabler, 1955, page 24).

What we have called pure mathematics is, therefore a *hypothetico-deductive system*. Its axioms serve as hypotheses or assumptions, which are entertained or considered for the propositions they imply.

(Cohen and Nagel, 1963, page 133).

[W]e can only describe arithmetic, namely, find its rules, not give a basis for them. Such a basis could not satisfy us, for the very reason that it must end sometime and then refer to something which can no longer be founded. Only the convention is the ultimate. Anything that looks like a foundation is, strictly speaking, already adulterated and must not satisfy us.

(Waismann, 1951, page 122).

Statements or propositions or theories may be formulated in assertions which may be true and their truth may be reduced, by way of derivations to that of primitive propositions. The attempt to establish (rather than reduce) by these means their truth leads to an infinite regress.

(Popper, 1979, extract from table on page 124).

The above criticism is decisive for the absolutist view of mathematics. However, it is possible to accept the criticism without adopting a fallibilist philosophy of mathematics. For it is possible to accept a form of hypothetico-deductivism which denies the corrigibility and the possibility of deep-seated error in mathematics. Such a position views axioms simply as hypotheses from which the theorems of mathematics are logically deduced, and relative to which the theorems are certain. In other words, although the axioms of mathematics are tentative, logic and the use of logic to derive theorems from the axioms guarantee a secure development of mathematics, albeit from an assumed basis. This weakened form of the absolutist position resembles Russell's 'if thenism' in its strategy of adopting axioms without either proof or cost to the system's security. However this weakened absolutist position is based on assumptions which leave it open to a fallibilist critique.

6. The Fallibilist Critique of Absolutism

The central argument against the absolutist view of mathematical knowledge can be circumvented by a hypothetico-deductive approach. However, beyond the problem of the assumed truth of the axioms, the absolutist view suffers from further major weaknesses.

The first of these concerns the underlying logic on which mathematical proof rests. The establishment of mathematical truths, that is the deduction of theorems from a set of axioms, requires further assumptions, namely the axioms and rules of inference of logic itself. These are non-trivial and non-eliminable assumptions, and the above argument (the ultimate irreducibility of assumptions on pain of a vicious circle) applies equally to logic. Thus mathematical truth depends on essential logical as well as mathematical assumptions.

It is not possible to simply append all the assumptions of logic to the set of mathematical assumptions, following the 'if-thenist' hypothetico-deductive strategy. For logic provides the canons of correct inference with which the theorems of mathematics are derived. Loading all logical and mathematical assumptions into the 'hypothetico' part leaves no basis for the 'deductive' part of the method. Deduction concerns 'correct inference', and this in turn is based on the notion of truth (the preservation of truth value). But what then is the foundation of logical truth? It cannot rest on proof, on pain of a vicious circle, so it must be assumed. But any assumption without a firm foundation, whether it be derived through intuition, convention, meaning or whatever, is fallible.

In summary, mathematical truth and proof rest on deduction and logic. But logic itself lacks certain foundations. It too rests on irreducible assumptions. Thus the dependence on logical deduction increases the set of assumptions on which mathematical truth rests, and these cannot be neutralized by the 'if-thenist' strategy.

A further presumption of the absolutist view is that mathematics is fundamentally free from error. For inconsistency and absolutism are clearly incompatible. But this cannot be demonstrated. Mathematics consists of theories (e.g., group theory, category theory) which are studied within mathematical systems, based on sets of assumptions (axioms). To establish that mathematical systems are safe (i.e., consistent), for any but the simplest systems we are forced to expand the set of assumptions of the system (Godel's Second Incompleteness Theorem, 1931). We have therefore to assume the consistency of a stronger system to demonstrate that of a weaker. We cannot therefore know that any but the most trivial mathematical systems are secure, and the possibility of error and inconsistency must always remain. Belief in the safety of mathematics must be based either on empirical grounds (no contradictions have yet been found in our current mathematical systems) or on faith, neither providing the certain basis that absolutism requires.

Beyond this criticism, there are further problems attendant on the use of proof as a basis for certainty in mathematics. Nothing but a fully formal deductive proof can serve as a warrant for certainty in mathematics. But such proofs scarcely exist. Thus absolutism requires the recasting of informal mathematics into formal deductive systems, which introduces further assumptions. Each of the following assumptions is a necessary condition for such certainty in mathematics. Each, it is argued, is an unwarranted absolutist assumption.

Assumption A

The proofs that mathematicians publish as warrants for asserting theorems can, in principle, be translated into fully rigorous formal proofs.

The informal proofs that mathematicians publish are commonly flawed, and are by no means wholly reliable (Davis, 1972). Translating them into fully rigorous formal proofs is a major, non-mechanical task. It requires human ingenuity to bridge gaps and to remedy errors. Since the total formalization of mathematics is unlikely to be carried out, what is the value of the claim that informal proofs can be translated into formal proofs 'in principle'? It is an unfulfilled promise, rather than grounds for certainty. Total rigor is an unattained ideal and not a practical reality. Therefore certainty cannot be claimed for mathematical proofs, *even if the preceding criticisms are discounted.*

Assumption B

Rigorous formal proofs can be checked for correctness.

There are now humanly uncheckable informal proofs, such as the Appel-Haken (1978) proof of the four colour theorem (Tymoczko, 1979). Translated into fully rigorous formal proofs these will be much longer. If these cannot possibly be surveyed by a mathematician, on what grounds can they be regarded as absolutely correct? If such proofs are checked by a computer what guarantees can be given that the software and hardware are designed absolutely flawlessly, and that the software runs perfectly in practice? Given the complexity of hardware and software it seems implausible that these can be checked by a single person. Furthermore, such checks involve an empirical element (i.e., does it run according to design?). If the checking of formal proofs cannot be carried out, or has an empirical element, then any claim of absolute certainty must be relinquished (Tymoczko, 1979).

Assumption C

Mathematical theories can be validly translated into formal axiom sets.

The formalization of intuitive mathematical theories in the past hundred years (e.g., mathematical logic, number theory, set theory, analysis) has led to unanticipated deep problems, as the concepts and proofs come under ever more piercing scrutiny, during attempts to explicate and reconstruct them. The satisfactory formalization of the rest of mathematics cannot be assumed to be unproblematic. Until this formalization is carried out it is not possible to assert with certainty that it can be carried out validly. But until mathematics is formalized, its rigour, which is a necessary condition for certainty, falls far short of the ideal.

Assumption D

The consistency of these representations (in assumption C) can be checked.

As we know from Godel's Incompleteness Theorem, this adds significantly to the burden of assumptions underpinning mathematical knowledge. Thus there are no *absolute* guarantees of safety.

Each of these four assumptions indicates where further problems in establishing certainty of mathematical knowledge may arise. These are not problems concerning the assumed truth of the basis of mathematical knowledge (i.e., the basic assumptions). Rather these are problems in trying to transmit the assumed truth of these assumptions to the rest of mathematical knowledge by means of deductive proof, and in establishing the reliability of the method.

7. The Fallibilist View

The absolutist view of mathematical knowledge has been subject to a severe, and in my view, irrefutable criticism.[6] Its rejection leads to the acceptance of the opposing fallibilist view of mathematical knowledge. This is the view that mathematical truth is fallible and corrigible, and can never be regarded as beyond revision and correction. The fallibilist thesis thus has two equivalent forms, one positive and one negative. The negative form concerns the rejection of absolutism: mathematical knowledge is not absolute truth, and does not have absolute validity. The positive form is that mathematical knowledge is corrigible and perpetually open to revision. In this section I wish to demonstrate that support for the fallibilist viewpoint, in one form or the other, is much broader than might have been supposed. The following is a selection from the range of logicians, mathematicians and philosophers who support this viewpoint:

In his paper 'A renaissance of empiricism in the philosophy of mathematics', Lakatos quotes from the later works of Russell, Fraenkel, Carnap, Weyl, von Neumann, Bernays, Church, Godel, Quine, Rosser, Curry, Mostowski and Kalmar (a list that includes many of the key logicians of the twentieth century) to demonstrate their common view concerning 'the impossibility of complete certainty' in mathematics, and in many cases, their agreement that mathematical knowledge has an empirical basis, entailing the rejection of absolutism. (Lakatos, 1978, page 25, quotation from R. Carnap)

> It is now apparent that the concept of a universally accepted, infallible body of reasoning — the majestic mathematics of 1800 and the pride of man — is a grand illusion. Uncertainty and doubt concerning the future of mathematics have replaced the certainties and complacency of the past . . . The present state of mathematics is a mockery of the hitherto deep-rooted and widely reputed truth and logical perfection of mathematics.
>
> (Kline, 1980, page 6)

> There are no authoritative sources of knowledge, and no 'source' is particularly reliable. Everything is welcome as a source of inspiration, including 'intuition'. . . But nothing is secure, and we are all fallible.
>
> (Popper, 1979, page 134)

> I should like to say that where surveyability is not present, i.e., where there is room for a doubt whether what we have really is the result of this substitution, the *proof* is destroyed. And not in some silly and unimportant way that has nothing to do with the *nature* of proof.

Or: logic as the foundation of mathematics does not work, and to show this it is enough that the cogency of logical proof stands and falls with its geometrical cogency. . . .

The logical certainty of proofs — I want to say — does not extend beyond their geometrical certainty.

(Wittgenstein, 1978, pages 174–5)

A Euclidean theory may be claimed to be true; a quasi-empirical theory — at best — to be well-corroborated, but always conjectural. Also, in a Euclidean theory the true basic statements at the 'top' of the deductive system (usually called 'axioms') *prove*, as it were, the rest of the system; in a quasi-empirical theory the (true) basic statements are *explained* by the rest of the system . . . Mathematics is quasi-empirical

(Lakatos, 1978, pages 28–29 & 30)

Tautologies are necessarily true, but mathematics is not. We cannot tell whether the axioms of arithmetic are consistent; and if they are not, any particular theorem of arithmetic may be false. Therefore these theorems are not tautologies. They are and must always remain tentative, while a tautology is an incontrovertible truism . . .

[T]he mathematician feels compelled to accept mathematics as true, even though he is today deprived of the belief in its logical necessity and doomed to admit forever the conceivable possibility that its whole fabric may suddenly collapse by revealing a decisive self-contradiction.

(Polanyi, 1958, pages 187 and 189)

The doctrine that mathematical knowledge is a priori *mathematical apriorism* has been articulated many different ways during the course of reflection about mathematics . . . I shall offer a picture of mathematical knowledge which rejects mathematical apriorism . . . the alternative to mathematical apriorism — *mathematical empiricism* — has never been given a detailed articulation. I shall try to give the missing account.

(Kitcher, 1984, pages 3–4)

[M]athematical knowledge resembles *empirical* knowledge — that is, the criterion of truth in mathematics just as much as in physics is success of our ideas in practice, and that mathematical knowledge is corrigible and not absolute.

(Putnam, 1975, page 51)

It is reasonable to propose a new task for mathematical philosophy: not to seek indubitable truth but to give an account of mathematical knowledge as it really is — fallible, corrigible, tentative and evolving, as is every other kind of human knowledge.

(Hersh, 1979, page 43)

Why not honestly admit mathematical fallibility, and try to defend the dignity of *fallible* knowledge from cynical scepticism, rather than delude ourselves that we shall be able to mend invisibly the latest tear in the fabric of our 'ultimate' intuitions.

(Lakatos, 1962, page 184)

8. Conclusion

The rejection of absolutism should not be seen as a banishment of mathematics from the Garden of Eden, the realm of certainty and truth. The 'loss of certainty' (Kline, 1980) does not represent a loss of knowledge.

There is an illuminating analogy with developments in modern physics. General Relativity Theory requires relinquishing absolute, universal frames of reference in favour of a relativistic perspective. In Quantum Theory, Heisenberg's Uncertainty Principle means that the notions of precisely determined measurements of position and momentum for particles also has had to be given up. But what we see here are not the loss of knowledge of absolute frames and certainty. Rather we see the growth of knowledge, bringing with it a realization of the limits of what can be known. Relativity and Uncertainty in physics represent major advances in knowledge, advances which take us to the limits of knowledge (for so long as the theories are retained).

Likewise in mathematics, as our knowledge has become better founded and we learn more about its basis, we have come to realize that the absolutist view is an idealization, a myth. This represents an advance in knowledge, not a retreat from past certainty. The absolutist Garden of Eden was nothing but a fool's paradise.

Notes

1 In this chapter, for simplicity, the definition of truth in mathematics is assumed to be unproblematic and unambiguous. Whilst justified as a simplifying assumption, since none of the arguments of the chapter hinge on the ambiguity of this notion, the meaning of the concept of truth in mathematics has changed over time. We can distinguish between three truth-related concepts used in mathematics:

(a) There is the traditional view of mathematical truth, namely that a mathematical truth is a general statement which not only correctly describes all its instances in the world (as does a true empirical generalisation), but is necessarily true of its instances. Implicit in this view is the assumption that mathematical theories have an intended interpretation, namely some idealization of the world.

(b) There is the modern view of the truth of a mathematical statement relative to a background mathematical theory: the statement is satisfied by some interpretation or model of the theory. According to this (and the following) view, mathematics is open to multiple interpretations, i.e., possible worlds. Truth consists merely in being true (i.e., satisfied, following Tarski, 1936) in one of these possible worlds.

(c) There is the modern view of the logical truth or validity of a mathematical statement relative to a background theory: the statement is satisfied by all interpretations or models of the theory. Thus the statement is true in all of these possible worlds.

Truth in sense (c) can be established by deduction from the background theory as an axiom set. For a given theory, truths in a sense (c) are a subset (usually a proper subset) of truths in sense (b).

Incompleteness arises (as Godel, 1931, proved) in most mathematical theories as there are sentences true in sense (b) (i.e., satisfiable) which are not true in sense (c).

Thus not only does the concept of truth have multiple meanings, but crucial mathematical issues hinge upon this ambiguity. Beyond this, the modern mathematical view of truth differs from the traditional mathematical view of truth (a), and the everyday sense of the term, which resembles it. For in a naive sense truths are statements which accurately describe a state of affairs — a relationship — in some realm of discourse. In this view, the terms which express the truth name objects in the realm of discourse, and the statement as a whole describes a true state of affairs, the relationship that holds between the denotations of the terms. This shows that the concept of truth employed in mathematics no longer has the same meaning as either the everyday, naive notion of truth, or its equivalent (a) as was used in mathematics, in the past (Richards, 1980, 1989).

The consequence of this is that the traditional problem of establishing the indubitable foundations of mathematical truth has changed, as the definition of truth employed has changed. In particular to claim that a statement is true in sense (b) is much weaker than senses (a) or (c). '1 + 1 = 1' is true in sense (b) (it is satisfied in Boolean algebra, but not in sense (a) which assumes the standard Peano interpretation).

2 For the proof to be rigorous, a formal language L for Peano Arithmetic should be specified in full. L is a first-order predicate calculus in universally quantified free variable form. The syntax of L will specify as usual the terms and formulas of L, the formula 'P(r)' in the term 'r', and the result 'P(t)' of substituting the term 't' for the occurrences of 'r' in 'P(r)' (sometimes written P(r)[r/t]). It should also be mentioned that a modernised form of the Peano Axioms is adopted above (see, for example, Bell and Machover, 1977), which is not literally that of Peano (Heijenoort, 1967).

3 Scholars believe that Euclid's fifth postulate was not considered to be as self-evident as the others. It is less terse, and more like a proposition (a theorem) than a postulate (it is the converse of proposition I 17). Euclid does not use it until proposition I 29. For this reason, over the ages, many attempts to prove the posulate were made including Sacchieri's attempt to prove it by *reductio ad absurdam* based on its denial (Eves, 1953).

4 It is worth remarking that the classical predicate calculus is translatable into intuitionist logic in a constructive way that preserves deducibility (see Bell and Machover, 1977). This means that all the theorems of classical mathematics expressible in the predicate calculus can be represented as intuitionistic theorems. Thus classical mathematics cannot easily be claimed to be intuitionistically unintelligible. (Note that the reverse translation procedure is intuitionistically unacceptable, since it replaces '-P' by 'P', and '-(x)-P' by '(Ex)P', reading -, (x), and (Ex) as 'not', 'for all x' and 'there exists x', respectively).

5 Some readers may feel that assertion requires justification. What valid warrant can there be for mathematical knowledge other than demonstration or proof? Clearly it is necessary to find other grounds for asserting that mathematical statements are true. The principal accounts of truth are the correspondence theory of truth, the coherence theory of truth (Woozley, 1949), the pragmatic theory of truth (Dewey, 1938) and truth as convention (Quine, 1936; Quinton, 1963). We can first dismiss the coherence and pragmatic theories of truth as irrelevant here, since these do not claim that truth can be warranted absolutely. The correspondence theory can be interpreted either empirically or non-empirically, to say that basic mathematical truths describe true states of affairs either in the world or in some abstract realm. But then the truths of mathematics are justified empirically or intuitively, respectively, and neither grounds serve as warrants for certain knowledge.

The conventional theory of truth asserts that basic mathematical statements are true by virtue of the meanings of the terms therein. But the fact that the axioms express what we want or believe terms to mean does not absolve us from having to assume them, even if we simply stipulate them by *fiat*. Rather it is an admission that we simply have to assume certain basic propositions. Beyond this, to say that complex axioms such as those of Zermelo-Fraenkel set theory are true by virtue of the meanings of the constituent terms is not supportable. (Maddy, 1984, gives an account of set-theoretic axioms in current use which by no stretch of the imagination are considered true). We must regard these axioms as implicit definitions of their constituent terms, and it is evident we must assume the axioms to proceed with set theory.

6 The critique of absolutism can be used to criticize this chapter, as follows. If no knowledge, including mathematics is certain how can the modestly founded assertions of this chapter be true? Is not the assertion that there is no truth self-defeating?

The answer is that the assertions and arguments of this chapter do not pretend to be the truth, but a plausible account. The grounds for accepting the truths of mathematics, imperfect as they are, are far firmer than the arguments of the chapter. (The argument can be defended analogously to the way Ayer, 1946, defends the Principle of Verification.)

2

The Philosophy of Mathematics Reconceptualized

1. The Scope of the Philosophy of Mathematics

In the previous chapter we entertained the hypothesis that mathematical knowledge is a set of truths, in the form of a set of propositions with proofs, and that the function of the philosophy of mathematics is to establish the certainty of this knowledge. Having found that this hypothesis is untenable we are forced to reconsider the nature of the philosophy of mathematics. What is the function and scope of the philosophy of mathematics?

> As the philosophy of law does not legislate, or the philosophy of science devise or test scientific hypotheses — the philosophy of mathematics does not add to the number of mathematical theorems and theories. It is not mathematics. It is reflection upon mathematics, giving rise to its own particular questions and answers.
>
> (Korner, 1960, page 9)
>
> The philosophy of mathematics begins when we ask for a general account of mathematics, a synoptic vision of the discipline that reveals its essential features and explains just how it is that human beings are able to do mathematics.
>
> (Tymoczko, 1986, page viii)

Priest (1973) boldly outlines the task as follows:

> All the problems concerning the philosophy of mathematics can neatly be summarized by the question:
>
> Question 0. What is pure mathematics? . . .
>
> Firstly, what is meant by 'mathematics'? The only answer we can give without begging the question is 'That which is done and has been done for the last four thousand years by mathematicians.' . . . Knowledge of the nature of mathematics lies in an ability to do it.
>
> [To answer question 0 we need to answer the following:]

> Question 1. Why are the truths of mathematics true?
> Any reasonable answer must also permit reasonable answers to the following questions
>
> Question 1(a). Why is it that such truths appear necessary and inviolable, and why are we unable to conceive of them being false?
>
> Question 1(b). How is it we come to know such truths?
>
> Question 1(c). Why is it that the truths of mathematics can be applied in practical matters e.g., surveying, building bridges, sending rockets to the moon, etc. In short, why are they useful? . . .
>
> Now the naive answer to question 1 is that mathematical truths are so because they are true of certain objects such as numbers, functions, propositions, points, groups, models etc., i.e., these are what mathematics is about.
>
> Hence we must be able to answer:
>
> Question 2. What exactly are the above objects, and in what sense do they exist? . . .
>
> Question 2 (cont.). And if they don't exist, why is it we have such a strong impression that they do?
>
> (Priest, 1973, pages 115–117)

According to these views, the role of the philosophy of mathematics is to reflect on, and give an account of the nature of mathematics. The key issue concerns how 'giving an account of' mathematics is conceived. Absolutist philosophies of mathematics such as logicism, formalism and intuitionism attempt to provide *prescriptive* accounts of the nature of mathematics. Such accounts, as we have seen, are programmatic, legislating how mathematics should be understood, rather than providing accurately *descriptive* accounts of the nature of mathematics. Thus they are failing to account for mathematics as it is, in the hope of fulfilling their vision of how it should be. But 'to confuse description and programme — to confuse 'is' with 'ought to be' or 'should be' — is just as harmful in the philosophy of mathematics as elsewhere.'

(Korner, 1960, page 12)

The inquiry can begin with the traditional questions of epistemology and ontology. What is the nature and basis of mathematical knowledge? What is the nature of, and how do we account for, the existence of mathematical objects (numbers, functions, sets, etc.)?

However, the answers to these questions will not provide a descriptive account of the nature of mathematics. For the narrow focus of these 'internal' questions concerning the philosophy of mathematics fails to locate mathematics within the broader context of human thought and history. Without such a context, according to Lakatos, the philosophy of mathematics loses its content.

> Under the present dominance of formalism (i.e., foundationism), one is tempted to paraphrase Kant: the history of mathematics, lacking the guidance of philosophy has become *blind*, while the philosophy of

> mathematics turning its back on the most intriguing phenomena in the history of mathematics, has become *empty*.
>
> (Lakatos, 1976, page 2)

Thus much more should fall within the scope of the philosophy of mathematics than merely the justification of mathematical knowledge, provided through its reconstruction by a foundationist programme. Mathematics is multi-faceted, and as well as a body of propositional knowledge, it can be described in terms of its concepts, characteristics, history and practices. The philosophy of mathematics must account for this complexity, and we also need to ask the following questions. What is the purpose of mathematics? What is the role of human beings in mathematics? How does the subjective knowledge of individuals become the objective knowledge of mathematics? How has mathematical knowledge evolved? How does its history illuminate the philosophy of mathematics? What is the relationship between mathematics and the other areas of human knowledge and experience? Why have the theories of pure mathematics proved to be so powerful and useful in their applications to science and to practical problems?

These questions represent a broadening of the scope of the philosophy of mathematics from the internal concerns of absolutism. Three issues may be selected as being of particular importance, philosophically and educationally. Each of these issues is expressed in terms of a dichotomy, and the absolutist and fallibilist perspectives on the issue are contrasted. The three issues are as follows.

First of all, there is the contrast between knowledge as a finished product, largely expressed as a body of propositions, and the activity of knowing or knowledge getting. This latter is concerned with the genesis of knowledge, and with the contribution of humans to its creation. As we have seen, absolutist views focus on the former, that is finished or published knowledge, and its foundations and justification. Absolutist views not only focus on knowledge as an objective product, they often deny the philosophical legitimacy of considering the genesis of knowledge at all, and consign this to psychology and the social sciences. One partial exception to this is constructivism, which admits the knowing agent in a stylized form.

In contrast, fallibilist views of the nature of mathematics, by acknowledging the role of error in mathematics cannot escape from considering theory replacement and the growth of knowledge. Beyond this, such views must be concerned with the human contexts of knowledge creation and the historical genesis of mathematics, if they are to account adequately for mathematics, in all its fullness.

Because of the importance of the issue, it is worth adding a further and more general argument for the necessity for considering the genesis of knowledge. This argument is based on the reality of knowledge growth. As history illustrates, knowledge is perpetually in a state of change in every discipline, including mathematics. Epistemology is not accounting adequately for knowledge if it concentrates only on a single static formulation, and ignores the dynamics of knowledge growth. It is like reviewing a film on the basis of a detailed scrutiny of a single key frame! Thus epistemology must concern itself with the basis of knowing, which underpins the dynamics of knowledge growth, as well as with the specific body of knowledge accepted at any one time. Traditional philosophers such as Locke and Kant admit the legitimacy and indeed the necessity of genetic considerations in epistemology. So do an increasing number of modern philosophers, such as Dewey (1950), Wittgenstein (1953), Ryle (1949), Lakatos (1970), Toulmin (1972), Polanyi (1958), Kuhn (1970) and Hamlyn (1978).

Secondly, there is the distinction between mathematics as an isolated and discrete discipline, which is strictly demarcated and separated from other realms of knowledge, as opposed to a view of mathematics which is connected with, and indissolubly a part of the whole fabric of human knowledge. Absolutist views of mathematics accord it a unique status, it being (with logic) the only certain realm of knowledge, which uniquely rests on rigorous proof. These conditions, together with the associated internalist denial of the relevance of history or genetic or human contexts, serve to demarcate mathematics as an isolated and discrete discipline.

Fallibilists include much more within the ambit of the philosophy of mathematics. Since mathematics is seen as fallible, it cannot be categorically divorced from the empirical (and hence fallible) knowledge of the physical and other sciences. Since fallibilism attends to the genesis of mathematical knowledge as well as its product, mathematics is seen as embedded in history and in human practice. Therefore mathematics cannot be divorced from the humanities and the social sciences, or from a consideration of human culture in general. Thus from a fallibilist perspective mathematics is seen as connected with, and indissolubly a part of the whole fabric of human knowledge.

The third distinction can be seen as a specialisation and further development of the second. It distinguishes between views of mathematics as objective and value free, being concerned only with its own inner logic, in contrast with mathematics seen as an integral part of human culture, and thus as fully imbued with human values as other realms of knowledge and endeavour. Absolutist views, with their internal concerns, see mathematics as objective and absolutely free of moral and human values. The fallibilist view, on the other hand, connects mathematics with the rest of human knowledge through its historical and social origins. Hence it sees mathematics as value-laden, imbued with moral and social values which play a significant role in the development and applications of mathematics.

What has been proposed is that the proper concern of the philosophy of mathematics should include external questions as to the historical origins and social context of mathematics, in addition to the internal questions concerning knowledge, existence, and their justification. For some years there has been a parallel debate over an internalist-externalist dichotomy in the philosophy of science (Losee, 1987). As in the philosophy of mathematics there has been a split between philosophers promoting an internalist view in the philosophy of science (such as the logical empiricists and Popper) and those espousing an externalist view. The latter include many of the most influential recent philosophers of science, such as Feyerabend, Hanson, Kuhn, Lakatos, Laudan and Toulmin. The contributions of these authors to the philosophy of science is a powerful testimony to the necessity of considering 'external' questions in the philosophy of the sciences. However in the philosophy of science, even philosophers espousing an internalist position, such as Popper, admit the importance of considering the development of scientific knowledge for epistemology.

Criteria for an Adequate Philosophy of Mathematics

It has been argued that the role of the philosophy of mathematics is to account for the nature of mathematics, where this task is conceived broadly to include 'external' issues such as the history, genesis and practice of mathematics, as well as 'internal' epistemological and ontological issues, such as the justification of mathematical knowledge. These criteria can be stated more explicitly: a proposed philosophy of mathematics should account for:

(i) Mathematical knowledge: its nature, justification and genesis.
(ii) The objects of mathematics: their nature and origins.
(iii) The applications of mathematics: its effectiveness in science, technology and other realms.
(iv) Mathematical practice: the activities of mathematicians, both in the present and the past.

It is proposed, therefore, to adopt these as adequacy criteria for any proposed philosophy of mathematics. These criteria represent a reconceptualization of the role of the philosophy of mathematics. However, this role, it is argued, represents the proper task of the philosophy of mathematics, which was obscured by the mistaken identification of the philosophy of mathematics with the study of the logical foundations of mathematical knowledge.

2. A Further Examination of Philosophical Schools

The new criteria provide a means to assess the adequacy of schools of thought in the philosophy of mathematics.

A. The Absolutist Schools

In the previous chapter we saw that the logicist, formalist and intuitionist schools are absolutist. We have given an account of the failure of the programmes of these schools, and indeed refuted in general the possibility of absolutism in the philosophy of mathematics. On the basis of the above criteria we can further criticize these schools for their inadequacy as philosophies of mathematics. Their task should have included accounting for the nature of mathematics, including external social and historical factors, such as the utility of mathematics, and its genesis. Because of their narrow, exclusively internal preoccupations, these schools have made no contribution to a broadly conceived account of mathematics (with the possible exception of intuitionism, see below). Thus, not only have they failed in their self-chosen foundationist goals, but even had they succeeded they would remain inadequate philosophies of mathematics, in terms of the criteria adopted. Furthermore, this criticism is probably applicable to any absolutist philosophy with a foundationist programme.

B. *Progressive Absolutism*

Although the various forms of absolutism are grouped and criticized together, different forms of absolutism in mathematics can be distinguished. Drawing a parallel with the philosophy of science, Confrey (1981) separates formal absolutist and progressive absolutist philosophies of mathematics.[1] The formal absolutist view of mathematics is

> the epitome of certainty, immutable truths, and irrefutable methods... secure through the infallibility of its supreme method, deduction... Concepts in mathematics do not develop, they are discovered... the previous truths left unchanged by the discovery of a new truth... mathematics proceeding by an accumulation of mathematical truths and as having an inflexible, a priori structure.
>
> (Confrey, 1981, pages 246–247)

This is contrasted with the progressive absolutist view of mathematics, which whilst absolutist sees mathematics as resulting from human striving for truth, rather than its attainment. According to this view

> progress is a process of replacement of previous theories by superior theories which account for all the previous data and more. Each progressive theory approximates truth more and more precisely... progress consists of discovering mathematical truths which are not consistent with a theory or not accounted for in the theory, and then extending the theory to account for this larger realm of mathematical phenomenon.
>
> (Confrey, 1981, pages 247–248)

The key distinction is between static and dynamic absolutist conceptions of mathematical knowledge and theories, with human activity contributing the dynamic in progressive absolutism. Formalism and logicism are formal absolutist. They accept the discovery and proof of new theorems within a formal mathematical theory, building on its axioms. However, they do not treat the creation or change of mathematical theories nor informal mathematics, let alone human agency. According to such views, mathematics consists of nothing more than fixed, formal mathematical theories.

In contrast, progressive absolutist philosophies:

1 accommodate the creation and change of axiomatic theories;
2 acknowledge that more than purely formal mathematics exists, for mathematical intuition is needed as the basis for theory creation; and hence
3 acknowledge human activity and its outcomes, in the creation of new knowledge and theories.

Intuitionism (and constructivism, more generally) fit this description. For intuitionism is foundationist and absolutist, seeking secure foundations for mathematical knowledge through intuitionistic proofs and 'ur-intuition' (Kalmar, 1967). However, intuitionism (1) acknowledges human mathematical activity as fundamental in the construction of proofs or mathematical objects, the creation of new knowledge, and (2) acknowledges that the axioms of intuitionistic mathematical theory (and logic) are fundamentally incomplete, and need to be added to as more mathematical truth is revealed informally or by intuition (Brouwer, 1927; Dummett, 1977).

In consequence, intuitionism, and progressive absolutist philosophies in general, satisfy more of the adequacy criteria than formal absolutist philosophies, whilst nevertheless remaining refuted overall. For they give some place, although restricted, to the activities of mathematicians (criterion 4). They acknowledge human agency, albeit in stylized form, in the domain of informal mathematics. This partial fulfilment of the criteria deserves acknowledgment, for it means that not all absolutist philosophies are on a par. It also turns out to be significant for education.

C. *Platonism*

Platonism is the view that the objects of mathematics have a real, objective existence in some ideal realm. It originates with Plato, and can be discerned in the writings of the logicists Frege and Russell, and includes Cantor, Bernays (1934), Hardy (1967) and Godel (1964) among its distinguished supporters. Platonists maintain that the objects and structures of mathematics have a real existence independent of humanity, and that doing mathematics is the process of discovering their pre-existing relationships. According to platonism mathematical knowledge consists of descriptions of these objects and the relationships and structures connecting them.

Platonism evidently provides a solution to the problem of the objectivity of mathematics. It accounts both for its truths and the existence of its objects, as well as for the apparent autonomy of mathematics, which obeys its own inner laws and logic.

The problem of platonism, unlike that of the absolutist foundational school, is not entirely one of failure, for it offers no foundationist programme to reconstruct and safeguard mathematics. What is of more interest is to account for the fact that such an implausible philosophy provides aid and comfort to successful mathematicians of the stature of Cantor and Godel.

This interest notwithstanding, platonism suffers from two major weaknesses. First of all, it is not able to offer an adequate account of how mathematicians gain access to knowledge of the platonic realm. We can grant that platonism accounts for mathematical knowledge in the way that naive inductivist science accounted for its knowledge. That is as being based on observations of the real world (an ideal world, in the case of platonism), subsequently generalized. But if mathematics is the natural history of the crystalline platonic universe, how is it that mathematicians gain knowledge of it? It must surely be through intuition, or some such special mental faculty, and no account of this is given. If access is through intuition, then a reconciliation is needed between the facts that (i) different mathematicians' intuitions vary, in keeping with the subjectivity of intuition, and (ii) platonist intuition must be objective, and lead to agreement. Thus the platonist view is inadequate without an account of human access to the realm of platonic objects which overcomes these difficulties.

If, on the other hand, the platonist's access to the world of mathematical objects is not through intuition but through reason and logic, then further problems arise. How does the platonist know that his or her reasoning is correct? Either another form of intuition is needed, which allows the platonist to see which proofs correctly describe mathematical reality, or the platonist is in the same boat as everybody else with regard to proof. But in this second case, what is platonism but empty faith, since it provides no insight into truth or existence?

The second flaw in the platonist account is that it is not able to offer an adequate account of mathematics, neither internally nor externally. Internally, an important part of mathematics is its constructive, computational side. This depends vitally on the representation of dynamic mathematical processes, such as iteration, recursive functions, proof theory, and so on. Platonism accounts only for the static set-theoretic and structural aspects of mathematics. Thus it omits a central area of mathematics from its account. Externally, platonism fails to account adequately for the utility of mathematics, its relations with science, human activity or culture, and the genesis of knowledge. For platonists to say that mathematics advances as it is progressively uncovered, just as geography advanced with the voyages of the explorers, is not enough. Nor does it suffice to say its utility stems from the fact that mathematics describes the necessary structure of observable reality. For these explanations beg the very questions they are meant to settle.

Since it fails on all the above counts, platonism is rejected as a philosophy of mathematics.

D. Conventionalism

The conventionalist view of mathematics holds that mathematical knowledge and truth are based on linguistic conventions. In particular, that the truths of logic and mathematics are analytic, true by virtue of the meanings of the terms involved. A moderate form of conventionalism, such as that of Quine (1936) or Hempel (1945), uses linguistic convention as the source of basic mathematical truth on which the edifice of mathematics is constructed. According to this view linguistic conventions provide the basic, certain truths of mathematics and logic, and deductive logic (proofs) transmits this truth to the remainder of the body of mathematical knowledge, thus establishing its certainty. This form of conventionalism is more or less the same as 'if-thenism', discussed in Chapter 1 as a fall-back position for defeated foundationists. This view remains absolutist, and as such is subject to the same refutation.

The more interesting forms of conventionalism are not absolutist (and it is these that I shall be referring to with the term 'conventionalism'). Priest (1973) proposes to revive conventionalism, but the best known proponent of this view is Wittgenstein, who both laid the foundations of the moderate form by declaring the truth of mathematics to be tautologous (Wittgenstein, 1922), before making his extensive later contributions (Wittgenstein, 1953, 1978). Wittgenstein's later philosophy of mathematics is not clearly laid out because of his epigrammatic style, in which he eschews systematic exposition, and because most of his contributions to the philosophy of mathematics were published posthumously, in an unfinished state (Wittgenstein, 1953, 1978).

Wittgenstein criticizes the foundationist schools, and dwells at length upon knowing as a process in mathematics (Wittgenstein, 1953, 1978). In his conventionalism, Wittgenstein claims that mathematics is a 'motley', a collection of 'language games', and that the notions of truth, falsity and proof depend upon our accepting the conventional linguistic rules of these games; as the following quotations indicate.

> Of course, in one sense mathematics is a branch of knowledge — but still it is also an *activity*. And 'false moves' can only exist as the exception. For if what we now call by that name became the rule, the game in which they were false moves would have been abrogated.
>
> The word 'agreement' and the word 'rule' are related to one another, they are cousins. If I teach anyone the use of the one word, he learns the use of the other word with it.
>
> 'So you are saying that human agreement decides what is true and what is false?' — It is what human beings *say* that is true and false: and they *agree* in the language they use. That is not agreement in opinions but in form of life.
>
> (Wittgenstein, 1953, pages 227, 86 and 88, respectively)

> *What* is unshakably certain about what is proved? To accept a proposition as unshakably certain — I want to say — means to use it as a grammatical rule: this removes uncertainty from it.
>
> (Wittgenstein, 1978, page 170)

These quotations illustrate Wittgenstein's view that the uses of language (in various language games or meaning contexts) involve the acceptance of rules, which are a precondition, a *sine qua non*, for linguistic communication. The agreement he refers to is the sharing of 'a form of life', a group socio-linguistic practice based on the common following of rules, which is essential for any meaningful language use. Such agreement is not merely the voluntary assent to a practice, such as with the conventions of bridge.

Rather it is inbuilt in our communicative behaviour, which presupposes a common underlying language usage and rule following.

Thus according to Wittgenstein's conventionalist philosophy of mathematics the 'truths' of mathematics and logic depend for their acceptance on the linguistic rules of use of terms and grammar, as well on the rules governing proofs. These underlying rules confer certainty on the 'truths', for they cannot be false without breaking the rules, which would be flying in the face of accepted use. Thus it is the linguistic rules underlying the 'truths' of mathematics and logic that ensure that they cannot be falsified.

> I have not yet made the role of miscalculating clear. The role of the proposition: 'I must have miscalculated'. It is really the key to an understanding of the 'foundations' of mathematics.
>
> (Wittgenstein, 1978, p. 221)

What Wittgenstein is saying here is that if our results contradict the underlying rules of use, then we reject the results, we do not question the underlying rules.

In summary, Wittgenstein proposes that the logical necessity of mathematical (and logical) knowledge rests on linguistic conventions, embedded in our socio-linguistic practices.[2]

Conventionalism might appear, on the basis of the account given, to be absolutist, for it claims that mathematical axioms, for example, are absolutely true on the basis of linguistic conventions. But locating the foundations of mathematical knowledge in the rules governing natural language usage allows for the development of mathematical knowledge, and indeed for changes in the nature of mathematical truth and meaning, as its basis evolves. For language and its patterns of use develop all the while organically, and its sets of conventions and rules change. This is especially true of informal mathematical language, in which the rules governing the use of such terms as 'set', 'infinity', 'infinitesimal' and 'proof' have changed dramatically in the last hundred years, as a mathematical practice has developed. Likewise new conventions have warranted new truths (such as Hamilton's '$ij = -ji$' and, in logic '$1 = 2$ implies $1 = 1$'). Thus conventionalism is not absolutist, for it allows for the dethronement and replacement of basic mathematical truths (such as '$xy = yx$'). This form of conventionalism is therefore consistent with fallibilism.

The conventionalist philosophy of mathematics has been criticized by previous authors on two grounds. First of all, it is claimed to be uninformative: 'apart from pointing out the essentially social nature of mathematics, conventionalism tells us remarkably little'. (Machover, 1983, page 6). The force of this criticism is that to be an adequate philosophy of mathematics, a much more elaborated version of conventionalism is needed.

The second objection is due to Quine.

> Briefly the point is that the logical truths, being infinite in number, must be given by general conventions rather than singly; and logic is needed then to

> begin with, in the metatheory, in order to apply the general conventions to individual cases'.
>
> (Quine, 1966, page 108)

Thus according to Quine, our linguistic conventions must either include the infinite number of truths of the form '(Sentence 1) and (Sentence 2) implies (Sentence 2)', or this single, general convention, in which case we need logic in the metalanguage to derive all its instances.

But notice that the same objection applies to the possibility of grammatical conventions in language. We either need to know of the infinite number of grammatical instances of the form '(Subject) is a (Predicate)', or we would need meta-linguistic rules of substitution to derive its instances from the general grammatical convention. But we evidently do not need such additional rules to speak, because the very scheme is a 'production rule'. The sole function of such a rule in a natural language is to generate instances. Likewise, logical schemes are rules which guide the production of logical truths. Thus it is not the case that we need to presuppose logic in a meta-language to derive instances from our logical scheme. It is inappropriate to seek all the forms and distinctions of formal languages in natural languages, which, for example, already differ in being their own meta-languages.

In fact, truths of the form 'A&B implies B' are not likely to depend on the above sentential scheme, but on the rules governing the use of the word 'and'. These rules are likely to be semantic rules linking 'and' with 'combine', 'join', and 'put together', that is with the conjunctive meaning of 'and'. These semantic rules imply that the consequences of 'A&B' are the consequnces of 'A' combined with those of 'B'.

Quine's objection is therefore dismissed in that it does not apply to natural languages, and imposes an overly restrictive role on general conventions. On the other hand he is right to say that we will not find all the truths of mathematics and logic represented literally as linguistic rules and conventions.

Although Quine is critical of conventionalism in logic, he regards its potential as a philosophy of mathematics quite differently.

> For set theory the linguistic doctrine has seemed less empty; in set theory, moreover, convention in quite the ordinary sense seems to be pretty much what goes on. Conventionalism has a serious claim to attention in the philosophy of mathematics, if only because of set theory.
>
> (Quine, 1966, page 108)

Conventionalism offers the beginnings of a descriptive account of the nature of mathematics, formulated in terms of its linguistic basis. It accommodates a fallibilist view of mathematics, and may account for both the objectivity of mathematical knowledge, through our necessary acceptance of linguistic rules, and for at least part of its genesis, via the acquisition of language. Since language connects mathematics with other areas of knowledge, conventionalism has the potential to account for the applications of mathematics. Thus conventionalism is not refuted, and indeed may satisfy many of the adequacy criteria proposed earlier. Conventionalism is discussed further in the next chapter, as one of several contributors to a proposed social constructivist philosophy of mathematics.

E. Empiricism

The empiricist view of the nature of mathematics ('naive empiricism', to distinguish it from Lakatos' quasi-empiricism) holds that the truths of mathematics are empirical generalisations. We can distinguish two empiricist theses: (i) the concepts of mathematics have empirical origins, and (ii) the truths of mathematics have empirical justification, that is, are derived from observations of the physical world. The first thesis is unobjectionable, and is accepted by most philosophers of mathematics (given that many concepts are not directly formed from observations but are defined in terms of other concepts which lead, via definitional chains, to observational concepts). The second thesis is rejected by all but empiricists, since it leads to some absurdities. The initial objection is that most mathematical knowledge is accepted on theoretical, as opposed to empirical grounds. Thus I know that 999,999 + 1 = 1,000,000 not through having observed its truth in the world, but through my theoretical knowledge of number and numeration.

Mill (1961) partly anticipates this objection, suggesting that the principles and axioms of mathematics are induced from observations of the world, and that other truths are derived from these by deduction. However, empiricism is open to a number of further criticisms.

First of all, when our experience contradicts elementary mathematical truths, we do not give them up (Davis and Hersh, 1980). Rather we assume that some error has crept in to our reasoning, because there is shared agreement about mathematics, which precludes the rejection of mathematical truths (Wittgenstein, 1978). Thus, '1 + 1 = 3' is necessarily false, not because one rabbit added to another does not give three rabbits, but because by definition '1 + 1' means 'the successor of 1' and '2' is the successor of '1'.

Secondly, mathematics is largely abstract, and so many of its concepts do not have their origins in observations of the world. Rather they are based on previously formed concepts. Truths about such concepts, which form the bulk of mathematics, cannot therefore be said to be induced from observations of the external world.

Finally, empiricism can be criticized for focusing almost exclusively on foundationist issues, and failing to account adequately for the nature of mathematics. This, as has been argued above, is the major purpose of the philosophy of mathematics. On the basis of this criticism we can reject the naive empiricist view of mathematics as inadequate.

3. Quasi-empiricism

Quasi-empiricism is the name given to the philosophy of mathematics developed by Imre Lakatos (1976, 1978). It is the view that mathematics is what mathematicians do and have done, with all the imperfections inherent in any human activity or creation. Quasi-empiricism represents a 'new direction in the philosophy of mathematics' (Tymoczko, 1986), because of the primacy it accords to mathematical practice. Supporters of this view include Davis (1972), Hallett (1979), Hersh (1979), Tymoczko (1979) and at least in part, Putnam (1975). A preliminary sketch of the quasi-empiricist view of mathematics is as follows.

Mathematics is a dialogue between people tackling mathematical problems. Mathematicians are fallible and their products, including concepts and proofs, can never be considered final or perfect, but may require renegotiation as standards of rigour change, or as new challenges or meanings emerge. As a human activity, mathematics cannot be viewed in isolation from its history and its applications in the sciences and elsewhere. Quasi-empiricism represents 'a renaissance of empiricism in the recent philosophy of mathematics' (Lakatos, 1967).

A. Exposition of Lakatos' Quasi-empiricism

Five theses of quasi-empiricism can be identified, as follows.

1. Mathematical knowledge is fallible

In quasi-empiricism, the search for a basis for absolute certainty in mathematics is rejected, and mathematical knowledge is acknowledged to be fallible, corrigible, and without certain foundations. (See Lakatos quotations in Chapter 1.)

2. Mathematics is hypothetico-deductive

Mathematics is acknowledged to be a hypothetico-deductive system, like the widely accepted conception of empirical science due to Popper (1959). As in science, the emphasis in such a system is not on the transmission of truth from true premises to conclusions (the absolutist view), but on the re-transmission of falsity from falsified conclusions ('falsifiers') to hypothetical premises. Since axiomatic theories are formalizations of previously existing informal mathematical theories, their potential falsifiers are the informal theorems of the pre-existing theory, (in addition to formal contradictions). The existence of such a (informal theorem) falsifier shows that the axiomatization has not validly expressed the informal theory, i.e. its source (Lakatos, 1978).

3. History is central

The epistemological task of the philosophy of mathematics is not simply to answer the question 'how is (any) mathematical knowledge possible?', but to account for the actual mathematical knowledge that exists. Thus, the philosophy of mathematics is indissolubly linked with the history of mathematics, since the latter is the history of the evolution of mathematical knowledge.

4. The primacy of informal mathematics is asserted

Informal mathematics is of paramount importance, both as a practice and as a product. As a product, it is the source of all formal mathematics, since it is what is formalized. It is also, as we have seen, the source of potential falsifiers of formal mathematics. The importance of mathematical practice is that it is the 'stuff' of the history of mathematics, and the quasi-empirical source of mathematics. It provides individuals with the premises and conclusions of deductive mathematics (informal axioms, definitions and conjectures), and the informal proofs through which the premises and conclusions are connected.

5. A theory of knowledge creation is included

A central concern of the philosophy of mathematics is the logic of mathematical discovery, or 'heuristic'. This underlies 'the autonomous dialectic of mathematics' (Lakatos, 1976, page 146), the mechanism for the genesis of mathematical knowledge. In this process the productions of individual mathematicians (usually a constellation of definitions, conjectures and informal proofs) are exposed to criticism, and reformulated in response to the criticism, in an iterated dialectical cycle. This process, following its own autonomous logic, is necessary for new items (definitions, theorems, proofs) to become incorporated into the body of accepted mathematical knowledge.

> There is a simple pattern of mathematical discovery — or of the growth of informal mathematical theories. It consists of the following stages:
>
> 1 Primitive conjecture.
> 2 Proof (a rough thought-experiment or argument, decomposing the primitive conjecture into subconjectures or lemmas).
> 3 'Global' counterexamples (counterexamples to the primitive conjecture) emerge.
> 4 Proof re-examined: the 'guilty lemma'to which the global counterexample is a 'local' counterexample is spotted. This guilty lemma may have previously remained 'hidden' or may have been misidentified. Now it is made explicit, and built into the primitive conjecture as a condition. The theorem — the improved conjecture — supersedes the primitive conjecture with the new proof-generated concept as its paramount new feature.
>
> These four stages constitute the essential kernel of proof analysis. But there are some further standard stages which frequently occur:

5 Proofs of other theorems are examined to see if the newly found lemma or the new proof-generated concept occurs in them: this concept may be found lying at cross-roads of different proofs, and thus emerge as of basic importance.
6 The hitherto accepted consequences of the original and now refuted conjecture are checked.
7 Counterexamples are turned into new examples — new fields of inquiry open up.

(Lakatos, 1976, pages 127–128)

> Mathematical activity is human activity. Certain aspects of this activity — as of any human activity — can be studied by psychology, others by history. Heuristic is not primarily interested in these aspects. But mathematical activity produces mathematics. Mathematics, this product of human activity, 'alienates itself' from the human activity which has been producing it. It becomes a living, growing organism, that acquires a certain autonomy from the activity which has produced it; it develops its own autonomous laws of growth, its own dialectic. The genuine creative mathematician is just a personification, an incarnation of these laws which can only realise themselves in human action. Their incarnation, however, is rarely perfect. The activity of human mathematicians as it appears in history, is only a fumbling realisation of the wonderful dialectic of mathematical ideas. But any mathematician, if he has talent, spark, genius, communicates with, feels the sweep of, and obeys this dialectic of ideas.
>
> (Lakatos, 1976, page 146).

It can be seen that at the heart of Lakatos' philosophy of mathematics is a theory of the genesis of mathematical knowledge. This is a theory of mathematical practice, and hence a theory of the history of mathematics. Lakatos is not offering a psychological theory of mathematical creation or discovery, for he does not deal with the origins of the axioms, definitions and conjectures in the mind of the individual. His focus is instead on the process which transforms private creations into accepted public mathematical knowledge, a process which centrally involves criticism and reformulation. In this, his philosophy closely resembles the falsificationist philosophy of science of Karl Popper, a debt that Lakatos readily acknowledges. For Popper (1959) proposes a 'logic of scientific discovery', in which he argues that science advances through a process of conjectures and refutations. The difference is that Popper concerns himself only with rational reconstructions or idealizations of theories, and denies the philosophical validity of the application of his model of science to history. Lakatos on the other hand, refuses to separate the philosophical theory of knowledge growth from its historical realization.

Despite his avoidance of the pitfalls of psychologism, Lakatos may be accused of straying outside the boundaries of legitimate philosophical concern. For in contrast with most epistemology in the English-speaking world, which focus exclusively on objective knowledge or a single knowing subject, quasi-empiricism discusses knowing or knowledge generation as part of a social process. However, in the philosophy of mathematics, as we have seen, there is a dearth of theories offering an adequate account of mathematics. Thus the traditional limitations on what counts as legitimate in philosophy may in fact be an obstacle to an adequate philosophy of mathematics.

We thus turn to an evaluation of Lakatos' quasi-empiricist philosophy of mathematics. However it must be remarked that Lakatos' philosophy of mathematics is far from a complete or fully worked out system. This is due to two factors. First of all, his untimely death. Lakatos' wrote only one slim volume and five papers on the philosophy of mathematics, and a number of these were unfinished and published posthumously. Secondly, his style of presentation in his major work was indirect, utilizing a platonic dialogue to reconstruct an aspect of the history of mathematics. Thus Lakatos' has bequeathed us an exciting but incomplete philosophy of mathematics, far from fully worked or elaborated. Thus the potential, as well as its realization, must be born in mind in assessing his quasi-empiricism.

B. The Adequacy Criterion and Quasi-empiricism

To evaluate the quasi-empiricist philosophy we consider it in the light of the adequacy criteria.

Quasi-empiricism offers a partial account of the nature of mathematical knowledge, and its genesis and justification. In this Lakatos offers a more extensive account than the other philosophies of mathematics we considered, far exceeding them in scope. To the traditional stratum of formalized mathematical knowledge he adds a new, lower stratum, namely that of informal mathematical knowledge. To this extended system he adds a dynamic, which shows not only how knowledge in the lower stratum develops, but also the relationship between the two strata. In particular, how knowledge in the lower stratum is reflected upwards, by formalization, to form idealized images at the upper level, which are seen as the indubitable truths of mathematics. Lakatos accounts for the nature of mathematical knowledge as hypothetico-deductive and quasi-empirical, building a striking analogy with Popper's (1979) philosophy of science. He accounts for errors in mathematical knowledge, and provides an elaborate theory of the genesis of mathematical knowledge. This potentially accounts for much of mathematical practice, and for its history.

Since Lakatos' theory of the genesis of mathematical knowledge puts it on a par, in many respects, with scientific knowledge, the success of the applications of mathematics are potentially explicable by analogy with science and technology. Explaining the success of applied mathematics would be a significant strength, especially in the face of the neglect shown by other philosophies of mathematics (Korner, 1960). Finally, a key strength of Lakatos' philosophy of mathematics is that it is not prescriptive, but descriptive, and attempts to describe mathematics as it is and not as it ought to be practised.

In terms of the four adequacy criteria, quasi-empiricism partly satisfies those concerning mathematical knowledge (i), applications (iii) and practice (iv).

C. Weaknesses of Lakatos' Quasi-empiricism

Quasi-empiricism can be criticized on a number of counts. First of all, there is no account of mathematical certainty. Lakatos fails to explain why mathematical truth is seen to be the most certain of all knowledge. Likewise, he fails to account for the seeming certainty of deductive logic. Thus Lakatos fails to account for the apparent certainty of mathematical knowledge, and the weight that is attached to its justification.

Second, Lakatos gives no account of the nature of the objects of mathematics, or of their origins. There is no indication in his account of the plausibility of platonism. He does not even offer grounds for the objectivity of heuristic, the 'autonomous laws of growth' of mathematical knowledge, assumed in his account.

Third, Lakatos does not account for either the nature or the success of the applications of mathematics, or its effectiveness in science, technology, and in other realms. It may be argued, as I have done, that there is a potential justification of the utility of mathematics in quasi-empiricism. The fact remains it is neither hinted at nor given. Furthermore, Lakatos deals and refers exclusively to pure mathematics, so an account of the nature of applied mathematics is missing from his philosophy.

Fourth, Lakatos does not sufficiently establish the legitimacy of bringing the history of mathematics into the heart of his philosophy of mathematics. In this, he is flying in the face of a philosophical tradition (although, as we have seen, an increasing number of philosophers bring genetic issues into epistemology). His approach is potentially fruitful, perhaps more so than previous philosophies of mathematics. Nevertheless his historical approach needs explicit philosophical justification (beyond the implicit justification that he is trying to account for mathematics descriptively).

Fifth, there is the problem of the contrasting status of his philosophy and the historical thesis. Lakatos fails to provide justification for introducing an empirical (i.e. conjectural) historical thesis into an analytic philosophical approach, on an equal footing with logical methodology. This, I believe, is due to his failure to distinguish the general dialectical logic of mathematical discovery from a specific empirical thesis concerning the stages of development and elaboration of mathematical knowledge. The general logic of mathematical discovery, like Popper's (1968) logic of scientific discovery, is hypothetico-deductive. It involves the general dialectical method of conjectures (including proofs) and refutations. It is a purely logical methodology, describing the general form and conditions for the improvement, and ultimately, the acceptance of mathematical knowledge.

In contrast, Lakatos' empirical hypothesis concerns the actual historical stages through which mathematical knowledge passes during its development from initial conjecture to accepted mathematical knowledge (the seven stage heuristic quoted above). This is an empirical conjecture, and does not constitute a necessary condition for the validity of the general quasi-empiricist philosophy of mathematics. Thus the details of this specific empirical thesis may be modified, without calling into question the underlying (and general) philosophy.[3] In fact the philosophy and the historical thesis are logically independent, in that the rejection of one has no logical implications for the other. Lakatos seems to be unaware of this distinction.

Sixth, Lakatos' quasi-empiricist philosophy of mathematics provides grounds that are necessary but not sufficient for establishing mathematical knowledge. Examples can be found of mathematical knowledge that after development and reformulation, following the general pattern of Lakatos' heuristic, is still not incorporated into the body of accepted mathematical knowledge. Consider, as a fictional counterexample, the idiosyncratic mathematics that may be developed by a group of mystics, sharing a set of conventions and norms, including the basis of their critical methodology, which is peculiar to themselves. The fact that this group's mathematical creations survive their process of proofs and refutations does not give them general acceptance.

To rule out such examples, quasi-empiricism requires the assumption of a shared basis for its critical methodology, if there is to be universal agreement on its outcomes. In effect, this is the assumption of the use of a standard logic, and of its validity.

Lastly, there is no systematic exposition of quasi-empiricism, putting forward its theses in detail, and anticipating and rebutting objections to it. Lakatos' publications on the philosophy of mathematics comprise reconstructed historical case studies and polemical writings.

Overall, it can be seen that the major defects of quasi-empiricism are sins of omission, rather than of commission. The above critique, admittedly from a sympathetic viewpoint, discloses no fundamental flaw or defect. Rather it indicates a needed research programme, namely to develop quasi-empiricism systematically and fill the gaps.

D. Quasi-empiricism and the Philosophy of Mathematics

Quasi-empiricism offers a partial account of the nature of mathematical knowledge, and its genesis and justification. In this Lakatos offers potentially a far more extensive account than the other philosophies of mathematics which have been considered. In large part, this depends on Lakatos' reconceptualization and redefinition of the philosophy of mathematics. He has called into question the dominant orthodoxy in the philosophy of mathematics with regard to both its foundationism and absolutism (the arguments in the previous chapter are unlikely to have been formulated, but for Lakatos). Thus he has freed the philosophy of mathematics to reconsider its function in the terms described in this chapter, as well as to question the hitherto unchallenged status of mathematical truth. Although these are critical (i.e. essentially negative) achievements of quasi-empiricism their importance cannot be overestimated.

In positive terms quasi-empiricism has the potential to offer solutions to many of the new problems Lakatos posed for the philosophy of mathematics. In the next chapter I shall attempt to build a novel social constructivist philosophy of mathematics on the basis of Lakatos' quasi-empiricism.

Notes

1 'Formal absolutism' is my term, in place of Confrey's 'absolutism', to avoid ambiguity. Confrey also has a third category of 'conceptual change' theories, corresponding to fallibilism and drawing heavily on Lakatos (1976).

2 Wittgenstein's conventionalism is more radical than the account given in the chapter, although the selective account presented there can be considered independently of some of his other views. Dummett (1959) terms it a 'full blooded conventionalism', because all of the 'truths' of mathematics and logic, not just the axioms, are direct expressions of linguistic conventions. Consequently, Wittgenstein denies that mathematics has any logical foundations, it rests instead on the rules of actual practice, both linguistic and mathematical. He claims that every time we accept a new theorem, we accept a new rule of language. He also holds a very strict constructivist view of mathematics known as 'strict finitism', and thus rejects even more of classical mathematics than the intuitionists. Dummett (1959) criticizes some of these views.

3 Considered as an empirical hypothesis about the development of mathematics, Lakatos' thesis has its limitations. It is based on a single case study from nineteenth century mathematics, the Euler conjecture (perhaps two, counting uniform continuity). In other domains, such as number theory, conjectures concerning well defined concepts (not needing redefinition, 'monster barring', etc) may be proposed in final form, simply needing proofs (for example, Ramanujan's conjectures proved by Hardy and Littlewood). In yet further domains, such as axiomatic set theory, it may not be possible to distinguish sharply between substance (concepts) and form (proofs), as Lakatos does in his case studies. See my review of Lakatos (1976) in *Mathematical Reviews* for further elaboration of this point.

3

Social Constructivism as a Philosophy of Mathematics

1. Social Constructivism[1]

In this chapter I propose a new philosophy of mathematics called 'social constructivism'. Naturally, as it concerns a novel philosophy of mathematics, this chapter is more tentative than those preceding it, which were largely concerned with the exposition of well-established ideas. On the other hand, not too much novelty should be claimed, since social constructivism is largely an elaboration and synthesis of pre-existing views of mathematics, notably those of conventionalism and quasi-empiricism.[2]

Social constructivism views mathematics as a social construction. It draws on conventionalism, in accepting that human language, rules and agreement play a key role in establishing and justifying the truths of mathematics. It takes from quasi-empiricism its fallibilist epistemology, including the view that mathematical knowledge and concepts develop and change. It also adopts Lakatos' philosophical thesis that mathematical knowledge grows through conjectures and refutations, utilizing a logic of mathematical discovery. Social constructivism is a *descriptive* as opposed to a *prescriptive* philosophy of mathematics, aiming to account for the nature of mathematics understood broadly, as in the adequacy criteria.

The grounds for describing mathematical knowledge as a social construction and for adopting this name are threefold:

(i) The basis of mathematical knowledge is linguistic knowledge, conventions and rules, and language is a social construction.

(ii) Interpersonal social processes are required to turn an individual's subjective mathematical knowledge, after publication, into accepted objective mathematical knowledge.

(iii) Objectivity itself will be understood to be social.

A. Overview of Social Constructivism

Like quasi-empiricism, a central focus of social constructivism is the genesis of mathematical knowledge, rather than just its justification. Newly generated mathematical knowledge can be either subjective or objective knowledge, and a unique feature of social constructivism is that it considers both these forms of knowledge, and links them in a creative cycle. It is not uncommon to see subjective knowledge and objective knowledge treated together in philosophy, as in Popper (1979). What is less common is for their links to be treated, since this admits the genesis of knowledge into philosophy.

Social constructivism links subjective and objective knowledge in a cycle in which each contributes to the renewal of the other. In this cycle, the path followed by new mathematical knowledge is from subjective knowledge (the personal creation of an individual), via publication to objective knowledge (by intersubjective scrutiny, reformulation and acceptance). Objective knowledge is internalized and reconstructed by individuals, during the learning of mathematics, to become the individuals' subjective knowledge. Using this knowledge, individuals create and publish new mathematical knowledge, thereby completing the cycle.[3] Thus subjective and objective knowledge of mathematics each contributes to the creation and re-creation of the other. The assumptions underpinning the social constructivist account of knowledge creation are as follows.

1. An individual possesses subjective knowledge of mathematics

A key distinction is that between subjective and objective knowledge. The mathematical thought of an individual (both the process and its product, mathematical knowledge) is subjective thought. This is largely learned (i.e. reconstructed objective) knowledge, but, subject to certain powerful constraints, the process of re-creation results in unique subjective representations of mathematical knowledge. Furthermore, individuals use this knowledge to construct their own, unique mathematical productions, the creation of new subjective mathematical knowledge.

2. Publication is necessary (but not sufficient) for subjective knowledge to become objective mathematical knowledge

When an individual's subjective mathematical knowledge production enters the public domain through publication, it is eligible to become objective knowledge. This will depend on its acceptance, but first it must be physically represented (in print, electronically, in writing, or as the spoken word). (Here knowledge is understood to include not only statements, but also their justification, typically in the form of informal proofs).

3. Through Lakatos' heuristic published knowledge becomes objective knowledge of mathematics

Published mathematics is subject to scrutiny and criticism by others, following Lakatos' (1976) heuristic, which may result in its reformulation and acceptance as objective (i.e., socially accepted) knowledge of mathematics. The successful application of this heuristic is sufficient for acceptance as (tentative) objective mathematical knowledge, although the knowledge always remains open to challenge.

4. This heuristic depends on objective criteria

During the genesis of mathematical knowledge, objective criteria play an essential part (Lakatos' autonomous logic of mathematical discovery, understood philosophically, not historically). These criteria are used in the critical scrutiny of mathematical knowledge, and include shared ideas of valid inference and other basic methodological assumptions.

5. The objective criteria for criticizing published mathematical knowledge are based on objective knowledge of language, as well as mathematics

The criteria depend to a large extent on shared mathematical knowledge, but ultimately they rest on common knowledge of language, that is, on linguistic conventions (the conventionalist view of the basis of knowledge). These too are socially accepted, and hence objective. Thus both published mathematical knowledge and the lingusitic conventions on which its justification rests are objective knowledge.

6. Subjective knowledge of mathematics is largely internalized, reconstructed objective knowledge

A key stage in the cycle of mathematical creation is the internalization, that is the inner subjective representation, of objective mathematical and linguistic knowledge. Through the learning of language and mathematics inner representations of this knowledge, including the corresponding rules, constraints and criteria are constructed. These permit both subjective mathematical creation, and participation in the process of criticizing and reformulating proposed (i.e., public) mathematical knowledge.

7. Individual contributions can add to, restructure or reproduce mathematical knowledge

On the basis of their subjective knowledge of mathematics individuals make potential contributions to the pool of objective knowledge. These can add to, restructure, or simply reproduce existing knowledge of mathematics (subject to the heuristic).

Additions can be new conjectures or proofs, which may include new concepts or definitions. They can also be new applications of existing mathematics. Restructuring contributions may be new concepts or theorems that generalize or otherwise link two or more previously existing parts of mathematical knowledge. Contributions that reproduce existing mathematics are typically textbooks or advanced expositions.

B. *Immediate Problems of Social Constructivism*

Two immediate problem areas arise from this brief account. First of all, there is the identification of objectivity with the social or socially accepted. To identify the immutable and enduring objectivity of the objects and truths of mathematics with something as mutable and arbitrary as socially accepted knowledge does, initially, seem problematic. However we have already established that all mathematical knowledge is fallible and mutable. Thus many of the traditional attributes of objectivity, such as its enduring and immutable nature, are already dismissed. With them go many of the traditional arguments for objectivity as a super-human ideal. Following Bloor (1984) we shall adopt a necessary condition for objectivity, social acceptance, to be its sufficient condition as well. It remains to show that this identification preserves the properties that we expect of objectivity.

Secondly, there is the problem of the proximity of social constructivism to sociological or other empirical accounts of mathematics. Since it is quasi-empirical, and has the task of accounting for the nature of mathematics including mathematical practice, in a fully descriptive fashion, the boundary between mathematics and other disciplines is weakened. By removing traditional philosophical barriers these consequences bring the philosophy of mathematics closer to the history and sociology of mathematics (and psychology too, concerning subjective knowledge). Thus there is the danger of social constructivism straying into the provinces of history, sociology or psychology. We saw that Lakatos (1976) conflates his theory of the historical evolution of mathematical knowledge with his philosophical account of the genesis of mathematical knowledge. Thus there is a real danger of conflating empirical with philosophical accounts of mathematics, which social constructivism must avoid.

2. Objective and Subjective Knowledge

A. *The Nature of Objective and Subjective Knowledge*

Before proceeding further with the exposition and development of social constructivism it is necessary to establish some philosophical preliminaries. A key distinction that is employed is that between subjective and objective knowledge. This is clarified by a consideration of Popper's (1979) definition of three distinct worlds, and the associated types of knowledge.

> We can call the physical world 'world 1', the world of our conscious experiences 'world 2', and the world of the logical *contents* of books, libraries, computer memories, and suchlike 'world 3'.
>
> (Popper, 1979, p. 74)

Subjective knowledge is world 2 knowledge, objective knowledge is world 3 knowledge, and according to Popper includes products of the human mind, such as published theories, discussions of such theories, related problems, proofs; and is human-made and changing.

I shall use the term 'objective knowledge', in a way that differs from Popper, to refer to all knowledge that is intersubjective and social. I wish to count all that Popper does as objective knowlege, including mathematical theories, axioms, conjectures, proofs, both formal and informal. One difference is that I also want to include additional 'products of the human mind' as objective knowledge, notably the shared (but possible implicit) conventions and rules of language usage. Thus I am referring to publicly shared, intersubjective knowledge as objective, even if it is implicit knowledge, which has not been fully articulated. This extension is very likely unacceptable to Popper.

In fact, I wish to adopt the social theory of objectivity proposed by Bloor.

> Here is the theory: it is that objectivity is social. What I mean by saying that objectivity is social is that the impersonal and stable character that attaches to some of our beliefs, and the sense of reality that attaches to their reference, derives from these beliefs being social institutions.
>
> I am taking it that a belief that is objective is one that does not belong to any individual. It does not fluctuate like a subjective state or a personal preference. It is not mine or yours, but can be shared. It has an external thing-like aspect to it.
>
> (Bloor, 1984, page 229).

Bloor argues that Popper's world 3 can defensibly and fruitfully be identified with the social world. He also argues that not only is the three-fold structure of Popper's theory preserved under this transformation, but so are the connections between the three worlds. Naturally, the social interpretation does not preserve the meaning that Popper attaches to objectivity, who regards the logical character of theories, proofs and arguments sufficient to guarantee objectivity in an idealistic sense. Despite this, the social view is able to account for most, if not all, features of objectivity: the autonomy of objective knowledge, its external thing-like character (presumably the original meaning of 'object'-ivity), and its independence from any knowing subject's subjective knowledge. For the social view sees objective knowledge, like culture, developing autonomously in keeping with tacitly accepted rules, and not subject to the arbitrary dictates of individuals. Since objective knowledge and rules exist outside individuals (in the community), they seem to have an object-like and independent existence.

Thus it can be seen that the social view accounts for many of the necessary characteristics of objectivity. Beyond this, it is worth remarking that Bloor's social view of objectivity explains and accounts for objectivity. In contrast traditional views (including Popper's) elaborate on, or at best *define* objectivity (intensively or extensively), but never account for or *explain* objectivity. For the autonomous, independent existence of objective knowledge is traditionally shown to be necessary, without any explanation of what objectivity is, or how objective knowledge can emerge from subjective human knowledge. In contrast, the social view of objectivity is able to offer an account of the basis and nature of objectivity and objective knowledge.

One immediate problem the social view must face is that of accounting for the necessity of logical and mathematical truth. The answer given by Bloor (1983, 1984), and adopted here, is that this necessity (understood in a fallibilist sense) rests on linguistic conventions and rules, as Wittgenstein proposes. This is the full conventionalist account of the basis of logic and mathematical knowledge.

B. The Role of Objective Knowledge in Mathematics

Having clarified the sense in which objectivity is understood to be social, it is worth reiterating the social constructivist account of objective mathematical knowledge. According to social constructivism, published mathematics, that is mathematics that is represented symbolically in the public domain, has the potential to become objective knowledge. The application of Lakatos' logic of mathematical discovery to this published mathematics is the process that leads to social acceptance, and thus to objectivity. Once mathematical axioms, theories, conjectures, and proofs are formulated and presented publicly, even if only in conversation, the autonomous (i.e., socially accepted) heuristic begins to work. Both the process and its product are objective, being socially accepted. Likewise, both the implicit and explicit conventions and rules of language and logic on which this heuristic rests are objective, also being socially accepted. It is these conventions and rules which it is claimed, following conventionalism, underpin mathematical knowledge (including logic). For they provide the basis of logical and mathematical definitions, as well as the basis for the rules and axioms of logic and mathematics.

C. The Role of Subjective Knowledge in Mathematics

Given the centrality of the role of objective knowledge, I wish to argue that the role of subjective mathematical knowledge must also be acknowledged, or else the overall account of mathematics will be incomplete. For subjective knowledge is needed to account for the origins of new mathematical knowledge, as well as, according to the theory proposed, the re-creation and sustainment of existing knowledge. Since objective knowledge is social, and not a self-subsistent entity existing in some ideal realm, then like all aspects of culture this knowledge must be reproduced and transmitted from generation to generation (admittedly with the aid of artefacts, such as text books). According to the social constructivist account, subjective knowledge is what sustains and renews objective knowledge, whether it be of mathematics, logic or language. Thus subjective knowledge plays a central part in the proposed philosophy of mathematics.

Having said this, it must be acknowledged that the treatment of subjective as well as objective knowledge, in the proposed theory, is at odds with much modern thought in philosophy, and in the philosophy of mathematics, as we have seen (barring intuitionism, which we have rejected). For example, Popper (1959) has distinguished very carefully between the 'context of discovery' and the 'context of justification' in science. He regards the latter context as being subject to logical analysis, and thus being the proper concern of philosophy. The former context, however, concerns empirical matters, and therefore is the proper concern of psychology, and not of logic or philosophy.

Anti-psychologism, the view that subjective knowledge — or at least its psychological aspects — is unfitted for philosophical treatment, rests on the following argument. Philosophy consists of logical analysis, including methodological problems such as the general conditions for the possibility of knowledge. Such inquiry is *a priori*, and is wholly independent of any particular empirical knowledge. Subjective issues are of necessity psychological issues, since they of necessity refer to the contents of individual minds. But such matters, and psychology in general, are empirical. Therefore, because of this category difference (the *a priori* versus the empirical realm) subjective knowledge cannot be the concern of philosophy.[4]

This argument is rejected here on two grounds. First of all, a powerful critique of absolutism, and hence of the possibility of certain *a priori* knowledge has been mounted (Chapter 1). On this basis all so-called *a priori* knowledge, including logic and mathematics, depends for its justification on quasi-empirical grounds. But this effectively destroys the unique categorical distinction between *a priori* knowledge and empirical knowledge. Thus this distinction cannot be used to deny the applicability of the *a priori* philosophical methods of objective knowledge to subjective knowledge, on the grounds that the latter is empirically tainted. For now we see that all knowledge, including objective knowledge, is empirically (or rather quasi-empirically) tainted.

The second argument, which is independent of the first, is as follows. In discussing subjective knowledge it is not proposed to discuss the specific contents of individuals' minds, nor specific empirical psychological theories of the mind under the guise of philosophy. Rather the intention is to discuss the possibility of subjective knowledge in general, and what can be concluded about its possible nature on the basis of logical reasoning alone (given a number of theoretical assumptions). This is a legitimate philosophical activity, just as the philosophy of science can legitimately reflect on an empirical realm, namely science, without therefore being an empirical realm itself. Thus subjective knowledge is a proper matter for philosophical inquiry. Indeed, in discussing belief or the knowing subject, this is precisely what epistemologists such as Sheffler (1965), Woozley (1949), Chisholm (1966) and indeed Popper (1979), are considering. Further back, epistemology has traditionally considered subjective knowledge, at least from the time of Descartes (and probably further back to Plato), through the British empiricists Locke, Berkeley and Hume, via Kant to the present day. Thus subjective knowledge is a legitimate area of philosophical enquiry, based on a substantial philosophical tradition.

Although the claim that the consideration of subjective knowledge is psychologistic is thus refuted, it is acknowledged that there are real dangers and legitimate concerns arising from the philosophical treatment of subjective knowledge. For it makes it easier to commit the error of using psychologistic reasoning in philosophy, that is reasoning based on psychological belief of necessity as opposed to logical argument. Furthermore, the distinction between subjective and objective knowledge is a vital one to maintain, both for social constructivism, and for philosophy in general. These are two genuinely distinct domains of knowledge.

For these reasons, in the explication of the social constructivist philosophy of mathematics, the domains of objective and subjective knowledge will be treated separately. The objective aspect of this philosophy is independent of the subjective aspect in terms of its justification. So the reader wary of psychologism can follow the objective aspect of social constructivism without qualms (concerning this issue, at least).

3. Social Constructivism: Objective Knowledge

In providing a social constructivist account of objective knowledge in mathematics we need to establish a number of claims. We need to justify the account of objective mathematical knowledge by demonstrating both the objectivity of what is referred to, as well as the fact that it is indeed warranted knowledge. Having established these minimal conditions for an account of objective mathematical knowledge, we next need to establish that social constructivism provides an adequate philosophical account of mathematics. This involves satisfying the adequacy criteria for the philosophy of mathematics formulated in the last chapter.

A. Objectivity in Mathematics

In Chapter 1, on the basis of a powerful critique of absolutism, we accepted the fallibility of mathematical knowledge. Whilst this is a central assumption of social constructivism, the fact remains that the objectivity of both mathematical knowledge and the objects of mathematics is a widely accepted feature of mathematics, and must be accounted for by any philosophy of mathematics. It has been established that objectivity is understood to reside in public, intersubjective agreement, that is, it is social. Thus the objectivity of mathematics means that both the knowledge and objects of mathematics have an autonomous existence upon which there is intersubjective agreement, and which is independent of the subjective knowledge of any individual. We therefore need to establish the shared basis for this knowledge, which allows public access to it, and guarantees intersubjective agreement on it. Subsequently, we extend the discussion to the objectivity of the ontology of mathematics, that is the basis for the autonomous existence of the objects of mathematics. We consider first the substratum which provides the foundation for objectivity in mathematics, namely language.

The linguistic basis of objectivity in mathematics

The claim is that the objectivity of mathematical knowledge is based on shared knowledge of natural language. Such an account has already been sketched in the treatment of conventionalism in the previous chapter. There Wittgenstein's account of the lingusitic basis for mathematics and logic was presented and judged promising. Here we develop the argument further. It will be argued that acquiring competence in natural language necessarily involves the acquisition of a large, implicit, body of knowledge. Part of this knowledge is elementary knowledge of mathematics and logical reasoning, and their applications. Linguistic communication requires that the rules and conventions of this language, which embody its meaning, are presupposed. These shared presuppositions, without which communication breaks down, are the basis for the objectivity of mathematical knowledge (and objects). This is the essence of the argument.

This is not argued on the basis of psychological or empirical facts, but on logical and philosophical grounds. For it is a truism that any logical system of knowledge, be it deductive or definitional, depends ultimately on a set of primitive propositions or terms. For objective mathematical knowledge these primitive propositions and terms are to be found in the objective knowledge of natural language.

To flesh out the bare bones of this argument, we notice first that traditionally, objective knowledge is identified with a set of *propositions* or *statements*, (or the contents thereof), that is as a linguistically expressed body of knowledge. In previous chapters it has been argued that knowledge includes in addition to propositional knowledge such things as processes and procedures. However, these too can be represented as propositions. Thus the fundamental assumption that knowledge has a lingusitic basis is not in question. This means that the understanding of such knowledge depends essentially on linguistic competence, as does most of human cognitive and social activity.

Linguistic competence consists of the ability to communicate linguistically. This in turn depends on the shared use of grammatical forms, of the relationships between terms, and of the applicability of terms and descriptions to situations, including shared meanings of terms, at least in publicly observable, behavioural uses. It also depends on the ability to interrelate social contexts and certain forms of discourse. In brief, linguistic competence depends on the common following of rules, in conformity with public usage. As we have seen in the discussion of Wittgenstein's conventionalism, the participation in our shared social 'language games' requires the acceptance and following of language rules and conventions. Through being woven into the whole complex of human action and communication (Wittgenstein's 'forms of life'), these rules acquire a necessity, that cannot be questioned without threatening to unravel the whole enterprise.

We cannot question the fact that 'A and B' entails 'A' or that $1+1=2$ without withdrawing some of the possibility of communication. We can only get around this temporarily, by circumscribing a small domain of language use, and exposing and questioning some of the rules governing its use. We may 'freeze' and thus suspend some of these rules to dissect them. But in our other language games, including our meta-language, these rules remain in force. And when our inquiry moves on, the rules become reanimated, and reassume their living certitude. Like a boat in mid voyage, we may tentatively remove a plank from the hull and question its role. But unless we reinstate it before we continue our inspection, the whole enterprise may founder.

This is the general argument for the necessity of the rules associated with language use. These rules codify the shared linguistic behaviour which allows the possibility of communication. In detail, these rules depend on the particular terms and rules of mathematics and logic embedded in our language. We consider these next.

Our natural language contains informal mathematics as a subset, including such terms as 'square', 'circle', 'shape', 'zero', 'one', 'two', 'number', 'add', 'less', 'greater', 'equals', 'set', 'element', 'infinite' and so on. Some of these terms are directly applicable to the shared world of our experience, and natural language includes rules and conventions on how to apply these terms. In this sense, these terms resemble those of science, for their basic terms are learned together. Such terms allow us to describe events and objects in the world by classification and quantification. The intended interpretations of informal mathematics, such as these, are implicit in the semantics of natural language (which often provides multiple meanings for these terms). In addition, the inter-relations between terms are established by linguistic conventions and rules. Thus, for example, 'one is less than two' and 'an infinite set has more than two elements' are both warranted on the basis of the semantic rules of language. As was stated, the elementary applications of mathematics are also built into the rules of linguistic use. The presence of these two types of rule, those concerning the interconnections of terms and their intended applications in the world, account for much of the implicit mathematical knowledge we unconsciously acquire with linguistic competence.

This account is grossly oversimplified in one sense. For it appears to assume a single external world. In fact, there are many overlapping domains of linguistic discourse, many language games, each with their own shared worlds of reference. Some relate to what is socially accepted by the majority as objective reality, others less so, and some are wholly fictional or mythological. Each contains an informal theory, a set of relationships between the entities that inhabit them. What they all share is social agreement on the rules relating to discourse about them.

Many of our linguistic utterances, whatever 'language game' we are engaged in, are laden with mathematical concepts, or highly 'mathematized' (Davis and Hersh, 1986). For an example of the embeddedness of mathematics in every day language use, consider the Zen koan 'What is the sound of one hand clapping?'. This is based on the linguistic knowledge that it takes two hands to clap, one is half of two, but half the number of hands does not give half the amount of sound (I focus here on mathematical content, and not on the purpose of the koan which is through cognitive challenge to induce *satori*). Overall, I wish to claim that natural language such as English (and Japanese, apparently), and even more so informal mathematical language, is rich with implicit mathematical rules, meanings and conventions. These rules, such as 'two is the successor of one', necessitate the acceptance of truths, such as '1 + 1 = 2'.

The linguistic basis of logic

The same can be said for logic in language. Our use of the key logical terms 'not', 'and', 'or', 'implies', 'if, and only if', 'entails', 'there exists', 'for all', 'is a', and so on, strictly follows linguistic rules. (We ignore inconsistent colloquial variations such as 'not not = not', which are rejected by mathematics and logic.) These rules fix as true basic statements such as 'if A, then A or B', and rules of inference such as 'A' and 'A implies B' together entail 'B'. For these rules reflect the use of these terms, and hence (according to Wittgenstein) their meaning. The rules and conventions of logic underpin more than just the 'truths' of logic. As we have seen, they also underpin logical relationships, including implication and contradiction. Thus reasoning, and indeed, the whole basis of rational argument rests on the shared rules of language.

The rather more abstract and powerful forms of logic used in mathematics also rest on the logic embedded in natural language use. However, the rules and meanings of mathematical logic represent a formalized and refined version of this logic. They make up a tighter set of language games overlapping with those of natural language logic.

The linguistic basis accommodates conceptual change

It has been argued, on conventionalist grounds, that everyday mathematical knowledge is linguistic knowledge, and that it derives its security, indeed its apparent necessity, from the regularity and agreed use of language. But while linguistic convention provides everyday mathematical knowledge with its secure foundation, so too it provides the grounds for change in mathematics, as linguistic conventions and usage develop over time.

Consider the most unassailable of arithmetical truths. Since time immemorial it has been inconceivable to question the elementary fact '1 + 1 = 2' (but see Restivo, 1984, on '2 + 2 = 4'). However, since the time of George Boole we can assert the contradictory fact '1 + 1 = 1'. It can be rejoined that this is only because Boole has invented a formal system which assigns different meanings to the symbols. This is true, but the fact remains that '1 + 1 = 1' is no longer false, and that '1 + 1 = 2' is no longer absolutely true. It is true given certain presuppositions (admittedly embedded in our natural language) which, when conflict arises, need to be made explicit. Prior to Boolean algebra the questioning of '1 + 1 = 2' was simply not coherently possible. The real change, therefore, lies behind the scenes. It resides in the fact that we can suspend our everyday rules for portions of language, and consider the consequences of hypothetical conventions, i.e. those contrary or distinct from those embedded in natural language usage. This is the change that led Russell to claim that Boole originated pure mathematics. Does this mean that the unique meanings of mathematics have been lost? On the contrary, it means that we have added new, more abstract language games, to those associated with the mathematical part of natural language.

This notion of a range of language games encompassing the mathematical part of natural language allows a possible objection to be addressed. This concerns the claim that since the basis of mathematical and logical knowledge is inherent in natural language usage, then all of mathematical knowledge must be inherent in natural language. But this is patently false, the only legitimate conclusion from these premises about the sum of all mathematical knowledge is that its basis, and not the whole itself, is inherent in language usage. Given this basis, more and more new language games embodying mathematical meaning and knowledge can be (and are) developed, without necessitating the corresponding enlargement of the linguistic basis. For specialist formal and informal mathematical discourses may be enlarged, resting on the same natural language basis.

The mathematical knowledge embedded in language usage provides a basis for informal (and ultimately formal) mathematical knowledge. The meanings and rules embodied in this knowledge can be described in terms of a series of language games. These games provide the basis of further, more refined language games, which abstract, refine, extend and develop their rules and meanings. Thus a loose hierarchy can be posited, with the mathematical knowledge embedded in natural language making up the base. On this is built a series of language games embodying informal and ultimately formal mathematical knowledge. In the upper reaches of the hierarchy informal mathematical systems become formalized into axiomatized theories. At this level the rules of the games or systems become almost completely explicit. In this way the knowledge of mathematics implicit in language provides a basis for all mathematical knowledge. Truths embodied in and vouchsafed by linguistic usage are reflected up the hierarchy to justify elementary assumptions adopted in mathematics. The same is true for the assumptions and rules of logic. In the next section we will explore the role of such assumptions, in the justification of mathematical knowledge.

In this section we have seen that linguistic convention and usage provides mathematical knowledge with its secure foundations. Similarly, it provides grounds for change in mathematics, as linguistic conventions and usage develop over time. Mathematics, like every other realm of knowledge depends essentially on tacit linguistic assumptions. Fallibilism forces us to admit their presence, as well as their changing nature, over the passage of time.

B. The Conventionalist Warrant for Mathematical Knowledge

According to the social constructivist view, mathematical knowledge is fallible, in that it is open to revision, and it is objective in that it is socially accepted and publicly available for scrutiny. Valid mathematical knowledge is knowledge which is accepted on the basis of there being a public justification of the knowledge (a published proof) which has survived (or been reformulated in the light of) public scrutiny and criticism.[5]

The justification for a particular item of mathematical knowledge consists of a valid informal or formal deductive proof. The analysis of a proof justifying an item of knowledge must consider two aspects: the explicit starting assumptions, and the sequence of steps leading from them to the conclusion. We consider first the starting assumptions. These consist of (i) hypothetical statements or assumed axioms (e.g., the continuum hypothesis), (ii) definitions (e.g., Peano's inductive definition of '+'), (iii) previously accepted mathematical knowledge, typically previously established theorems, (iv) the 'truths' of accepted informal mathematical knowledge, which are embedded in mathematical language, or their formalization (e.g., the Peano Axioms), or (v) logical axioms. Of these types, (iii) is reducible to others (via proofs). The remaining assumptions are either hypothetical assumptions (case i and case iv in some instances), or rest on the conventions and rules of mathematical language. Definitions of type (ii) are conventions by *fiat*, they are simply laid down as such. The remaining two types of assumptions are either informal mathematical rules, or their formalization (case iv), or logical axioms (case v). The justification for these two types of assumptions is conventionalist, and is offered below.

Secondly, a mathematical proof consists of a finite sequence of steps leading from the initial assumptions of the proof, to its conclusion. The key feature of such a sequence is the means of proceeding from one step to the next, that is the justification for concluding a step from its predecessors. The justification for such a step consists of (i) the use of a logical rule of inference (e.g. the rule Modus Ponens), (ii) the use of a mathematical principle of inference (e.g. the Pigeon Hole Principle), (iii) the introduction of a new assumption (this is like the cases treated in the preceding paragraph), (iv) the claim that the step is justified by an elementary combination of the previous types of steps, and (v) by analogy with a similar proof given elsewhere. Assuming that any claims under cases (iv) and (v) are verified, this leaves (i) and (ii) to consider. These depend on the assumption of a rule or principle of mathematics or logic. These will either be reducible to simple assumptions (as is the Pigeon Hole Principle) or are themselves basic mathematical and logical rules. Such rules do not in principle differ from the basic mathematical and logical assumptions discussed above. In fact assumptions and rules are inter-translatable, so rules can be replaced by assumptions in sentence form, although at least one logical rule of inference is necessary. If, for simplicity, we thus dispose of the mathematical rules (replacing them by assumptions in propositional form), the assumptions on which the inferential steps in a mathematical proof are based reduce simply to some basic logical rules of inference. These rules of inference will be justified by the conventionalist argument.

We have seen that the warrant for asserting mathematical knowledge consists of a mathematical proof (of one step only, in the case of a basic assumption). The basis for such a warrant resides in a number of basic assumptions (excluding genuinely hypothetical axioms, such as 'V = L' of Godel, 1940, or the axioms of Tensor Theory). These basic assumptions consist of informal mathematical 'truths', and logical axioms and rules of inference. These were justified, in the previous section, as linguistic conventions, which are a part of the rules of meaning and use inherent in our grasp of language. Thus, it is argued, the whole body of mathematical knowledge is warranted by proofs, whose basis and security rests on linguistic knowledge and rules.[6]

C. The Objects of Mathematics

The objectivity of mathematical knowledge is social, based on the acceptance of linguistic rules, which are necessary for communication as we know it. Social acceptance also provides the basis for the independent existence of the objects of mathematics. For embedded in the rules and truths of mathematics is the assumption, the assertion even, that the concepts and objects of mathematics have an objective existence.

In natural language, each set of language games may be regarded as a discourse, including a set of linguistic entities, and rules and truths, together making up a naive theory. Associated with such a discourse and its function is a semantic realm, a universe of discourse. This is an informally delineated set of entities, with certain properties and relationships as specified by the associated naive theory. Thus the existence of a shared set of language games entails a realm of entities with an existence independent of any individual. In particular, a mathematical theory or discourse brings with it a commitment to the objective existence of a set of entities.

> Classical mathematics, as the example of primes greater than a million clearly illustrates, is up to its neck in commitments to an ontology of abstract entities. Thus it is that the great medieval controversy over universals has flared up anew in the modern philosophy of mathematics. The issue is clearer now than of old, because we now have a more explicit standard whereby to decide what ontology a given theory or form of discourse is committed to: a theory is committed to those, and only those entities to which the bound variables of the theory must be capable of referring in order that the affirmations made in the theory be true.
>
> (Quine, 1948, pages 13–14)

Objective mathematical definitions and truths specify the rules and properties determining the objects of mathematics. This confers on them as much of an objective existence as that of any social concepts. Just as universal linguistic terms, such as 'noun', 'sentence', or 'translation' have a social existence, so too the terms and objects of mathematics have the properties of autonomous, self-subsistent objects. The objects of mathematics inherit a fixity (i.e., a stability of definition) from the objectivity of mathematical knowledge, entailing in turn their own permanency and objective existence. Their objectivity is the ontological commitment that inevitably accompanies the acceptance of certain forms of discourse.

Of course, this is not the end of the matter, for discourses commit us to all manner of entities, from chairs, tables and cars, to ghosts, angels and souls. It cannot be claimed that these are all on a par. But likewise, the objects of mathematics vary from the relatively concrete, embedded in the natural language descriptions of the sensible world, to the abstract theoretical entities of mathematics e.g., the least inaccessible cardinal (Jech, 1971), many steps removed from this basis. However, most of the objects of mathematics have more reality than the objects in some discourses, such as the fantasy creatures of Tolkien's (1954) Middle Earth. For they are the result of social negotiation, not just the product of a single individual's imagination.

Many of the elementary terms and concepts of mathematics have concrete applications and examples in the world. For they are part of a language developed to describe the physical (and social) world. Thus such terms as 'one', 'two', 'ten', 'line', 'angle', 'square', 'triangle', and so on, describe properties of objects or sets of objects, in the world. Other terms such as 'add', 'subtract', 'divide', 'measure', 'rotate', and so on, describe actions that can be performed on concrete objects. The denotations of these terms, gain 'objecthood' by their concrete applications in objective reality.[7] Yet more terms, such as 'equation', 'identity', and 'inequality' refer to linguistic entities. Each of these sets of terms describes aspects of objective reality, whether external or linguistic, and thus provides a concrete basis for a 'mathematical reality'. On this basis further mathematical terms, such as 'number', 'operation', 'shape', and 'transformation', are defined, one level removed from concrete reference. At higher and higher levels, further mathematical terms, increasingly abstract, apply to those below them. Thus through such an hierarchy virtually all mathematical terms have definitions and denote objects at lower levels. These denotations behave exactly like objectively existing, autonomous objects. Thus the objects of mathematics are objective in the same way as the knowledge of mathematics. They are public linguistic objects, some concrete but most abstract.

An example is provided by algorithms. These denote precisely specified sequences of actions, procedures which are as concrete as the terms they operate on. They establish connections between the objects they operate on, and their products. They are a part of the rich structure that interconnects, and thus helps to implicitly define, the terms, and hence the objects of mathematics.

This account may seem to fall short of providing all that is required for objective existence. However, the analogy between the above conceptual hierarchy of mathematics and an empirical scientific theory should be noted. For although defined analogously, the theoretical entities of theoretical science are understood to have an autonomous existence. Hempel (1952) likens a scientific theory to a net. Knots represent terms, and threads represent sentences of the theory (definitions, theoretical statements, or interpretative links) which both bind together the net and anchor it to the bedrock of observation. The theoretical terms of science, such as 'neutrino', 'gravitational field', 'quark', 'strangeness' and 'big bang' correspond to the abstract entities of mathematics, in the analogy. The difference being that only concrete mathematical terms have an empirical reference, whereas the theoretical terms of science are taken to denote the physical entities whose empirical existence is posited by the current theory.

Both these types of entities exist in the 'world 3' of objective knowledge. Whether all such objects, especially those of mathematics 'really' exist or not, is the fundamental question of ontology, and is the subject of the traditional debate between realism and nominalism (see, for example, Putnam 1972). The social constructivist view is that the objects of mathematics are social constructs or cultural artefacts. They exist objectively in the sense that they are public and there is intersubjective agreeement about their properties and existence. The social constructivist view is that mathematical entities have no more permanent and enduring self-subsistence than any other universal concepts such as truth, beauty, justice, good, evil, or even such obvious constructs as 'money' or 'value'. Thus if all humans and their products ceased to exist, then so too would the concepts of truth, money and the objects of mathematics. Social constructivism therefore involves the rejection of platonism.[8]

D. The Genesis of Mathematical Knowledge

In accepting that mathematics is a social construct, it is implied that objective mathematical knowledge is the product of human beings. To defend this thesis, we need to be able to account for the addition of the mathematical creations of individuals (or groups) to the body of accepted mathematical knowledge. But the growth of mathematical knowledge is not exclusively incremental. Thus we must also account for the way that as a result of new contributions, the body of existing mathematical knowledge develops and changes.

Although he does not explicitly address both of these issues, we have seen that Lakatos' quasi-empiricism offers a potentially fruitful account of the genesis of mathematical knowledge, and we shall build on his account.

According to the usage adopted, the mathematical thought of an individual is subjective thought. In order for it to become objective thought it must be linguistically represented, typically in written form. The key act which transforms this published subjective thought into objective thought is social acceptance, following critical public scrutiny. Then it can be said to be a contribution to mathematical knowledge, even if, like Fermat's famous conjecture written in his copy of *Diophantus*, it is not scrutinized in the author's lifetime. Objectivity is conferred on the mathematical thought through its social acceptance, following publication. Here no restriction to written publication is intended. Thus communicating mathematical thought via a lecture to colleagues also constitutes publication, and can equally be a contribution to objective thought, providing it is socially accepted.

A crucial feature in the genesis of mathematical knowledge is transformation from publicly represented (subjective) knowledge in mathematics to objective, that is socially accepted mathematical knowledge. This transformation depends on surviving a process of public scrutiny and criticism. During this process, which is Lakatos' autonomous logic of mathematical discovery, objective criteria play an essential part.

They are used to judge the correctness of inferences, the consistency of assumptions, the consequences of definitions, the validity of formalizations in expressing informal notions, and so on. The shared criteria used in such a process of critical scrutiny include the ideas of logic and correct inference and other basic methodological notions and procedures, which depend to a large extent on shared mathematical and logical knowledge.

The fact that objective criteria exist, however, does not mean that all criticism is rational.[9] However, this account represents a discussion of the philosophical features of objective knowledge growth, and not the empirical factors that may arise in practice. The account is based on that of Lakatos, although elaborated in several respects. The original insight as to the essential role of public criticism in the growth of knowledge, as Lakatos acknowledges, is that of Popper (1959).

The varieties of mathematical creation

What has not yet been accounted for is how some additions to knowledge are incremental, whereas others result in a restructuring or reformulation of existing knowledge. Like science, mathematics is acknowledged to be hypothetico-deductive. Thus mathematicians work within an established mathematical theory. Much of this work consists of the development of new consequences of existing aspects of the theory, or the application of existing methods from within the theory to a range of problems. When fruitful, such work results in incremental additions to the body of mathematical knowledge.

Mathematicians also utilize the concepts and methods from one mathematical theory in another, or manage to establish links between two previously separate theories. Such work causes new structural links to be formed between separate parts of mathematics. This constitutes a restructuring of mathematics, which can be quite considerable if under the influence of the new links the two theories are reworked, reformulated and drawn closer together. Finally, work in some theories, often directed at the solution of some problem, may generate a new mathematical theory. This may simply be an additional theory, or it may subsume previous theories in a larger, more general theory. The move towards increasing abstraction and generality, such as in this case, is a major factor in the restructuring of mathematical knowledge. For increasingly general theories are more widely applicable, and several more specialized, pre-existing theories may fall within their more general structual patterns. An example is provided by Cantor's theory of sets, which initially seemed very specialized and recondite. Since its introduction, because of its wide generality, it has come to encompass most other mathematical theories, and provide them with reformulated and unified foundations.

This account of the genesis of mathematical knowledge provides an idealized account of the mechanism underlying the historical development of mathematics. At any given time, the existing body of objective mathematical knowledge is being reformulated and developed as the result of new contributions, which may either restructure existing knowledge, or simply add to it.

E. The Applicability of Mathematics

For adequacy, social constructivism must account for the 'unreasonable effectiveness of mathematics in science' (Wigner, 1960). It is able to account for the applicability of mathematics on two grounds: (1) mathematics is founded on our empirical natural language; and (2) the quasi-empiricism of mathematics means that it is not so very different from empirical science anyway.

First of all, we have already argued that mathematical knowledge rests on the rules and conventions of natural language. We have seen that there is a rich mathematical vocabulary directly applicable to the world of our experience, and natural language includes rules and conventions on how to apply these terms. Many of these belong both to mathematics and to science, and allow us to use classification and quantification in describing events and objects in the world (via conjectured explanations). Everyday and scientific uses of natural language are a key feature of its role, and in such uses the embedded mathematical concepts play an essential part. Thus the linguistic basis for mathematics, as well as the other functions language performs for mathematics, provides interpretative links with real world phenomena. In this way its linguistic roots provide mathematics with applications.

Secondly, we have accepted Lakatos' argument that mathematics is a quasi-empirical hypothetico-deductive system. In acknowledging this, we are admitting a much closer link between mathematics and empirical science than the traditional absolutist philosphies allow. This is reflected in the close resemblance between mathematical theories and scientific theories, which we have observed. Both types of theory contain relatively concrete exemplifiable or observational terms, and theoretical terms, interconnected by a 'net' of links and relationships. Quine (1960) even sees them both as interwoven in a single, connected fabric. In view of this striking structural analogy, it is not surprising that some of the general structures and methods of mathematics are imported into physical theories. Indeed, much of empirical theory is wholly expressed in the language of mathematics. Likewise, it is not surprising that many scientific problems, formulated in the language of mathematics, become the stimulus for mathematical creation. The need for ever better models of the world, as science advances, provides mathematics growth points for development. The consequent cross-fertilization and interpenetration of science and mathematics is a fact that the absolutist philosophical separation between *a priori* and empirical knowledge has masked and mystified. In its origins and throughout its development, mathematics has maintained contact with the physical world by modelling it, often in conjunction with empirical science. In addition, the forces that lead to generalization and integration in mathematical knowledge, described above, ensure that the contact and influence of the empirical world on mathematics are not merely marginal. The applicable theories within mathematics are subsumed into more general theories, as mathematics is restructured and remade. By these means, the applicability of mathematics extends to the central abstract theories of mathematics, and not merely those on its periphery.

Overall, the applicability of mathematical knowledge is sustained by the close relationships between mathematics and science both as bodies of knowledge and as fields of enquiry, sharing methods and problems. Mathematics and science are both social constructs, and like all human knowledge they are connected by a shared function, the explanation of human experience in the context of a physical (and a social) world.

4. A Critical Examination of the Proposals

The social constructivist account of objective mathematical knowledge potentially satisfies the adequacy criteria for the philosophy of mathematics, since it treats knowledge, ontology, applications and practice. However, a number of telling criticisms can be directed at the account, and these need to be anticipated and answered.

A. Mathematics is arbitrary and relative

First of all, there is the problem of the relativism of mathematical knowledge and truth. If, as is claimed, mathematical truth rests on social conventions, then it is both arbitrary and relative. It is arbitrary since it rests on arbitrary beliefs, practices and conventions. It is relative since it rests on the beliefs of one group of humans. Consequently there is no need for other groups of humans, let alone other intelligent creatures in the universe, to accept the necessity of mathematical knowledge, which only holds relative to a particular culture at a particular period.

To answer this, I wish to question two presuppositions. These are first, the notion that linguistic and mathematical conventions are arbitrary and mutable; and second, the misconception that mathematical and logical knowledge are necessary and immutable (although this was largely refuted in Chapter 1).

Arbitrariness

The arbitrariness of mathematics, in the account given, arises from the fact that mathematical knowledge is founded on linguistic conventions and rules. There is no necessity behind these rules, and they could have developed differently. This is undeniable. But the fact remains that language operates within the tight constraints imposed by physical reality and interpersonal communication. The conventions of language can be formulated differently, but the purpose of language in providing a functioning social description of the world remains constant. The shared rules and conventions of language are a part of a naive empirical theory of the physical world and social life. Thus, although every symbol in natural language is arbitrary, as the choice of any sign must be, the relationship between reality and the overall model of it, provided by language is not.

Although such modelling may not be the whole function of language, it provides crucial external anchors for language that keep it functioning viably. To maintain this viability, some of the logical rules of language appear to be necessary. For example, White (1982) argues that the principle of contradiction is necessary for any assertion to be made by means of language. For without the principle in operation there would be simultaneous assertion and denial. But assertions are ruled out by denial. Now in some language uses the principle might be relaxed for certain purposes, such as describing a deity. However it is difficult to argue that a language could function viably without some such rules. Thus although much of the formulation of language rules and conventions may be arbitrary in detail, the need for viability reduces much of the scope of the arbitrariness of language to inessential details. For example, the differences between natural languages indicate areas of arbitrariness in their formulation.

Relativism

The definition of objectivity adopted opens social constructivism to the charge of relativism. That is, it is just the knowledge of a particular group at a particular time. This is true, but there are two mitigating circumstances which remove much of the force of this criticism. As we have seen, mathematics through language must provide a viable description of aspects of empirical and social reality. Thus the relativism of mathematics is reduced by its anchoring via these applications. In other words, both mathematics and language are highly constrained by the need to describe, quantify and predict events in the physical and human worlds effectively. In addition, mathematics is constrained by its growth and development through the inner logic of conjectures, proofs and refutations, described above. Thus mathematics not only has its feet rooted in reality, but its upper parts have to survive the rigorous public procedures of justification and criticism, based on the thorough-going application of a small number of principles. Thus mathematical knowledge is relativistic knowledge in that its objectivity is based in social agreement. But its relativism does not make it equal or interchangeable with other social belief systems, unless they satisfy these same two criteria.

Critics of the possibility of relativism in mathematics claim that alternative mathematics or logic is inconceivable, which confirms the necessity and unique status of mathematics and logic. This raises the question: what would an alternative mathematics (or logic) look like? Bloor (1976) asks this question, and illustrates his answer with alternative notions of number, calculus, and so on from the history of mathematics. A critic's reply to this might be that although our conceptions have evolved and changed throughout history, they were just steps on the path to the necessary modern notions. If the questionable teleological aspect of this claim is ignored, then it is necessary to exhibit simultaneous competing alternatives for mathematics, to answer the criticism. However, a further question needs to be asked: how different does an alternative mathematics need to be to count as alternative (and hence to refute the uniqueness claim)?

The answer I propose is that an alternative mathematics (or logic) should be based on concepts defined differently, with different means of establishing truths, and which results in a very different body of truths. Furthermore, if the alternative is to be taken seriously, there should be a respectable body of mathematicians who adhere to the alternative, and who reject standard mathematics. This, in my view, is an adequately strong characterization of an alternative form of academic (as opposed to culturally embedded) mathematics. Strong as it is, it is not difficult to satisfy this requirement. Intuitionist mathematics fits the requirement perfectly. Intuitionist concepts, from the logical connectives 'not', 'there exists', to the concepts of 'set', 'spread' and the 'continuum' are quite different in meaning and in logical and mathematical outcomes, from the corresponding classical concepts, where they exist. Intuitionist axioms and principles of proof are also different, with the rejection of the classical Law of the Excluded Middle, '--P ↔ P', and '-(x)-A ↔ (Ex)A'. Intuitionist mathematics has its own body of truths including the countability of the continuum, the Fan Theorem and the Bar Theorem, which do not appear in classical mathematics, as well as rejecting the bulk of classical mathematics. Finally, since the time of Brouwer, intuitionism has always had a cadre of respected adherent mathematicians, committed to intuitionism (or constructivism) and who reject classical mathematics (e.g. A. Heyting, H. Weyl, E. Bishop, A. Troelstra). Thus there *is* an alternative mathematics, which includes an alternative logic.

This century there has been an explosion of other alternative or 'deviant' logics, including many-valued logics, Boolean-valued logic, modal logic, deontic logic and quantum logic. These show that further alternatives to logic are not only possible, but exist. (However these deviant logics may not satisfy the last criterion given above, i.e. the adherence of a group of mathematicians, who reject classical logic).

The example of intuitionism shows that classical mathematics is neither necessary nor unique, for an alternative is not only possible, but it exists. It also shows that there are alternatives to classical logic. The example also demonstrates the relativism of mathematics, subject to the constraints discussed above, since there are two mathematical communities (classical and intuitionist) with their own, opposing notions and standards of mathematical truth and proof. In previous chapters the absolutist view of mathematics as a body of immutable and necessary truth was refuted, and a fallibilist view argued in its place. This weakened the claim of necessity for mathematics. This has now been supplemented with an example of a genuine alternative, dispelling any possible claims of uniqueness or necessity for mathematics.

B. Social Constructivism Fails to Specify any Social Group

The account of social constructivism given refers to 'social acceptance', 'social construction' and 'objectivity as social'. However it fails to specify in any way which social groups are involved, and for the term social to have meaning, it must refer to a specific group. There are also hidden secondary problems such as how does one tell when something is accepted by the mathematical community? What happens when there is conflict in this community? Does this mean that a new piece of mathematics can hover on the boundary between subjective and objective knowledge?

To answer the main point first: it would be inappropriate in a *philosophical* account to specify any social groups or social dynamics, even as they impinge upon the acceptance of objective knowledge. For this is the business of history and sociology, and in particular, the history of mathematics and the sociology of its knowledge. The claim that there is a social mechanism involved in objectivity and in the acceptance of mathematical knowledge, and a conceptual analysis and elaboration of it remains within the province of philosophy. The importation of concepts from history and sociology to develop this theory, valuable as this might be, takes the discussion outside the philosophy of mathematics. Thus this is not a valid criticism.

The secondary criticisms do hold some problems for social constructivism. If there is simultaneous social acceptance of different sets of mathematical knowledge, as was explored in section A above, then they both constitute objective mathematical knowledge.

The transition of mathematical knowledge from subjective to objective knowledge may benefit from further clarification. It needs to be made clear that there is an intermediate state, which is neither. Subjective mathematical knowledge resides in the mind of an individual, possibly supported by external representation. For individuals developing subjective knowledge often do so with the aid of visual, oral or other representations. Already such representations mean that there is a public aspect supporting the individual's subjective knowledge. When fully represented in the public domain, it is no longer subjective knowledge as such, although the originating individual may have corresponding subjective knowledge. Public representations of knowledge are just that. They are not subjective knowledge, and need not be (or represent, to be precise) objective knowledge either. However, they have the potential to lead to the latter, when they are socially accepted.

Strictly speaking, public representations of knowledge are not knowledge at all, for they consist only of symbols, and meanings and assertions have to be projected into them by understanding subjects. Whereas knowledge is meaningful. This is consistent with the view adopted in communications theory, that signals have to be coded, transmitted and then decoded. During the transmission phase, that is when coded, signals have no meaning. This has to be constructed during decoding.

It is convenient to adopt the current (but strictly speaking false) usage of identifying public representations of objective knowledge (coded signals) with the knowledge itself, and speaking as if the representation embodied information and meaning. Such an attribution of meaning only works if it is assumed that the appropriate community share the decoding knowledge. In the case of mathematical knowledge this consists of knowledge of natural language and additional knowledge of mathematics.

These then are some to the essential presuppositions about the social groups on which social constructivism depends.

C. Social Constructivism Assumes a Unique Natural Language

Social constructivism employs a conventionalist justification for mathematical knowledge. This assumes that mathematical knowledge rests on a unique natural language, contradicting the fact that over 700 different natural languages are known, many with very different bases from English.

Whilst it can be said that mathematical concepts and truths do depend on certain structural features of English, these are found equally in European and some other languages, but not necessarily in all natural languages. This has two major consequences, neither of which is critical for social constructivism. First, if mathematics were based on languages with significantly different logic and structural features, then an alternative (i.e., different) mathematics could result. This is not a problem for social constructivism. Second, native language speakers whose language differs significantly from English, French, etc., in logic and structural features either have to acquire a second language, or restructure their understanding of their own, in order to learn academic Western mathematics. This again seems plausible, and in fact there is some evidence to support this.[10] In fact such evidence of cultural relativism strengthens rather than weakens the case in favour of social constructivism.

D. Previously Raised Objections

1. Social acceptance is not the same as objectivity.

An account has been given of objective mathematical knowledge, but objectivity has been re-interpreted to mean socially accepted, in the manner of Bloor (1984). Thus it is true to say that objectivity (understood socially) is used to mean something different. In defence of the social interpretation the following has been argued. First, the important properties of objectivity, such as impersonality and verifiability, are preserved. Second, objective existence in mathematics means consistently postulable. The immense ontological consequences of this definition for mathematics distort the meaning of 'objectivity' far beyond the sense of 'existing like an object'. Third, the social interpretation uniquely provides an explanation of the nature of objectivity in mathematics.

2. Social constructivism attends insufficiently to the warranting of mathematical knowledge.

It is true that the account given focuses on the genesis of mathematical knowledge, but it does not neglect to account for the justification of mathematical knowledge, although in doing so it challenges absolutist accounts. Mathematical knowledge is justified as hypothetico-deductive knowledge, which, in the case of derived knowledge, involves proofs. Some elementary truths and the basis of logic and proof are justified in terms of natural language, using the arguments of conventionalism. A problem with this latter justification is that natural language does not literally contain all the basic truths and rules of mathematics and logic. Rather it embodies the basic meanings, rules and conventions, which *in refined and elaborated form*, provide the elementary truths and rules of mathematics and logic. The account offered is superior in scope to that of traditional philosophies of mathematics, because it provides an objective basis, warranting these elementary assumptions. At best, other philosophies offer intuition (intuitionism, formalism, platonism) or induction (empiricism), for these assumptions, if they offer any basis at all.

3. Social constructivism conflates the contexts of discovery and justification, and commits the error of psychologism.

By challenging the widespread assumption that the business of philosophy is only with the context of justification, and not that of discovery, social constructivism may seem to lay itself open to this charge. The account given acknowledges the importance of these concepts, and distinguishes carefully between the two contexts, as well as between the different proper concerns of philosophy, history, psychology and sociology. However it is argued that on adequacy grounds the philosophy of mathematics must account for the development and genesis of mathematical knowledge, albeit from a philosophical perspective, as is accepted analogously in the philosophy of science. It is also argued that subjective knowledge is a legitimate area of philosophical enquiry, and need not lead to psychologism. Subjective thought and knowledge must be included in a social constructivist account because it is the fount of new mathematical knowledge. Naturally it must be treated philosophically, and not psychologically, to avoid psychologism.

Notes

1 Other authors refer to mathematics as a social construct, notably Sal Restivo (1985, 1988), in developing a sociology of mathematical knowledge. Although approaching mathematics from another perspective, he offers a range of insights compatible with the philosophy proposed here. This is considered further in chapter 5.

2 Ideas from constructivist epistemology and learning theory, due to Glasersfeld, Piaget and others, have also contributed to social constructivism.

3 Knowledge reproducing cycles occur also in sociology and education, but are concerned with the genesis and reproduction of knowledge, not its justification (see Chapters 11 and 12). Such a cycle is relatively novel in philosophical studies of mathematics, because it treats the genesis and social origins of knowledge, as well as its justification. The approach adopted may be seen as part of a new naturalistic approach to the philosophy of mathematics, typified by Kitcher (1984), and other authors.

4 'Twentieth-century epistemology has been characterized by an attitude of explicit distaste for theories of knowledge which describe the psychological capacities and activities of the subject. This attitude has fostered an *apsychologistic* approach to knowledge . . . present in the writings of Russell, Moore, Ayer, C. I. Lewis, R. Chisholm, R. Firth, W. Sellars and K. Lehrer, and is presupposed by the discussions of science offered by Carnap, Hempel and Nagel.' (Kitcher, 1984, page 14)

5 Notice the analogy with the 'replicability' criterion for experimentation in science, which demands

that results, for acceptance, should not be peculiar to a unique scientist, but should be replicable. Likewise a mathematical proof not only has to be surveyable to others, but this survey must result each time in acceptance.

6 One possible objection to this account is that there seems to be a gap between the implicit 'truths' embedded in natural language, and the more abstract and sophisticated logical and mathematical assumptions required to warrant mathematical knowledge. The answer to this is that linguistic competence is not defined by a static level of performance and knowledge. Linguistic competence presupposes competence in performing certain functions in certain social situations. The mastery of these different linguistic functions, which representing the mastery of different 'language games' in different social contexts, represents different types of linguistic competence. This range of contexts brings with it a range of ever sharper linguistic conventions and rules, and those interested in mathematical knowledge will of necessity have mastered a range of sophisticated mathematical and logical language games.

In addition, the relatively elementary notions of rationality (implication, and contradiction) underpin the more refined notions of logic. This ensures coherence and means that the more refined notions are extrapolations of, and not discontinuous with, the simpler notions. Thus I am claiming that the 'gap' is self closing. Those with knowledge enough to participate in the language games of warranting mathematical knowledge will have extended their linguistic rules and truths to the required point. Should they fall short, they will have their usage, and rules, extended.

7 By objective reality I am referring to the socially agreed features of the external world. I acknowledge the basic ontological assumption implicit in using this phrase, that is the assumption of the existence of a physical world (Popper's world 1). I also acknowledge the existence of human beings, as the basis for a social view of objective knowledge. However, I do not concede that this commits me to ontological assumptions about the particular, conventionally labelled objects in the world, beyond the fact that there is social agreement on their existence and objectivity.

8 In case it seems a weakness of the social constructivist view that it denies the absolute existence of mathematical objects, it is worth noting some of consequences of the mathematical notion of existence. The criterion for existence in mathematics is consistency. If a mathematical theory is consistent, then there is a set of objects (a model) providing denotations for all its terms and satisfying all the conditions of the theory. That is, these objects have *mathematical existence*. Thus, for example, provided the theories of Peano Arithmetic, the Real Numbers and Zermelo-Fraenkel (ZF) Set Theory are consistent (which is accepted), there are models satisfying them. Thus the Natural Numbers, Real Numbers and the universe of sets can be said to exist, which already allows for an unbelievable richness of entities. But worse is to come. The Generalized Lowenheim-Skolem Theorem (Bell and Slomson, 1971) establishes that these theories have models of *every* infinite cardinality. Thus countable models of both Real Number Theory and ZF set theory exist, as do models of the natural numbers of every (infinite) size. These all exist in world 3. Social constructivism denies the existence this unpredictable and undreamed of multitude of mathematical entities. In contrast, the outcome of the platonist view of existence in mathematics would be to populate the universe far more densely than merely putting infinitely many angels to dance on the head of every pin. The question is, what does it mean to say that the objects in the platonist ontology actually exist? Surely this is using the term 'exist' in a novel way. The social constructivist response to these ontological problems is basically to adopt the traditional conceptualist solution, but in a new social guise (Quine, 1948).

9 Russell's (1902) criticism of Frege's system was rational, i.e. logical, being based on logical features of the system (namely its inconsistency). Kronecker's criticism of Cantor's set theory evidently was not purely rational or logical, since it had strongly moral and religious undertones.

The Kuhn-Popper debate in the philosophy of science hinged on the issue of rational versus irrational criticism of scientific theories, which is analogous to the point made here concerning mathematics. Popper's position is prescriptive, he posits falsification as a rational criterion for the rejection of a scientific theory. Kuhn, on the other hand, proposes a more descriptive philosophy of science, which while treating the growth of objective knowledge acknowledges that rational features are neither necessary nor sufficient to account for theory acceptance or rejection.

10 Alan Bishop has presented evidence that in some of the 350 languages to be found in Papua and New Guinea the conceptual structure is markedly different from that presupposed by elementary mathematics. For example in one language the same word is used to denote 'up', 'top', 'surface' and 'area', indicating a conceptual structure very different from English and comparable languages.

4

Social Constructivism and Subjective Knowledge

1. Prologue

This chapter faces a difficult task: that of showing the relationship between subjective and objective knowledge of mathematics in social constructivism. The task is difficult for a number of reasons. It skirts the edge of psychologism, and it needs to conjoin two different languages, theories and modes of thought that apply to two different realms, the subjective and the objective. Beyond this, the epistemology underpinning social constructivism is quite slippery to grasp, since it is claimed that there is no realm where a determinate entity 'knowledge' basks in tranquillity. Knowledge, perhaps analogous to consciousness, is seen as an immensely complex and ultimately irreducible process of humankind dependent on the contributions of a myriad of centres of activity, but also transcending them. Science fiction authors (Stapledon, 1937) and mystical philosophers (Chardin, 1966) have groped for a vision of how the consciousnesses of individual human beings can meld into a greater whole. But these provide too simplistic a vision to account for knowledge and culture as dynamic, cooperative dances uniting millions of thinking and acting but separate human beings. The seduction of idealism is great: to say that knowledge exists somewhere in an ultimate form, possibly growing and changing, but that all our representations of knowledge are but imperfect reflections. The pull to view human knowledge attempts as parts of a convergent sequence that tends to a limit in another realm, is almost irresistible.

Once these simplifying myths are rejected, as they are by social constructivism, there is the complex task of accounting for knowledge. It is social, but where is the social? Is it a moving dance, a cloud of pirouetting butterflies, which when caught is no more? Books do not contain knowledge, according to this account. They may contain sequences of symbols, carefully and intentionally arranged, but they do not contain meaning. This has to be created by the reader, although books may guide the reader to create new meanings. This is subjective knowledge, the unique creation of each individual. Yet by some miracle of interaction, the way human beings use this knowlege in their transactions fits together.

The concept of the individual, the knowing subject, in Western thought is another problem. Since Locke, or earlier, the subject is a *tabula rasa*, and gradually knowledge is inscribed on its blank page by experience and education. But the form and content of mind cannot be separated like this, and there is no universal form of knowledge that can be written in our minds. The view that follows is that knowledge has to be created anew in the mind of every human being, and solely in response to their active efforts to know. Consequently, objective knowledge is all the while being born anew. Thus knowledge is more like a human body, with its every cell being replaced cyclically, or like the river: never the same twice! This is why I called the epistemology involved slippery!

These are some of the problems that the present chapter raises and tries to tackle.

2. The Genesis of Subjective Knowledge[1]

The fundamental problem to be accounted for in the growth of subjective knowledge concerns the acquisition of knowledge of the external world. How can an individual acquire knowledge of the external world by means of sensory organs alone? The external world includes other human beings, so acquisition of knowledge of the external world includes knowledge of human beings, their actions, and ultimately their speech. Only when we have accounted for the acquisition of speech can we begin to consider how the substantive structures of mathematics are acquired. We begin, therefore, by inquiring how any subjective knowledge is acquired.

A. The Construction of Subjective Knowledge

How does the individual acquire knowledge of the external world? Human beings have incoming sense impressions of the world, as well as being able to act physically on the world, and thus in some measure are able to control aspects of the environment. Clearly subjective knowledge is acquired on the basis of interaction with the external world, both through incoming sense data and through direct actions. What is also clear is that these interactions are necessary but not sufficient for the acquisition of knowlege of the external world. For the sense data are particulars. Whereas our knowledge is evidently general, since it includes general concepts (universals), and it allows for anticipation and the prediction of regularities in our experience. Therefore some further mechanism is required to account for the generation of general knowledge of the world of our experience, on the basis of particular items of information or experiences.

This is precisely the problem that the philosophy of science faces, but expressed at the subjective level. Namely, how can we account for (and justify) theoretical scientific knowledge on the basis of observations and experiment alone? The solution proposed is the same. The development of subjective knowledge, like that of science, is hypothetico-deductive. The answer proposed is that the minds of individuals are active, conjecturing and predicting patterns in the flow of experience, and thus building theories of the nature of the world, although these may be unconsciously made theories. These conjectures or theories serve as guides for action, and when they prove inadequate, as inevitably they do, they are elaborated or replaced by new theories that overcome the inadequacy or failing of the previous theory. Thus our subjective knowledge of the external world consists of conjectures, which are continually used, tested and replaced when falsified.

Thus the account of the formation of subjective knowledge is a recursive one. Our knowledge of the world of our experience consists of private conjectures or theories, which order the world of our experience. These theories are based on two factors. First, our immediate experiences of the world, including interactions with it, as perceived and filtered through our theories. Second, our previously existing theories. Thus the formation of our subjective theories is recursive in that it depends essentially on these theories, albeit in an earlier state.

This account mirrors that of Popper (1959), but at the level of subjective as opposed to objective knowledge.[2] However, it is clear that Popper intends his account of science to apply only to objective knowledge, and furthermore, he has nothing to say on the genesis of scientific theories. As a purely subjective view of knowledge, this view is elaborated by Glasersfeld (1983, 1984, 1989) as 'radical constructivism'.

> 'The world we live in' can be understood also as the world of our experience, the world as we see, hear and feel it. This world does not consist of 'objective facts' or 'things-in-themselves' but of such invariants and constancies as we are able to compute on the basis of our individual experience. To adopt this reading, however, is tantamount to adopting a radically different scenario for the activity of knowing. From an explorer who is condemned to seek 'structural properties' of an inaccessible reality, the experiencing organism now turns into a builder of cognitive structures intended to solve such problems as the organism perceives or conceives. Fifty years ago, Piaget characterised this scenario as one could wish: 'Intelligence organises the world by organising itself' (Piaget, 1937). What determines the value of the conceptual structures is their experimental adequacy, their goodness of fit with experience, their viability as means for the solving of problems, among which is, of course, the never-ending problem of consistent organisation that we call understanding.
>
> The world we live in, from the vantage point of this new perspective, is always and necessarily the world as we conceptualize it. 'Facts', as Vico saw long ago, are made by us and our way of experiencing, rather than given by an independently existing objective world. But that does not mean that we can make them as we like. They are viable facts as long as they do not clash with experience, as long as they remain tenable in the sense that they continue to do what we expect them to do.
>
> (Glasersfeld, 1983, p. 50–51)

> Constructivism is a theory of knowledge with roots in philosophy, psychology and cybernetics. It asserts two main principles . . . (a)

> knowledge is not passively received but actively built up by the cognizing subject; (b) the function of cognition is adaptive and serves the organization of the experiential world, not the discovery of ontological reality.
>
> (Glasersfeld, 1989, page 162)

This view accounts for the development of subjective knowledge of the external world. It explains how an individual constructs subjective knowledge, notably a theoretical model of a portion of the external world which fits that portion, and how this knowledge or model develops, improving the fit. It does this without presupposing that we construct true knowledge matching[3] the given portion of the world, which would contradict much modern thought, especially in the philosophy of science. Thus the theory provides an account of how external reality serves as a constraint in the construction of subjective knowledge, a constraint that ensures the continued viability of the knowledge. What the theory does not yet do, is to account for the possibility of communication and agreement between individuals. For the sole constraint of fitting the external world does not of itself prevent individuals from having wholly different, incompatible even, subjective models of the world.

Such differences would seem inescapable. However, this is not the case. Suitably elaborated, the social constructivist view also provides an account of the development of knowledge of the world of people and social interaction, and the acquisition of language. The very mechanism which improves the fit of subjective knowledge with the world also accounts for the fit with the social world, including patterns of linguistic use and behaviour. Indeed, the experiential world of the cognizing subject which Glasersfeld refers to, does not differentiate between physical or social reality. Thus the generation and adaptation of personal theories on the basis of sense data and interactions equally applies to the social world, as the following account shows.

Individuals, from the moment of birth, receive sense impressions from, and interact with, the external and social worlds. They also formulate subjective theories to account for, and hence guide, their interactions with these realms. These theories are continually tested through interaction with the environment, animate and inanimate. Part of this mental activity relates to other persons and speech. Heard speech leads to theories concerning word (and sentence) meaning and use. As these theories are conjectured, they are tested out through actions and utterances. The patterns of responses of other individuals (chiefly the mother or guardian, initially) lead to the correction of usage. This leads to the generation of an ever growing set of personal rules of language use. These rules are part of a subjective theory (or family of theories) of language use. But the growth of this theory is not monotonic. The correction of use leads to the abandonment of aspects of it, the adaptation of the theory and hence the refinement of use. This subjective knowledge of language is likely to be more procedural than propositional knowledge. That is, it will be more a matter of 'knowing how' than 'knowing that' (Ryle, 1949).

The acquisition of language involves the exchange of utterances with other individuals in shared social and physical contexts. Such interaction provides encounters with rule governed linguistic behaviour. In other words, it represents the confrontation with, and accommodation to, socially accepted or objective features of language. The acquisition of linguistic competence results from a prolonged period of social interaction. During this period, by dint of repeated utterances and correction, individuals construct subjective theories or personal representations of the rules and conventions underpinning shared language use. The viability of these theories is a function of their mode of development. Quine refers to the 'objective pull', which brings about adequate levels of agreement between individuals utterances and behaviour:

> Society, acting solely on overt manifestations, has been able to train the individual to say the socially proper thing in response even to socially undetectable stimulations.
>
> (Quine, 1960, pages 5–6)

Halliday (1978) describes linguistic competency in terms of mastery of three interlocking systems, namely the forms, the meanings and the (social) functions of language. Of these, language forms and functions are publicly manifested systems, which thus lend themselves to correction and agreement. Whilst the system of meanings is private, the other systems ensure that where they impact on public behaviour, there is a pull towards agreement.

> The uniformity that unites us in communication and belief is a uniformity of resultant patterns overlying a chaotic subjective diversity of connections between words and experience. Uniformity comes where it matters socially . . . Different persons growing up in the same language are like different bushes trimmed and trained to take the shape of identical elephants. The anatomical details of twigs and branches will fulfil the elephantine form differently from bush to bush, but the overall outward results are alike.
>
> (Quine, 1960, page 8)

What has been provided is an account of how individuals acquire (or rather construct) subjective knowledge, including knowledge of language. The two key features of the account are as follows. First of all, there is the active construction of knowledge, typically concepts and hypotheses, on the basis of experiences and previous knowledge. These provide a basis for understanding and serve the purpose of guiding future actions. Secondly, there is the essential role played by experience and interaction with the physical and social worlds, in both the physical actions and speech modes. This experience constitutes the intended use of the knowledge, but it provides the conflicts between intended and perceived outcomes which lead to the restructuring of knowledge, to improve its fit with experience. The shaping effect of experience, to use Quine's metaphor, must not be underestimated. For this is where the full impact of human culture occurs, and where the rules and conventions of language use are constructed by individuals, with the extensive functional outcomes manifested around us in human society.

Bauersfeld describes this theory as the *triadic nature of human knowledge*:

> the subjective structures of knowledge, therefore, are subjective constructions functioning as viable models, which have been formed through adaptations to the resistance of 'the world' and through negotiations in social interactions'
>
> (Grouws *et al.*, 1988, page 39)

The theory has a number of implications for communication, for the representation of information, and for the basis and location of objective knowledge. With regard to communication, the theory imposes severe limits on the possibility of communicating meanings by linguistic or other means. Since the subjective meanings of individuals are uniquely constructed (with certain constraints accommodated in view of their genesis), it is clear that communication cannot be correctly described as the transfer of meanings. Signals can be transmitted and received, but it is impossible to match the meanings that the sender and recipient of the signals attach to them, or even talk of such a match. However, the ways in which linguistic competence is acquired, mean that a fit between sender and receiver meanings can be achieved and sustained, as evidenced by satisfactory participation in shared language games. This view of communication is fully consistent with the Communication Theory of Shannon (cited in Glasersfeld, 1989).

However, there is something I want to call objective, which enters into communication. This is not the informational content of messages, but the pre-existing norms, rules and conventions of linguistic behaviour that every speaker meets (in some form) when entering into a linguistic community. These, in Wittgenstein's (1953) term, are a 'form of life', the enacted rules of linguistic behaviour shared (at least in approximation) by speakers. These rules, represent the constraints of the world of interpersonal communication, which permit the possibility of a fit between senders' and receivers' meanings. Such a fit will depend on the extent to which the actors are drawn from communities which share the same norms of linguistic competence, as well as on the success of the individuals in reconstructing these norms for themselves. These norms or rules are objective, in the sense that they are social, and transcend individuals. However, at any one time, they are located in the regularities of the linguistic behaviour of the group, sustained by the subjective representations of them, in the minds of the individual group members.

A further consequence of the view of subjective knowledge growth concerns the extent to which meanings are inherent in symbolic representations of information, such as a book or a mathematical proof. According to the view proposed, such meanings are the constructions of the reader. (This view is in essence, Derrida's deconstructive approach to textual meanings; Anderson *et al.*, 1986). The linguistic rules, conventions and norms reconstructed by a reader during their acquisition of language constrain the reader to a possible interpretation whose consequences fit with those of other readers. In other words, there is no meaning *per se* in books and proofs. The meanings have to be created by readers, or rather, constructed on the basis of their existing subjective meanings. Within a given linguistic community, the readers' private meaning structures are constructed to fit the constraints of publicly manifested linguistic rules. Thus it is the fit between the readers' subjective theories of language, brought about by a common context of acquisition, including shared constraints, rather than an inherent property of text that brings about a fit between interpretations. However, the social agreement within a community as to how a symbolism is to be decoded constrains individuals' meaning constuctions, giving the sense that there is informational content in the text itself.

This is consistent with the account given of objective knowledge in the previous chapter. For it was stated that public representations of subjective knowledge are just that. Knowledge, truth and meaning cannot be attributed to sets of marks or symbols. Only the assignment of meanings to a set of marks, or a symbol system, which ultimately has to be done by an individual, results in the knowledge or meaning of a published document. As in communications theory, decoding is essential if meaning is to be attributed to a set of broadcast codes.

The social constructivist account of subjective knowledge is also consistent with the conventionalist account of the basis of mathematical, logical and linguistic knowledge given in previous chapters. For according to the constructivist view, the growth of the subjective knowledge of an individual is shaped by interactions with others (and the world). This shaping takes place throughout a linguistic community, so that the constraints accommodated by all of its individuals allow shared participation in language games and activities. These constraints are the objective, publicly manifested rules and conventions of language. On the basis of these constraints individuals construct their own subjective rules and conventions of language. It can be said that these 'fit' (but do not necessarily match) since they allow for shared purposes and interchange which satisfy the participants to any degree of refinement desired.

One problem that arises from the account of the constructivist epistemology that has been given is that it seems to necessitate cumbersome circumlocutions. Knowledge is no longer 'acquired', 'learned', or 'transmitted', but 'constructed' or 'reconstructed' as the creative subjective response of an individual to certain stimuli, based on the individual's pre-existing knowledge which has been shaped to accommodate rules and constraints inferred (or rather induced) from interactions with others. Whilst the latter account is the accurate one from the constructivist viewpoint, it is convenient to retain the former usage on the understanding that it is merely a *façon de parler*, and an abbreviation for the latter.

An analogy for such usage is provided by the language of analysis in mathematics. To say that a function $f(x)$ (defined on the reals) approaches infinity is acceptable, provided this is understood to have the following more refined meaning (provided in the nineteenth century). Namely, that for every real number r there is another s such if $x > s$, then $f(x) > r$. This reformulation no longer says that the function literally approaches infinity, but that for every finite value, there is some point such that thereafter all values of the function exceed it. The two meanings expressed are quite different, but the convention is adopted that the first denotes the second. The rationale for this is that an abbreviated and historically prior mode of speech is retained, which in all contexts can be replaced by a more precise definition. Likewise, we may retain the use of terminology of knowledge transmission in situations where there is no danger of confusion, on the understanding that it has constructivist meaning, which can be unpacked when needed.

In summary, it has been argued that: (a) subjective knowledge is not passively received but actively built up by the cognizing subject, and that the function of cognition is adaptive and serves the organization of the experiential world of the individual (Glasersfeld, 1989). (b) This process accounts for subjective knowledge of the world and language (including mathematics). (c) Objective constraints, both physical and social, have a shaping effect on subjective knowledge, which allows for a 'fit' between aspects of subjective knowledge and the external world, including social and physical features, and other individuals' knowledge. (d) Meanings can only be attributed by individuals, and are not intrinsic to any symbolic system.

B. The Construction of Mathematical Knowledge

It has been argued that linguistic knowledge provides the foundation (genetic and justificatory) for objective mathematical knowledge, both in defending the conventionalist thesis, and subsequently as part of the social constructivist philosophy of mathematics. What is proposed here is the parallel but distinct claim that linguistic knowledge also provides the foundation, both genetic and justificatory, for the subjective knowledge of mathematics. In a previous section we saw how social (i.e. objective) rules of language, logic, etc., circumscribe the acceptance of published mathematical creations, allowing them to become part of the body of objective mathematical knowledge. Thus we were concerned with the subjective origins of objective knowledge. In this section the focus is on the genesis of subjective mathematical knowledge, and it will be argued that the origins of this knowledge lie firmly rooted in linguistic knowledge and competence.

Mathematical knowledge begins, it can be said, with the acquisition of linguistic knowledge. Natural language includes the basis of mathematics through its register of elementary mathematical terms, through everyday knowledge of the uses and interconnections of these terms, and through the rules and conventions which provide the foundation for logic and logical truth. Thus the foundation of mathematical knowledge, both genetic and justificatory, is acquired with language. For both the genetic basis of mathematical concepts and propositions, and the justificatory foundation of propositional mathematical knowledge, are found in this linguistic knowledge. In addition, the structure of subjective mathematical knowledge, particularly its conceptual structure, results from its acquisition through language.

One of the characteristics of mathematical knowledge is its stratified and hierarchical nature, particularly among terms and concepts. This is a logical property of mathematical knowledge, which is manifested both in public expositions of objective mathematical knowledge and, as will be claimed here, in subjective mathematical knowledge. We consider first the hierarchical nature of objective mathematical knowledge.

It is acknowledged that concepts and terms, both in science and mathematics, can be divided into those that are defined and those taken as primitive and undefined, in any theory (see, for example, Popper, 1979; Hempel, 1966; Barker, 1964). The defined terms are defined using other terms. Ultimately, after a finite number of defining links, chains of definition can be chased back to primitive terms, or else the definitions would be based on, and lead to, an infinite regress[4]. On the basis of the division of terms into primitive and defined, a simple inductive definition of the level of every term within an hierarchical structure can be given. Assuming that each concept is named by a term, this provides an hierarchy of both terms and concepts. Let the terms of level *1* be the primitive terms of the theory. Assuming that the terms of level n are defined, we define the terms of level $n + 1$ to be those whose definitions include terms of level n, but none of any higher level (although terms of lower level may be included). This definition unambiguously assigns each term of an objective mathematical theory to a level, and hence determines an hierarchy of terms and concepts (relative to a given theory).[5]

In the domain of subjective knowledge, we can, at least theoretically, divide concepts similarly, into primative observational concepts, and abstract concepts defined in terms of other concepts. Given such a division, an hierarchical structure may be imposed on the terms and concepts of a subjective mathematical theory precisely as above. Indeed Skemp (1971) offers an analysis of this sort. He terms observational and defined concepts primary and secondary concepts, respectively. He bases the notion of conceptual hierarchy upon this distinction in much the same way as above, without assigning numbers to levels. His proposals are based on a logical analysis of the nature of concepts, and their relationships. Thus the notion of a conceptual hierarchy can be utilized in a philosophical theory of subjective knowledge without introducing any empirical conjecture concerning the nature of concepts.

To illustrate the hierarchical nature of subjective mathematical knowledge, consider the following sample contents, which exemplify its linguistic origins. At the lowest level of the hierarchy are basic terms with direct empirical applications, such as 'line', 'triangle', 'cube', 'one', and 'nine'. At higher levels there are terms defined by means of those at lower levels, such as 'shape', 'number', 'addition' and 'collection'. At higher levels still, there are yet more abstract concepts such as 'function', 'set', 'number system', based on those at lower levels, and so on. In this way, the concepts of mathematics are stratified into a hierarchy of many levels. Concepts on succeeding levels are defined implicitly or explicitly in terms of those on lower levels. An implicit definition may take the following form: numbers consist of 'one', 'two', 'three', and other objects with the same properties as these. 'Shape' applies to circles, squares, triangles, and other objects of similar type. Thus new concepts are defined in terms of the implicit properties of a finite set of exemplars, whose membership implicitly includes (explicitly includes, under the new concept) further exemplars of the properties.

It is not the intention to claim here that there is a uniquely defined hierarchy of concepts in either objective or subjective mathematical knowledge. Nor is it claimed than an individual will have but one conceptual hierarchy. Different individuals may construct distinct hierarchies for themselves depending on their unique situations, learning histories, and for particular learning contexts. We saw in the previous section that different individuals' use of the same terms in ways that conform to the social rules of use does not mean that the terms denote identical concepts or meanings (such an assertion would be unverifiable, except negatively). Similarly, such conformity does not mean that individuals' conceptual structures are isomorphic, with corresponding connections. All that can be claimed is that the subjective conceptual knowledge of mathematics of an individual is ordered hierarchically.

It is conjectured that the generation of a hierarchy of increasingly abstract concepts reflects a particular tendency in the genesis of human mathematical knowledge. Namely, to generalize and abstract the shared structural features of previously existing knowledge in the formation of new concepts and knowledge. We conjecture the existence of some such mechanism to account for the genesis of abstract concepts and knowledge (as was noted above). At each succeeding level of the conceptual hierarchy described, we see the results of this process. That is the appearance of new concepts implicitly defined in terms of a finite set of lower level terms or concepts.

This abstractive, vertical process contrasts with a second mode of mathematical knowledge generation: the refinement, elaboration or combination of existing knowledge, without necessarily moving to a higher level of abstraction. Thus the genesis of mathematical knowledge and ideas within individual minds is conjectured to involve both vertical and horizontal processes, relative to an individual's conceptual hierarchy. These directions are analogous with those involved in inductive and deductive processes, respectively. We discuss both these modes of knowledge generation in turn, beginning with that described as vertical.

Before continuing with the exposition of the mechanisms underpinning the genesis of mathematical knowledge, a methodological remark is called for. It should be noted that the conjectures concerning the vertical and horizontal modes of thought in the genesis of subjective mathematical knowledge are inessential for social constructivism. It has been argued that some (mental) mechanism is needed to account for the generation of abstract knowledge from particular and concrete experience. This is central to social constructivism. But as a philosophy of mathematics it is not necessary to analyse this mechanism further, or to conjecture its properties. Thus the rejection of the following exploration of this mechanism need not entail the rejection of the social constructivist philosophy of mathematics.

The vertical processes of subjective knowledge generation involve generalization, abstraction and reification, and include concept formation. Typically, this process involves the transformation of properties, constructions, or collections of constructions into objects. Thus, for example, we can rationally reconstruct the creation of the number concept, beginning with ordination, to illustrate this process. The ordinal number '5' is associated with the 5th member of a counting sequence, ranging over 5 objects. This becomes abstracted from the particular order of counting, and a generalization '5', is applied as an adjective to the whole collection of 5 objects. The adjective '5' (applicable to a set), is *reified* into an object, '5', which is a noun, the name of a thing-in-itself. Later, the collection of such numbers is reified into the set 'number'. Thus we see how a path can be constructed from a concrete operation (using the ordinal number '5'), through the processes of abstraction and reification, which ultimately leads (via the cardinal number '5') to the abstract concept of 'number'. This account is not offered as a psychological hypothesis, but as a theoretical reconstruction of the genesis of subjective mathematical knowledge by abstraction.

What is proposed is that by a vertical process of abstraction or concept formation, a collection of objects or constructions at lower, pre-existing levels of a personal concept hierarchy become 'reified' into an object-like concept, or noun-like term. Skemp refers to this 'detachability', or 'the ability to isolate concepts from any of the examples which give rise to them' (Skemp, 1971, page 28) as an essential part of the process of abstraction in concept formation. Such a newly defined concept applies to those lower level concepts whose properties it abstracts, but it has a generality that goes beyond them. The term 'reification' is applied because such a newly formed concept acquires an integrity and the properties of a primitive mathematical object, which means that it can be treated as a unity, and at a subsequent stage it too can be abstracted from, in an iteration of the process.[6]

The increasing complexity of subjective mathematical knowledge can also be attributed to horizontal processes of concept and property elaboration and clarification. This horizontal process of object formation in mathematics is that described by Lakatos (1976), in his reconstruction of the evolution of the Euler formula and its justification. Namely, the reformulation (and 'stretching') of mathematical concepts or definitions to achieve consistency and coherence in their relationships within a broader context. This is essentially a process of elaboration and refinement, unlike the vertical process which lies behind 'objectification' or 'reification'.

Thus far, the account given has dwelt on the genesis and structure of the conceptual and terminological part of subjective mathematics. There is also the genesis of the propositions, relationships and conjectures of subjective mathematical knowledge to be considered. But this can be accommodated analogously. We have already discussed how the elementary truths of mathematics and logic are acquired during the learning of mathematical language. As new concepts are developed by individuals, following the hierarchical pattern described above, their definitions, properties and relationships underpin new mathematical propositions, which must be acquired with them, to permit their uses. New items of propositional knowledge are developed by the two modes of genesis described above, namely by informal inductive and deductive processes. Intuition being the name given to the facility of perceiving (i.e., conjecturing with belief) such propositions and relationships between mathematical concepts on the basis of their meaning and properties, prior to the production of warrants for justifying them. Overall, we see, therefore, that the general features of the account of the genesis of mathematical concepts also holds for propositional mathematical knowledge. That is we posit analogous inductive and deductive processes, albeit informal, to account for this genesis.

In summary, this section has dealt with the genesis of the concepts and propositions of subjective mathematical knowledge. The account given of this genesis involves four claims. First of all, the concepts and propositions of mathematics originate and are rooted in those of natural language, and are acquired (constructed) alongside linguistic competence. Secondly, that they can be divided into primitive and derived concepts and propositions. The concepts can be divided into those based on observation and direct sensory experience, and those defined linguistically by means of other terms and concepts, or abstracted from them. Likewise, the propositions consist of those acquired linguistically, and those derived from pre-existing mathematical propositions, although this distinction is not claimed to be clear cut. Thirdly, the division of concepts, coupled with the order of their definition, results in a subjective (and personal) hierarchical structure of concepts (with which the propositions are associated, according to their constituent concepts). Fourthly, the genesis of the concepts and propositions of subjective mathematics utilizes both vertical and horizontal processes of concept and proposition derivation, which take the form of inductive and deductive reasoning.

These claims comprise the social constructivist account of the genesis of subjective mathematical knowledge. However, in providing the accounts, examples have been given, especially concerning the third and fourth of these claims, which may have the status of empirical conjectures. The hierarchical nature of subjective mathematical knowledge can be accepted, without relying on such empirical conjectures. Likewise the existence of the horizontal process of subjective concept refinement or propositional deduction, by analogy with Lakatos' logic of mathematical discovery, can be accepted in principle. This leaves only the vertical processes of abstraction, reification or induction to account for, without assuming empirical grounds. But some such procedure is necessary, if subjective knowledge is to be constructed by individuals on the basis of primitive concepts derived from sense impressions and interactions, or elementary mathematical propositions embedded in language use, as we have assumed. For it is clear that relatively abstract knowledge must be constructed from relatively concrete knowledge, to account for the increasing abstraction of the subjective knowledge of mathematics. Hence, as with the horizontal process, the existence of this vertical process is needed in principle, irrespective of the fact that some of the details included in the account might be construed as empirical conjectures. For this reason, these details were characterized as inessential to the central thesis of social constructivism.

C. *Subjective Belief in the Existence of Mathematical Objects*

The account given above of the development of individuals' knowledge of the external world is that it is a free construction of the individual subject to the constraints of the physical and social worlds. The individual directly experiences these worlds and has his or her conjectural maps of these worlds confirmed as viable or demonstrated to be inadequate on the basis of the responses to their actions. The consequence of this is that the individual constructs personal representations of these worlds, which are unique and idiosyncratic to that individual, but whose consequences fit with what is socially accepted. Such a fit is due to the shared external constraints which all individuals accommodate (more or less), and in particular, the constraint of viable negotiation of meanings and purposes in social intercourse. Thus, according to this account, individuals' construct their own subjective knowledge and concepts of the external and social worlds, as well as that of mathematics, so that they fit with what is socially accepted.

These self-constructed worlds represent reality to the individuals who have made them, be it physical or social reality. Since the same mechanism lies behind the construction of mathematics as the other representations, it is not surprising that it too seems to have a measure of independent existence. For the objects of mathematics have objectivity, in that they are socially accepted constructs. Other socially constructed concepts are known to have a powerful impact upon our lives, such as 'money', 'time' (o'clock), 'the North Pole', 'the equator', 'England', 'gender', 'justice' and 'truth'. Each of these is, undeniably a social construct. Yet each of these concepts has as tangible an impact as many concretely existing objects.

Consider 'money'. This represents an organizing concept in modern social life of great power, and more to the point, of undeniable existence. Yet it is clearly an abstract human-made symbol of conventional, quantified value, as opposed to some aspect of the physical world. Let us explore 'money' further. What is it that gives money its existence? There are two features on which its ontological status is based. First, it is socially accepted, which gives it objectivity. Second, it is represented by tokens, which means that it has tangible concrete reference.

Now consider the analogy with the objects of mathematics. These have objectivity, being socially accepted. In addition, the primitive concepts of mathematics, such as 'square' and '7', have concrete examples in our perceptions of the physical world. So far, the analogy is good. The defined concepts of mathematics do not fit so well with the analogy, for they may only have concrete applications indirectly, via chains of definition. Although there is an analogy between these abstract objects of mathematics and the more abstracted applications of money (budgeting, financial forecasting, etc.), this is stretching it a bit far. What can be said is that the analogy between 'money' and mathematical objects lends some plausibility to subjective belief in the latter objects. They are both objective social constructions and have concrete manifestations.

Of course mathematics has a further feature supporting this belief. This is the necessary relationships between its objects, due to their strict logical relationships in deductive systems. Logical necessity attaches to the objects of mathematics through their defining relationships, their inter-relationships and their relationships with mathematical knowledge. This lends necessity to the objects of mathematics (a feature that money lacks).

In a nutshell, the argument is this. If an individual's knowledge of the real world, including its conventional components, is a mental construct constrained by social acceptance, then belief in such constructs evidently can be as strong as beliefs in anything. Subjective knowledge of mathematics, and acquaintance with its concepts and objects is also a mental construct. But like other socially determined constructs, it has an external objectivity arising from its social acceptance. The objects of mathematics also have (i) concrete exemplifications, either directly (for the primitive mathematical concepts), or indirectly (for the defined mathematical concepts); and (ii) logical necessity, through their logical foundations and deductive structure. These properties are what give rise to a belief in the objective existence of mathematics and its objects.

Traditionally, knowledge has been divided into the real and the ideal. It is common to accept the reality of the external world and our scientific knowledge of it (scientific realism). It is also common to accept the ideal existence of (objective) mathematics and mathematical objects (idealism or platonism). This dichotomy places physical and scientific objects in one realm (Popper's world 1) and mathematical objects in another (subjective knowledge of them in world 2, objective knowledge in world 3). Thus it places mathematical and physical objects in different categories. The social constructivist thesis is that we have no direct access to world 1, and that physical and scientific objects are only accessible when *represented* by constructs in world 3 (objective concepts) or in world 2 (subjective concepts). Thus our knowledge of physical and mathematical objects has the same status, contrary to traditional views. The difference resides only in the nature of the constraints physical reality imposes on scientific concepts, through the means of verification adopted for the two types of knowledge (scientific or mathematical). The similarity, including the social basis of the objectivity of both types of knowledge, accounts for the subjective belief in the existence of mathematical objects (almost) just as for theoretical physical objects.[7]

3. Relating Objective and Subjective Knowledge of Mathematics

The relationship between subjective and objective knowledge of mathematics is central to the social constructivist philosophy of mathematics. According to this philosophy, these realms are mutually dependent, and serve to recreate each other. First of all, objective mathematical knowledge is reconstructed as subjective knowledge by the individual, through interactions with teachers and other persons, and by interpreting texts and other inanimate sources. As has been stressed, interactions with other persons (and the environment), especially through negative feedback, provides the means for developing a fit between an individuals subjective knowledge of mathematics, and socially accepted, objective mathematics. The term 'reconstruction', as applied to the subjective representation of mathematical knowledge, must not be taken to imply that this representation matches objective mathematical knowledge. As has been said, it is rather that the subjective knowledge 'fits', to a greater or lesser extent, socially accepted knowledge of mathematics (in one or more of its manifestations).

Secondly, subjective mathematical knowledge has an impact on objective knowledge in two ways. The route through which individuals' mathematical creations become a part of objective mathematical knowledge, provided they survive criticism, has been described. This represents the avenue by means of which new creations (including the restructuring of pre-existing mathematics) are added to the body of objective mathematical knowledge. It also represents the way in which existing mathematical theories are reformulated, inter-related or unified. Thus it includes creation not only at the edges of mathematical knowledge, but also throughout the body of mathematical knowledge. This is the way that subjective knowledge of mathematics *explicitly* contributes to the creation of objective mathematical knowledge.[8] However, there is also a more far-reaching but *implicit* way in which subjective mathematical knowledge contributes to objective mathematical knowledge.

The social constructivist view is that objective knowledge of mathematics is social, and is not contained in texts or other recorded materials, nor in some ideal realm. Objective knowledge of mathematics resides in the shared rules, conventions, understandings and meanings of the individual members of society, and in their interactions (and consequently, their social institutions). Thus objective knowledge of mathematics is continually recreated and renewed by the growth of subjective knowledge of mathematics, in the minds of countless individuals. This provides the substratum which supports objective knowledge, for it is through subjective representations that the social, the rules and conventions of language and human interaction, is sustained. These mutually observed rules, in their turn, legitimate certain formulations of mathematics as accepted objective mathematical knowledge. Thus objective knowledge of mathematics survives through a social group enduring and reproducing itself. Through passing on their subjective knowledge of mathematics, including their knowledge of the meaning to be attributed to the symbolism in published mathematical texts, objective knowledge of mathematics passes from one generation to the next.

This process of transmission does not merely account for the genesis of mathematical knowledge. It is also the means by which both the justificatory canons for mathematical knowledge, and the warrants justifying mathematical knowledge itself are sustained. Kitcher (1984) likewise claims that the basis for the justification of objective mathematical knowledge is passed on in this way, from one generation of mathematicians to the next, starting with empirically warranted knowledge.

As a rational reconstruction of mathematical history to warrant mathematical knowledge, Kitcher's account has some plausibility. Like Kitcher, social constructivism sees as primary the social community whose acceptance confers objectivity on mathematical knowledge. However, unlike Kitcher, social constructivism sees the social as sustaining the full rational justification for objective mathematical knowledge, without the need for historical support for this justification. According to social constructivism, the social community which sustains mathematics endures smoothly over history, with all its functions intact, just as a biological organism smoothly survives the death and replacement of its cells. These functions include all that is needed for warranting mathematical knowledge.

It should be made clear that the claim that objective mathematical knowledge is sustained by the subjective knowledge of members and society does not imply the reducibility of the objective to the subjective. Objective knowledge of mathematics depends upon social institutions, including established 'forms of life' and patterns of social interaction. These are sustained, admittedly, by subjective knowledge and individual patterns of behaviour as is the social phenomenon of language. But this no more implies the reducibility of the objective to the subjective, than materialism implies that thought can be reduced to, and explained in terms of physics. The sum of all subjective knowledge is not objective knowledge. Subjective knowledge is essentially private, whereas objective knowledge is public and social. Thus although objective knowledge of mathematics rests on the substratum of subjective knowledge, which continually recreates it, it is not reducible to subjective knowledge.

As a thought experiment, imagine that all social institutions and personal interactions ceased to exist. Although this would leave subjective knowledge of mathematics intact, it would destroy objective mathematics. Not necessarily immediately, but certainly within one lifetime. For without social interaction there could be no acquisition of natural language, on which mathematics rests. Without interaction and the negotiation of meanings to ensure a continued fit, individual's subjective knowledge would begin to develop idiosyncratically, to grow apart, unchecked. The objective knowledge of mathematics, and all the implicit knowledge sustaining it, such as the justificatory canons, would cease to be passed on. Naturally no new mathematics could be socially accepted either. Thus the death of the social would spell the death of objective mathematics, irrespective of the survival of subjective knowledge.

The converse also holds true. If, as another thought experiment we imagine that all subjective knowledge of mathematics ceased to exist, then so too would objective knowledge of mathematics cease to exist. For no individual could legitimately assent to any symbolic representations as embodying acceptable mathematics, being deprived of the basis for such assent. Therefore there could be no acceptance of mathematics by any social group. This establishes the converse relationship, namely that the existence of subjective knowledge is necessary for there to be objective knowledge of mathematics.

Of course it is hard to follow through all the consequences of the second thought experiment, because of the impossibility of separating out an individual's subjective knowledge of language and mathematics. Knowledge of language depends heavily on the conceptual tools for classifying, categorising and quantifying our experience and for framing logical utterances. But according to social constructivism these form the basis for mathematical knowledge. If we delete these from subjective knowledge in the thought experiment, then virtually all knowledge of language and its conceptual hierarchy, would collapse. If we leave this informal knowledge and only debate explicit knowledge of mathematics (that learned as mathematics and not as language), then subjective knowledge of mathematics could be rebuilt, for we would have left its foundations intact.[9]

In summary, the social constructivist thesis is that objective knowledge of mathematics exists in and through the social world of human actions, interactions and rules, supported by individuals' subjective knowledge of mathematics (and language and social life), which need constant re-creation. Thus subjective knowledge re-creates objective knowledge, without the latter being reducible to the former. Such a view of knowledge is endorsed by a number of authors. Paul Cobb, argues from a radical constructivist perspective that:

> the view that cultural knowledge in general and mathematics in particular can be taken as solid bedrock upon which to anchor analyses of learning and teaching is also questioned. Instead it is argued that cultural knowledge (including mathematics) is continually recreated through the coordinated actions of the members of a community.
>
> (Cobb, 1988, page 13)

Paulo Freire has elaborated an epistemology and philosophy of education that places individual consciousness, in the context of the social, at the heart of objective knowledge. He 'recognize(s) the indisputable unity between subjectivity and objectivity in the act of knowing.' (Freire, 1972b, page 31) Freire argues, as we have done, that objective knowledge is continually created and re-created as people reflect and act on the world.

Even the received view of epistemology (see, for example, Sheffler, 1965) can be interpreted as logically founding objective knowledge on subjective knowledge. For this view defines knowledge (rather more narrowly than it has been used above) as justified true belief. Belief includes what has been termed subjective knowledge, in this chapter. In mathematics, justified true belief can be interpreted as consisting of assertions that have a justification necessitating their acceptance (in short, a proof). According to the social constructivist philosophy, such mathematical statements are socially accepted, on the basis of their justification, and thus constitute objective mathematical knowledge. Thus, in the terms of this chapter, 'knowledge is justified true belief' translates into 'objective knowlege of mathematics is socially acccepted subjective knowledge, expressed in the form of linguistic assertions'. According to this translation, objective knowledge of mathematics depends logically on subjective knowledge, because of the order of definition.

The social constructivist view of mathematics places subjective and objective knowledge in mutually supportive and dependent positions. Subjective knowledge leads to the creation of mathematical knowledge, via the medium of social interaction and acceptance. It also sustains and re-creates objective knowledge, which rests on the subjective knowledge of individuals. Representations of objective knowledge are what allow the genesis and re-creation of subjective knowledge. So we have a creative cycle, with subjective knowledge creating objective knowledge, which in turn leads to the creation of subjective knowledge. Figure 4.1 shows the links between the private realm of subjective knowledge and the social realm of objective knowledge each sustaining the creation of the other. Each must be publicly represented for this purpose. Thereupon there is an interactive social negotiation process leading to the reformulation of the knowledge and its incorporation into the other realm as new knowledge.

Of course there are powerful constraints at work throughout this creative cycle. These are the physical and social worlds, and in particular the linguistic and other rules embodied in social forms of life.

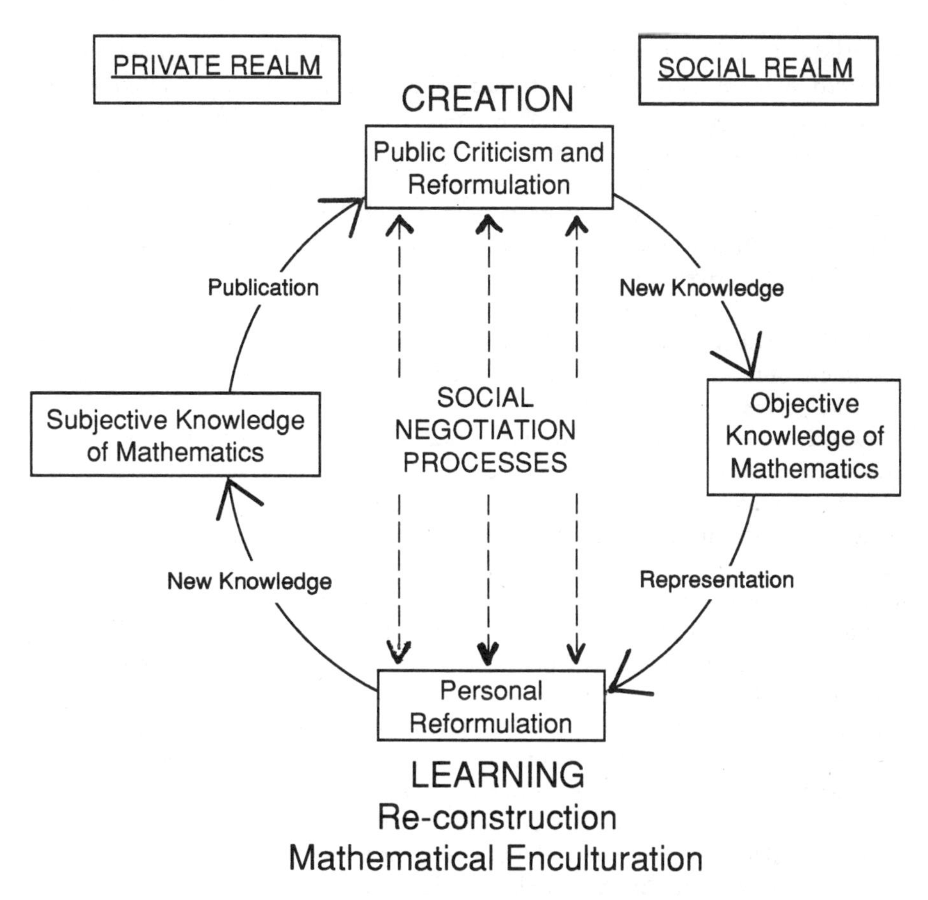

Figure 4.1: The Relationship between Objective and Subjective Knowledge of Mathematics

4. Criticism of Social Constructivism

The account of social constructivism has brought together three philosophical perspectives as a basis for a unified philosophy of mathematics. These are quasi-empiricism, conventionalism and radical constructivism. As a consquence, two types of criticism can be directed at social constructivism. First of all, there is criticism directed at the assumptions and philosophical stance adopted by one of these tributary philosophies. For example, there are the problems in accounting for logical necessity in mathematics from the social perspective of conventionalism. An attempt has been made to anticipate and answer such criticisms. To those who reject the conventionalist assumptions such arguments will not be convincing. However, no further criticisms of this type, that is, directed at the tributary philosophies, will be addressed. For in addition to having been treated above, they can also be found in the relevant literature.

Secondly, there are criticisms that can be directed at the novel synthesis that is provided by social constructivism. Two criticisms in this category will be considered. These concern the use of empirical assumptions in social constructivism, and the tension between the subsumed theories of conventionalism and radical constructivism.

Throughout the exposition, it has been argued that the genesis as well as the justification of knowledge is the proper concern of the philosophy of mathematics. Consequently both these contexts have been discussed, but an attempt has been made to distinguish between them carefully, and to avoid or to demarcate carefully any empirical assumptions, especially concerning the genesis of mathematical knowledge. This was a criticism directed at Lakatos in Chapter 2, that in his account of the conditions of the genesis of mathematical knowledge, he introduced an historical (i.e. empirical) conjecture. It may be felt that the social constructivist account errs similarly. However, I believe that this would indicate that a clearer exposition of social constructivism is needed, rather than necessitating a rejection of the entire philosophy.

A more substantial criticism arises from a possible tension, or even inconsistency, between the subsumed theories of conventionalism and radical constructivism. For the former gives primacy to the social, comprising accepted rules and conventions underpinning the use of language and objective knowledge of mathematics. This reflects a 'form of life', constituting accepted social and verbal behaviour patterns. The latter, gives primacy to the knowing subject, an unreachable monad constructing hypothetical world-pictures to represent experiences of an unknowable reality. To an adherent of one but not the other philosophy, their conjunction in social constructivism may seem to be an unholy alliance, for neither of the two foci is given precedence. Rather each is the centre of a separate realm. The knowing subject is at the centre of the private realm of individuals and subjective knowledge. This realm assumes a real but unknowable world, as well as the knowing subject. But this realm is not enough to account for objective knowledge, let alone for humanity. For humankind is a social animal, and depends essentially on interchange and language. The social realm takes this as its basis, including social institutions and social agreements (albeit tacit). This realm assumes the existence of social groups of human beings. However, this perspective seems weak in terms of the interior life and consciousness it ascribes to individuals.

Thus although the primacy of focus of each of conventionalism and radical constructivism is sacrificed in social constructivism, their conjunction in it serves to compensate for their individual weaknesses, yet this conjunction raises the question as to their mutual consistency. In answer it can be said that they treat different domains, and both involve social negotiation at their boundaries (as Figure 4.1 illustrates). Thus inconsistency seems unlikely, for it could only come about from their straying over the interface of social interaction, into each other's domains.

The separateness of the private and social realms, together with their separate theoretical accounts, has another consequence. It means that the parts of the social constructivist account could be modified (e.g., the account of subjective knowledge) without changing the whole philosophy. This suggests that the philosophy lacks a single overarching principle. However, there are unifying concepts (or metaphors) which unite the private and social realms, namely construction and negotiation. For both subjective and objective knowledge are deemed to be human constructions, built up from pre-existing knowledge components. The second unifying concept is that of social negotiation. This not only plays a central role in the shaping of subjective and objective knowledge. It also plays a key role in the justification of mathematical knowledge according to social constructivism, from the quasi-empiricist component.

Notes

1 Reference is made to *subjective knowledge*. Philosophically this usage is problematic, because knowlege has been defined as justified belief, where the justification is understood to be objective. Thus knowledge is socially verifiable and socially accepted belief. But subjective belief which is not publicly represented cannot be socially accepted, and hence cannot be knowledge. To follow this strict usage, all references to subjective knowledge should be replaced by references to subjective belief. However, there is a well known precedent for continuing with the current usage. Individuals are frequently described as possessing procedural knowledge (Sheffler, 1965) or knowledge as 'know how' (Ryle, 1949). Thus a precedent exists for not observing the strict usage described above. I shall continue to refer to 'subjective knowledge', in full consciousness of the possible transgression involved, but without assuming that subjective knowledge is objectively justified.

2 There is also a powerful analogy with Kuhn's (1970) theory of the structure of scientific revolutions. For in the proposed account we have an alternation between viable subjective theories used to guide action, which parallels Kuhn's periods of 'normal science', followed by conflicts between the predictions of the subjective theories and observations, paralleling Kuhn's periods of 'revolutionary science' (as well as Popper's falsification of theories), when the old subjective theories are rejected, and new theories are built from them and accepted, completing the cycle (see Ernest, 1990).

3 I follow Glasersfeld's (1984) use of the terms 'fit' and 'match' which make an important distinction. They both refer to the relationship between a representation and what it represents. Thus a map 'matches' the geographical region it represents because there is a morphism, a structure preserving relation, between them. In this way 'matching' resembles the correspondence theory of truth. In contrast, a key 'fits' a lock when it works and turns the lock. Such a 'fit' does not represent the structure of the lock, but merely a functional relationship between the two. When a theory, account or representation 'fits' it means that it satisfies a basic set of constraints, but that beyond that it can have any form. 'Fitting' resembles the pragmatic conception of truth. The crucial difference is that theories that 'fit' a portion of the world do not tell us about the structure of the world, whereas any that 'match' would do so.

4 If this assertion is considered to be controversial, we simply exclude from the discussion any terms which do not fit the description. Subsequent discussion of hierarchies therefore only refer to those terms included in them, and not any excluded circularly defined terms.

5 This hierarchy can also be used as a basis for imposing a hierarchical structure on the sentences and formulas of the theory. We simply define the level of a sentence or formula to be the maximum level of the terms occurring in it.

6 Such an account has previously been offered to explain the genesis of set-theoretic objects from mathematical constructions, within an objective philosophy of mathematics, by Machover (1983). The purpose being to found abstract, classical mathematics on the more intuitive (and presumably safer) realm of constructive mathematics, analogous to the formalist strategy. However, by proposing that the objects of mathematics are interpreted as 'reified constructions', Machover is opening the door for a genetic epistemology in mathematics, as is proposed by Piaget (1972) and Kitcher (1984), among others, albeit from a different viewpoint.

It is also worth mentioning that the vertical process of object formation or reification is part of set theory, both formal and informal, in the standard objective mathematical theory. Typically, such a process is permitted by a comprehension axiom, which allows the collection of objects satisfying a defined mathematical property to become a new mathematical object, a set:

For all P, there exists p, such that p = {x/*P*(x)}) (Where 'P(x)' is a first-order defined property with free variable 'x', and 'p' is the set comprising the extension of 'P(x)').

Of course such an unrestricted comprehension principle leads to contradictions (Russell, 1902). Indeed many of the innovations in logic and set theory in the early part of this century (Russell's theory of types, Zermelo-Fraenkel and Godel-Bernays-von Neumann set theories) were expressly motivated to permit the safe use of some form of the comprehension principle.

Thus the process of reification, i.e., the elevation of (the extension of) a property to objecthood, is an established principle in objective mathematics. Since, according to social constructivism, objective mathematics is a socially accepted reflection of subjective mathematics, this adds plausibility to the assumption of the vertical reifying process in the genesis of subjective mathematical knowledge.

7 There is a further explanation for the subjective belief in the objects of mathematics based on the propensity to objectify and reify the concepts of mathematics into objects, discussed above. Once objectified, such mentally constructed mathematical objects can be (nearly) as potent as physical objects. Indeed a platonist view of mathematical existence is often held by practising mathematicians (such as Frege, Hardy, Godel, Thom) whose extensive mathematical activities by constantly using their mathematical concepts and objects reinforce their subjective 'solidity' or 'object'-like qualities. Thus the reification of mathematical objects, coupled with other features, such as a propensity for belief in socially accepted constructs, gives them an apparent existence, accounting for some mathematicians' platonism.

8 Subjective knowledge of mathematics (*syntactical* knowledge of mathematics, Schwab, 1975) also plays a part in the process of acceptance of new mathematical knowledge, via the process of criticism.

9 Popper (1979, pages 107-108) uses this second interpretation of the thought experiment as a basis for an argument that books contain objective knowledge. For with our capacity to learn from books intact, Popper argues that we can re-acquire objective knowledge from libraries. However, the social constructivist view is that only the explicit, more advanced part of subjective (and objective) knowledge is eradicated in this interpretation of the thought experiment. The reconstruction of these forms of knowledge with the aid of libraries shows that the foundations of this knowledge has survived. (However, this conclusion does not contradict Popper's, since his sense of 'objectivity' does not include implicit, socially accepted linguistic knowledge).

5

The Parallels of Social Constructivism

1. Introduction

Today's burgeoning intellectual climate views all human knowledge as a social construction. Mathematics, the last bastion of certainty in knowledge, has been trying to resist this current of thought. But as we have seen, more and more authors are joining the stream and viewing mathematics as a social construction. Of course social constructivism, the particular philosophy expounded above, is just one account of mathematics from this perspective. Not many alternatives are yet formulated in the *philosophy* of mathematics, but beyond, parallel views of mathematics and knowledge abound in other disciplines. This chapter explores some of these parallels, showing how overwhelming the intellectual current is becoming.

2. Philosophical Parallels

A. Sceptical Philosophy

The most central claim of social constructivism is that no certain knowledge is possible, and in particular no certain knowledge of mathematics is possible. Concerning empirical knowledge, this thesis is one that is subscribed to by many philosophers. These include continental sceptics beginning with Descartes; British empiricists such as Hume; American pragmatists such as James and Dewey; some modern American philosophers such as Goodman, Putnam, Quine and Rorty; and modern philosophers of science, including Popper, Kuhn, Feyerabend, Lakatos and Laudan.

Throughout a number of these strands of thought there is agreement that empirical knowledge of the world is a human construction. Beyond those cited, this view is shared by Kant and his followers, who see any knowledge of the world as shaped by innate mental categories of thought.

Scepticism concerning absolute empirical knowledge has grown to be the dominant view. However, until Lakatos (1962) the full extension of scepticism to mathematics was not made. Since then, it has gained partial acceptance, whilst remaining controversial. Social constructivism is an attempt to extend Lakatos' sceptical approach systematically to a philosophy of mathematics.

However, social constructivism does not entail the fullest form of scepticism, such as cartesian doubt. For it accepts the existence of the physical world (whilst denying *certain* knowledge of it) and accepts the existence of language and the social group. Both the physical and the social worlds play an essential part in the social constructivist epistemology. As a commentator on Wittgenstein says: 'Doubt presupposes mastery of a language-game.' Kenny (1973, page 206) Social constructivism is sceptical about the possibility of any certain knowledge, particularly in mathematics, but it is not sceptical about the necessary pre-conditions for knowledge.

B. The Private Language Problem

One challenge for the social constructivist account of subjective knowledge is the 'private language' problem. If an individual's concepts are personal constructions, how are they able to communicate using a shared language? Why should different mathematicians understand the same thing by a concept or proposition, when their meanings are personally unique? May not each have a private language, to refer to his or her own private meanings?

Social constructivism overcomes this problem through the interpersonal negotiation of meanings to achieve a 'fit'. Support for this position, if not the precise form of argument, is widespread. Wittgenstein (1953) first answered the problem, arguing that private languages cannot exist. A number of philosophers commenting on his work, such as Kenny (1973) and Bloor (1983), support the rejection of private languages, as do others including Ayer (1956) and Quine (1960). With regard to mathematics, the private language problem is also considered soluble, for example by Tymoczko (1985) and Lerman (1989), both arguing from a position close to social constructivism.

The solution of the private language problem by social constructivism reflects a substantial body of philosophical opinion. Generally, it is argued that the shared rules and 'objective pull' of inter-personal language use makes it public, consistent with social constructivism.

C. Knowing and the Development of Knowledge

The social constructivist philosophy of mathematics treats knowledge as the result of a process of coming to know, including the social processes leading to the justification of mathematical knowledge. Thus it attaches great weight to knowing and the *development* of knowledge, in addition to its product, knowledge. This emphasis, although far from universal, is to be found in the works of a number of philosophers, including Dewey (1950), Polanyi (1958), Rorty (1979), Toulmin (1972), Wittgenstein (1953) and Haack (1979).

Other authors have looked to an evolutionary model to account for the growth and development of knowledge. This includes the genetic epistemology of Piaget (1972, 1977), and the evolutionary epistemologies of Popper (1979), Toulmin (1972) and Lorenz (1977).

The majority of modern philosophers of science view it as a growing and developing body of knowledge either detached from history (Popper, 1979) or embedded in human history (Kuhn, Feyerabend, Lakatos, Toulmin and Laudan).

Educational thinkers have also stressed the processes and means of knowledge acquisition, as a basis for the curriculum, including, most notably, Schwab (1975) and Bruner (1960).

The process of coming to know relates to practical knowledge and the applications of knowledge. Ryle (1949) established that practical knowledge ('knowing how') belongs to epistemology as well as declarative knowledge ('knowing that'). Sneed (1971) proposes a model of scientific knowledge which incorporates the range of intended applications (models) as well as the core theory. This model has been extended to mathematics by Jahnke (Steiner, 1987). Such approaches admitting practical knowledge or its applications into the traditional domain of knowledge thus parallel aspects of the social constructivist proposals.

The social constructivist account of the nature and genesis of subjective knowledge of mathematics is to a large extent based on the radical constructivism of Glasersfeld (1984, 1989). This has parallels in the thought of Kant, and even more so, Vico, as well as with the American pragmatists and modern philosophers of science cited above.

Thus there is a growing current of thought in modern philosophy which gives a central place in epistemology to considerations of the human activity of knowing and the evolution of knowledge, as in social constructivism.

D. The Divisions of Knowledge

A key tenet of social constructivism, following Lakatos, is that mathematical knowledge is quasi-empirical. This leads to the rejection of the categorical distinction between *a priori* knowledge of mathematics, and empirical knowledge. Other philosophers have also rejected this distinction, most notably Duhem and Quine (1951), who hold that because the assertions of mathematics and science are all part of a continuous body of knowledge, the distinction between them is one of degree, and not of kind or category. White (1950) and Wittgenstein (1953) also reject the absoluteness of this distinction, and a growing number of other philosophers also reject the watertight division between knowledge and its empirical applications (Ryle, 1949; Sneed, 1971; Jahnke).

A further parallel is found in 'post-structuralist' and 'post-modernist' philosophers, such as Foucault (1972) and Lyotard (1984), who take the existence of human culture as their starting point. Foucault claims that the divisions of knowledge are modern constructs, defined from certain social perspectives. Throughout history, he argues, the different disciplines have changed. Their objects, concepts, accepted rules of thought and aims have evolved and changed, even amounting, in extreme cases, to discontinuities. Knowledge, in his view, is but one component of 'discursive practice', which includes language, social context and social relations. In evidence, he documents how certain socially privileged groups, such as doctors and lawyers, have established discourses creating new objects of thought, grouping together hitherto unconnected phenomena defined as delinquent behaviour or crime. Elsewhere, Foucault (1981) shows how a new area of knowledge, the discourse of human sexuality, was defined by church and state, to serve their own interests.

Lyotard (1984) considers all human knowledge to consist of narratives, whether literary or scientific. Each disciplined narrative has its own legitimation criteria, which are internal, and which develop to overcome or engulf contradictions. He describes how mathematics overcame crises in its axiomatic foundations due to Godel's Theorem by incorporating meta-mathematics into an enlarged research paradigm. He also claims that continuous differentiable functions are losing their pre-eminence as paradigms of knowledge and prediction, as mathematics incorporates undecidability, incompleteness, Catastrophe theory and chaos. Thus a static system of logic and rationality does not underpin mathematics, or any discipline. Rather they rest on narratives and language games, which shift with organic changes of culture.

These thinkers exemplify a move to view the traditional objective criteria of knowledge and truth within the disciplines as internal myths, which attempt to deny the social basis of all knowing. This new intellectual tradition affirms that all human knowledge is interconnected through a shared cultural substratum, as social constructivism asserts.

Another post-structuralist is Derrida, who as well as supporting this view, argues for the 'deconstructive' reading of texts:

> In writing, the text is set free from the writer. It is released to the public who find meaning in it as they read it. These readings are the product of circumstance. The same holds true even for philosophy. There can be no way of fixing readings . . .
>
> Anderson *et al*. (1986, page 124)

This offers a parallel to the social constructivist thesis that mathematical texts are empty of meaning. Meanings must be constructed for them by individuals or groups on the basis of their knowledge (and context).

E. The Philosophy of Mathematics

Various modern philosophers of mathematics have views consistent with some if not all of the theses of social constructivism. Here we draw together some of the points of contact between them and social constructivism.

Some philosophers emphasize the significance of the history and empirical aspects of mathematics for philosophy. Kitcher (1984) erects a system basing the justification of mathematical knowledge on its empirical basis, with the justification transmitted from generation to generation by the mathematical community.[1] An empirical or quasi-empirical justification of mathematical knowledge, drawing on mathematical practice, is also adopted by N. D. Goodman, Wang, P. Davis, Hersh, Wilder, Grabiner, Tymoczko (all in Tymoczko, 1986), Tymoczko (1986a), Stolzenberg (1984), MacLane (1981), McCleary and McKinney (1986), and Davis (1974). Thus a move away from the traditional *aprioristic* view or justification of mathematics, as advocated by social constructivism, is widespread.

A number of other contributory theses of social constructivism are espoused by philosophers of mathematics. The conventionalist viewpoint is implicit in several of these authors' work. Those who make it explicit include Stolzenberg (1984), as well as Bloor, Quine and Wittgenstein, cited above, and others mentioned in Chapter 2. In addition, the thesis that the objects of mathematics are reified constructions is proposed by both Davis (1974) and Machover (1983).

Beyond such piecemeal comparisons, two philosophers who have anticipated much of the social constructivist philosophy of mathematics are Bloor (1973, 1976, 1978, 1983, 1984) and Tymoczko (1985, 1986, 1986a). Both argue that objectivity in mathematics can best be understood in terms of social acceptance, and draw upon the seminal work of Wittgenstein and Lakatos.

Although no new paradigm is yet fully accepted, social constructivism sits comfortably in a growing quasi-empiricist tradition. Beyond this, a few contemporary philosophers are beginning to propose approaches to the philosophy of mathematics similar to and coherent with social constructivism.

3. Sociological Perspectives of Mathematics

A. Cultural and Historical Approaches

Several authors offer historico-cultural accounts of the nature of mathematics, treating the relationships between the social and cultural groups involved in mathematics, and the genesis and nature of mathematical knowledge. These include Crowe (1975), Mehrtens (1976), Restivo (1985), Richards (1980, 1989), Szabo (1967), Wilder (1974, 1981) and Lakatos (1976). These authors have offered theories of the development of mathematical knowledge, which relate it to its social, historical and cultural context. In particular, they theorize how the micro social context (i.e. interactions in small groups), in the case of Szabo and Lakatos, or the macro social context, in the case of Crowe, Mehrtens, Restivo, Richards and Wilder, influence the development of mathematical knowledge.

Studies of the micro social context concern negotiation within groups of individuals, leading to the acceptance of logical argument or mathematical knowledge, as well as concepts. Such theories reflect the quasi-empiricist account of the acceptance of knowledge, on the empirical level. Lakatos (1976) offers an account of this type with his conjectured 7 stage pattern of mathematical discovery. As an empirical conjecture this fits here, because it represents an historical parallel with aspects of quasi-empiricism and social constructivism, at the micro social level. Szabo (1967) argues that the deductive logic of Euclid derives from pre-socratic dialectics, with conversation serving as the model. Again, this fits with the social constructivist account.

Studies of the macro social context offer theories of the structural patterns, social relationships or 'laws' in the development of mathematical knowledge in history and culture. Many of these are social constructionist accounts, consistent with conventionalism, and hence social constructivism, albeit in a different realm. In this bracket can be included a new breed of histories of mathematics acknowledging its fallibility (Kline, 1980) and its multi-cultural social construction (Joseph, 1990).

Historical and cultural studies of mathematics with a bearing on the philosophy of mathematics draw strength and inspiration from the comparable 'externalist' approaches to the philosophy of science, such as those of Kuhn (1970) and Toulmin (1972). Such historical approaches, as well as the philosophy of science, provide parallels and support for social constructivism. Likewise, when the social constructivist account is supplemented with empirical hypotheses, a theory of the history of mathematics results, as in the quasi-empiricism of Lakatos (1976).

B. The Sociology of Knowledge

A number of sociologial theses offer parallels to social constructivism.

Knowledge as a social construction

First of all, there is the 'social constructionist' thesis that all knowledge is a social construction. There is a tradition in the sociology of knowledge supporting and elaborating this thesis, including such theorists as Marx, Mannheim, Durkheim, Mead, Schutz, Berger and Luckman, and Barnes (although the first few named in this list assert that some knowledge, notably mathematics, can be free from social bias). This is the dominant view in the sociology of knowledge, contrasting with the main traditions in philosophy which claim that there is certain knowledge of the world from observations (empiricism) or through abstract thought (idealism).

In the sociology of knowledge there are variations in the degree of relativism ascribed to knowledge. In the extreme case, all human knowledge is seen as relative to social groups and their interests, and physical reality itself is regarded as a social construction. More moderate positions regard knowledge (and not reality) as a social construction, and accept an enduring world as a constraint on the possible forms of knowledge. For example, Restivo (1988a) argues that although the new sociology of science regards knowledge as a social construction, it is better aligned with realism than with simple relativism, with which it has no necessary connection. Such positions are parallel to social constructivism in the assumptions they adopt, although they remain sociological as opposed to philosophical theories. Their existence suggests the potential fruitfulness of a sociological version of social constructivism, to account for the social structures and development of mathematics.

The 'strong programme' in the sociology of knowledge

Bloor (1976) has laid down criteria (the tenets of the 'strong programme') that a sociology of knowledge should satisfy if it is to provide a *sociologically* acceptable account of knowledge. Briefly, these require that for adequacy a theory of knowledge should account for: (i) the social genesis of knowledge; (ii and iii) both true and false knowledge and beliefs symmetrically; and (iv) itself (reflexivity).

Although designed for the sociology of knowledge, it is interesting to apply these criteria to social constructivism. Re (i): the account given evidently accounts for the social genesis of mathematical knowledge. Re (ii): it can be said that social constructivism accounts for beliefs and knowledge without regard to their truth or falsity. For the generation of knowledge by the hypothetico-deductive method has no implications concerning its truth. Social constructivism accounts both for the adoption of new, and for the rejection of old beliefs and knowledge when falsified, or for other reasons, denied acceptance. Like sociological accounts of knowledge, social constructivism is symmetrical in these explanations, in terms of social acceptance, and not in terms of a 'match' with a transcendent reality.

Re (iv): Although social constructivism is primarily a philosophy of *mathematics*, it can be extended to account for itself, at least in part. For it is based on a number of epistemological and ontological assumptions, from which conclusions are inferred. As such it has similar status to that which it ascribes to mathematics, namely an hypothetico-deductive theory, except for differences in subject matter and rigour. Both start with a set of plausible but conjectural assumptions (albeit concerning different realms of knowledge), from which consequences are inferred. In addition, any justification for social constructivism must reside in its social acceptance, directly paralleling its account of mathematics. Finally, social constructivism rejects the analytic-empirical distinction, and views all knowledge as inter-related. Consequently, it is legitimately applicable throughout realms of human knowledge, including to itself. Thus social constructivism may be said to be reflexive, since a parallel account can be applied to itself.

Overall, social constructivism largely satisfies the criteria of the 'strong programme'. This contrasts with absolutist philosophies, which treat truth quite differently from falsehood, failing to satisfy (ii) and (iii), as well as being unable to satisfy (iv). Whilst in terms of traditional philosophy, this is of limited significance, it suggests that a sociological parallel will satisfy the criteria, auguring well for an over-arching social constructivist theory.

Knowledge is value laden

Third, there is the value laden-ness of knowledge. Values are the basis for choice, and so to be value-laden is to represent the preference or interest of a social group. Values can be manifested explicitly, as in a conscious act of choice, or tacitly, as in unconscious compliance or acceptance. For example, Polanyi (1958) argues that many of the shared values of the scientific community, such as the endorsement of the scientific consensus, are tacit. However, the traditional view of science and scientific knowledge is that it is logical, rational, objective, and hence value-free.[2] Both social constructivism and the sociology of knowledge reject this view, for different reasons. The sociology of knowledge asserts that all knowledge is value-laden, because it is the product of social groups, and embodies their purposes and interests.

Social constructivism denies that mathematical knowledge is value-free. First, because it rejects the categorical distinction between mathematics and science, and as is increasingly accepted by philosophers of science, science is value-laden. Second, because it posits a shared linguistic basis for all knowledge, which since it serves all human purposes, it is imbued with human values. The mathematical use of language, formal and informal, attempts to eradicate values, by adhering to objective logical rules for the definition and justification of mathematical knowledge. However, the use of the hypothetico-deductive (i.e. axiomatic) method means that values are involved in the choice of hypotheses (and definitions). Beyond this, there are values implicit in logic and the scientific method.

Although mathematics was thought to epitomize value-free objectivity, alongside the sociology of knowledge social constructivism rejects this belief, arguing that objectivity itself is social, and that consequently mathematical knowledge is laden with human and cultural values.

The reification of concepts

Fourth, there is the reification of concepts, in which they become autonomous, objective things-in-themselves. In sociology, this mechanism was first proposed by Marx, by analogy with the genesis of religion.

> . . . the productions of the human brain appear as independent beings endowed with life, and entering into a relation both with one another and with the human race.
>
> (Marx, 1967, page 72)

He argues that the form of products becomes reified and fetishized into an abstract thing: money, value or commodity (Lefebvre, 1972). Subsequent theorists in this tradition of thought, such as Lukacs, have extended the range of operation of reification to a much broader range of concepts.

Evidently the social constructivist thesis concerning the reification of newly defined concepts has a strong parallel in Marxist sociology. This parallel has been extended to mathematics by Davis (1974) and others such as Sohn-Rethel, as Restivo (1985) reports.

C. *The Sociology of Mathematics*

The sociology of mathematics is a substantial field of study concerning the social development and organization of mathematics, as the survey in Restivo (1985) indicates. In contrast with the philosophy of mathematics, it is concerned to offer empirical theories of the growth, development and organisation of mathematical knowledge. To achieve this, it tries to account for mathematics and mathematical knowledge as a social construction[3] (unlike the traditional philosophies of mathematics). Consequently, the social constructivist philosophy of mathematics offers a parallel with sociologial accounts, but whereas the former is concerned with a logical and conceptual analysis of the conditions of knowledge, the latter is concerned with the social determinants of the actual body of knowledge.

One of the aims of social constructivism is to offer a descriptive philosophy of mathematics, as opposed to the prescription of the traditional philosophies. Thus parallel accounts of mathematics from sociological, as well as historical and psychological perspectives should be possible. Therefore this section offers a preliminary sociology of mathematical knowledge.

A social constructivist sociology of mathematics

From a sociological perspective, mathematics is the name given to the activities of, and knowledge produced by a social group of persons called mathematicians. When linked to social history by a definition like this, the term 'mathematics' has an organic, changing denotation, as does the set of mathematicians.

> '[M]athematics' by 1960, consisted of various subgroups working, to some extent, within different cognitive and technical norms, on different orders of phenomena and different types of problems. What had changed, with some exceptions — such as computing — was the relative numerical strength and status within the overall discipline of groups carrying particular norms.
>
> (Cooper, 1985, page 7)

> Subjects (e.g. mathematics) will be regarded not as monoliths, that is as groups of individuals sharing a consensus both on cognitive norms and on perceived interests, but rather as constantly shifting coalitions of individuals and variously sized groups whose members may have, at any specific moment, different and possibly conflicting missions and interests. These groups may, nevertheless, in some arenas, all successfully claim allegiance to a common name, such as 'mathematics'.
>
> (Cooper, 1985, page 10)

These complexities form a backdrop of the brief, conjectural sociological account of mathematics that follows, in line with social constructivism.

(i) *Mathematicians*. At any one time, the nature of mathematics is determined primarily by a fuzzy set of persons: mathematicians. The set is partially ordered by the relations of power and status. The set and the relations on it are continually changing, and thus mathematics is continuously evolving. The set of mathematicians has different strengths of membership (which could in theory be quantified from 0 to 1). This includes 'strong' members (institutionally powerful or active research mathematicians) and 'weak' members (teachers of mathematics). The 'weakest' members could simply be numerate citizens. The notion of a fuzzy set usefully models the varying strengths of individuals' contribution to the institution of mathematics. Mathematical knowledge is legitimated through acceptance by the 'strongest' members of the set. In practice the set of mathematicians is made up of many sub-sets pursuing research in sub-fields, each with a similar sub-structure, but loosely interconnected through various social institutions (journals, conferences, universities, funding agencies).

(ii) *Joining the set*. Membership of the set of mathematicians results from an extended period of training (to acquire the necessary knowledge and skills) followed by participation in the institutions of mathematics, and presumably the adoption of (at least some) of the values of the mathematics community (Davis and Hersh, 1980; Tymoczko, 1985). The training requires interaction with other mathematicians, and with information technology artefacts (books, papers, software, etc.). Over a period of time this results in personal knowledge of mathematics. To the extent that it exists, the shared knowledge of mathematics results from this period of training in which students are indoctrinated with a 'standard' body of mathematical knowledge. This is achieved through common learning experiences and the use of key texts, which have included Euclid, Van der Waerden, Bourbaki, Birkhoff and MacLane, and Rudin, in the past. Many, probably most students fall away during this process. Those that remain have successfully learned part of the official body of mathematical knowledge and have been 'socialized' into mathematics. This is a necessary, but not sufficient condition for entry into the set of mathematicians (with a membership value significantly greater than 0). The 'standard' body of knowledge will have a shared basis, but will vary according to which subfields the mathematician contributes.

(iii) *Mathematicians' culture*. Mathematicians form a community with a mathematical culture, with sets of concepts and prior knowledge, methods, problems, criteria of truth and validity, methodology and rules, and values, which are shared to a varying degree. A number of authors have explored the culture and values of mathematics, including Bishop (1988), Davis and Hersh (1980) and Wilder (1974, 1981). Here we will undertake a more limited inquiry, restricted to the different realms of discourse and knowledge of mathematicians, and their associated values. The analysis given here is three-fold, proposing that mathematicians operate with knowledge on the three levels of the syntax, semantics and pragmatics of mathematics. This is based on the classificatory system of Charles Morris (1945) who distinguishes these three levels in language use. In his sytem the syntax, semantics and pragmatics of a language refer to the formal rule system (grammar and proof), the system of meanings and interpretations, and the nexus of human rules, purposes and decisions concerning language use, respectively. In constructing this system, Morris added to the formal logical levels of syntax and semantics a further level of pragmatics, inspired by pragmatism.

There is also a parallel with the three interlocking systems of language distinguished by Halliday (1978), consisting the forms, meanings and functions of language. In the sociology of mathematics, Restivo (1985) distinguishes the syntactical and semantic properties of an object (following Hofstadter), paralleling the syntax-semantics distinction. Hersh (1988) makes an analogous distinction between the 'front' and 'back' of mathematics. Restivo (1988) also distinguishes between 'social' and 'technical' talk of mathematics, paralleling the distinction between the third level of pragmatic considerations and the first two levels taken as one, respectively. Thus precursors of these three levels, in various forms, are to be found in the literature.

The three levels of mathematical discourse proposed are as follows. First of all, there is the level of syntax or formal mathematics. This consists of rigorous formulations of mathematics, consisting of the formal statement and proof of results, comprising such things as axioms, definitions, lemmas, theorems and proofs, in pure mathematics, and problems, boundary conditions and values, theorems, methods, derivations, models, predictions and results in applied mathematics. This level includes the mathematics in articles and papers accepted for conferences and journal publication, and constitutes what is accepted as official mathematics. It is considered to be objective and impersonal, the so-called 'real' mathematics. This is the level of high status knowledge in mathematics, what Hersh (1988) terms 'the front' of mathematics. This level is not that of total rigour, which would require exclusive use of one of the logical calculi, but of what passes in the profession for acceptable rigour.

Secondly, there is the level of informal or semantic mathematics. This includes heuristic formulations of problems, informal or unverified conjectures, proof attempts, historical and informal discussion. This is the level of unofficial mathematics, concerned with meanings, relationships and heuristics. Mathematicians refer to remarks on this level as 'motivation' or 'background'. It consists of subjective and personal mathematics. It is considered to be low status knowledge in mathematics, what Hersh (1988) terms 'the back' of mathematics.

Third, there is the level of pragmatic or professional knowledge of mathematics and the professional mathematical community. It concerns the institutions of mathematics, including the conferences, places of work, journals, libraries, prizes, grants, and so on. It also concerns the professional lives of mathematicians, their specialisms, publications, position, status and power in the community, their work places and so on. This is not considered to be mathematical knowledge at all. The knowledge has no official status in mathematics, since it does not concern the cognitive content of mathematics, although aspects of it are reflected in journal announcements. This is the level of 'social talk' of mathematics (Restivo, 1988).

These three levels are the different domains of practice within which mathematicians operate. As languages and domains of discourse they form a hierarchy, from the more narrow, specialized and precise (the level of syntax), to the more inclusive, expressive and vague (the level of pragmatics). The more expressive systems can refer to the contents of the less expressive systems, but the relation is asymmetric.

The hierarchy also embodies some of the values of mathematicians. Namely, the more formal, abstract and impersonal that the mathematical knowledge is, the more highly it is valued. The more heuristic, concrete and personal mathematical knowledge is, the less it is prized. Restivo (1985) argues that the development of abstract mathematics follows from the economic and social separation of the 'hand' and 'brain'. For abstract mathematics is far removed from practical concerns. Since the 'brain' is associated with wealth and power in society, this division may be said to lead to the above values.

The values described above lead to the identification of mathematics with its formal representations (on the syntactical level). This is an identification which is made both by mathematicians, and philosophers of mathematics (at least those endorsing the absolutist philosophies). The valuing of abstraction in mathematics may also partly explain why mathematics is objectified. For the values emphasize the pure forms and rules of mathematics, facilitating their objectification and reification, as Davis (1974) suggests.[4] This valuation allows the objectified concepts and rules of mathematics to be depersonalized and reformulated with little concerns of ownership, unlike literary creations. Such changes are subject to strict and general mathematical rules and values, which are a part of the mathematical culture. This has the result of offsetting some of the effects of sectional interests exercised by those with power in the community of mathematicians. However, this in no way threatens the status of the most powerful mathematicians. For the objective rules of acceptable knowledge serve to legitimate the position of the elite in the mathematical community.

Restivo (1988) distinguishes between 'technical' and 'social' talk of mathematics, as we saw, and argues that unless the latter is included, mathematics as a social construction cannot be understood. Technical talk is identified here with the first and second levels (the levels of syntax and semantics), and social talk is identified with the third level (that of pragmatics and professional concerns).

Denied access to this last level, no sociology of mathematics is possible, including a social constructionist sociology of mathematics. However, social constructivism as a philosophy of mathematics does not need access to this level, although it requires the existence of the social and language, in general. An innovation of social constructivism is the acceptance of the second level (semantics) as central to the philosophy of mathematics, following Lakatos. For traditional philosophies of mathematics focus on the first level alone.

Sociologically, the three levels may be regarded as distinct but inter-related discursive practices, after Foucault. For each has its own symbol systems, knowledge base, social context and associated power relationships, although they may be hidden. For example, at the level of syntax, there are rigorous rules concerning acceptable forms, which are strictly maintained by the mathematics establishment (although they change over time). This can be seen as the exercise of power by a social group. In contrast, the absolutist mathematician's view is that nothing but logical reasoning and rational decision-making is relevant to this level. Thus a full sociological understanding of mathematics requires an understanding of each of these discursive practices, as well as their complex inter-relationships. Making these three levels explicit, as above, is a first step towards this understanding.

D. Sociological Parallels of Social Constructivism

The above suggests that social constructivism may offer a potentially fruitful parallel sociological account of mathematics. Such a parallel, highly compatible with social constructivism is already partly developed by Restivo (1984, 1985, 1988) and others. Although sociological parallels do not add weight to social constructivism in purely philosophical terms, they offer the prospect of an interdisciplinary social constructivist theory, offering a broader account of mathematics than a philosophy alone. Mathematics is a single phenomenon, and a single account applicable to each of the perspectives of philosophy, history, sociology and psychology is desirable, since it reflects the unity of mathematics. If successful, such an account would have the characteristics of unity, simplicity and generality, which are good grounds for theory choice.[5]

4. Psychological Parallels

A. Constructivism in Psychology

Constructivism in psychology can be understood in narrow and broad senses.[6] The narrow sense is the psychological theory of Piaget and his school. Piaget's epistemological starting point resembles that of social constructivism in its treatment of subjectve knowledge. His epistemological assumptions are developed into the philosophy of radical constructivism by von Glasersfeld, as we have seen. However, Piaget's psychological theory goes far beyond its epistemological starting point. Fully articulated, it is a specific empirical theory of conceptual development, with particular concepts and stages. It also assumes the narrow Bourbakiste structural view of mathematics, which is not compatible with social constructivism.

The Bourbaki group have been developing and publishing a unified axiomatic formulation of pure mathematics for about fifty years in *Elements de Mathematique* (see for example, Kneebone, 1963). Their formulation is structuralist, based on axiomatic set theory in which three 'mother-structures' are defined: algebraic, topological and ordinal, providing the foundation for pure mathematics. As a view of mathematics, the Bourbaki programme may be criticised as narrow. First, because it excludes constructive mathematical processes, and second, because it represents mathematics as fixed and static. Thus it reflects the state of mathematics during a single era (mid twentieth-century), although this is denied in Bourbaki (1948). It is incompatible with social constructivism because of this narrowness, and because it is a foundationist programme, and hence is implicitly absolutist.

However, the Bourbaki programme is *not* a philosophy of mathematics, and does not need to defend itself against this criticism. For it can be seen merely as a programme, albeit ambitious, to reformulate the structural part of mathematics. But Piaget views Bourbaki as revealing the nature of mathematics. Thus this criticism can be validly directed at Piaget's implicit philosophy of mathematics, rendering the details of his psychological theory incompatible with social constructivism. For he takes the three 'mother-structures' of Bourbaki as *a priori*, and assumes that they are an integral part of the psychological development of individuals. This is evidently an error, due to a misinterpretation of the significance of Bourbaki.

Other aspects of Piaget's theory do offer a parallel to social constructivism. For example, the notion of 'reflective abstraction', which allows mental operations to become objects of thought in their own right, accommodates the social constructivist thesis of mathematical objects as reifications. However, much of Piaget's developmental psychology, such as his stage theory, goes beyond any parallel with social constructivism, and is extensively criticized on both psychological (Bryant, 1974; Brown and Desforges, 1979; Donaldson, 1978), and mathematical grounds (Freudenthal, 1973).

The broad sense of constructivism in psychology is what Glasersfeld (1989) refers to as 'trivial constructivism', based on the principle that knowledge is not passively received but actively built up by the cognizing subject. This broad sense encompasses many different psychological theories including the personal construct theory of Kelly (1955), the information processing theory of Rumelhart and Norman (1978), the schema theory of Skemp (1979) and others, the social theory of mind of Vygotsky (1962), as well as the basis of the constructivism of Piaget and his followers. This list indicates some of the diversity of thought that falls under the broad sense of constructivism. What these authors share is a belief that the acquisition and development of knowledge by individuals involves the construction of mental structures (concepts and schemas), on the basis of experience and reflection, both on experience and on mental structures and operations. Many, but not all psychologists in this group accept that knowledge grows through the twin processes of assimilation and accommodation, first formulated by Piaget.

On the basis of their epistemological assumptions alone, both the broad and narrow senses of constructivism offer a psychological parallel to social constructivism. The auxiliary hypotheses of individual constructivist psychologies, such as Piaget's, may be incompatible with the social constructivist philosophy of mathematics. But the potential for a psychological theory of mathematics learning paralleling social constructivism clearly exists.

A number of researchers are developing a constructivist theory of mathematics learning, including Paul Cobb, Ernst von Glasersfeld and Les Steffe (see, for example, Cobb and Steffe, 1983; Glasersfeld, 1987; Steffe, Glasersfeld, Richards and Cobb, 1983). As they appear to have rejected many of the problematic aspects of Piaget's work, such as his stages, much of their theory can be seen as parallel to social constructivism, on the psychological plane. However it is not clear that all of their auxiliary assumptions, such as those involved in accounting for young children's number acquisition, are fully compatible with social constructivism.

No attempt will be made to develop a psychological parallel to social constructivism here, although in the next sections we consider briefly some of the key components of such a theory.

B. Knowledge Growth in Psychology

Following Piaget, schema theorists such as Rumelhart and Norman (1978), Skemp (1979) and others, have accepted the model of knowledge growth utilizing the twin processes of assimilation and accommodation. These offer parallels to the social constructivist accounts of subjective and objective knowledge growth. For knowledge, according to this account, is hypothetico-deductive. Theoretical models or systems are conjectured, and then have their consequences inferred. This can include the applications of known procedures or methods, as well as the elaboration, application, working out of consequences, or interpretation of new facts within a mathematical theory or framework. In subjective terms, this amounts to elaborating and enriching existing theories and structures. In terms of objective knowledge, it consists of reformulating existing knowledge or developing the consequences of accepted axiom systems or other mathematical theories. Overall, this corresponds to the psychological process of assimilation, in which experiences are interpreted in terms of, and incorporated into an existing schema. It also corresponds to Kuhn's (1970) concept of normal science, in which new knowledge is elaborated within an existing paradigm, which, in the case of mathematics, includes applying known (paradigmatic) procedures or proof methods to new problems, or working out new consequences of an established theory.

The comparison between assimilation, on the psychological plane, and Kuhn's notion of normal science, in philosophy, depends on the analogy between mental schemas and scientfic theories. Both schemas (Skemp, 1971; Resnick and Ford, 1981) and theories (Hempel, 1952; Quine 1960) can be described as interconnected structures of concepts and propositions, linked by their relationships. This analogy has been pointed out explicitly by Gregory (in Miller, 1983), Salner (1986), Skemp (1979) and Ernest (1990), who analyzes the parallel further.

The comparison may be extended to schema accommodation and revolutionary change in theories. In mathematics, novel developments may exceed the limits of 'normal' mathematical theory development, described above. Dramatic new methods can be constructed and applied, new axiom systems or mathematical theories developed, and old theories can be restructured or unified by novel concepts or approaches. Such periods of change can occur at both the subjective and objective knowledge levels. It corresponds directly to the psychological process of accommodation, in which schemas are restructured. It also corresponds to Kuhn's concept of revolutionary science, when existing theories and paradigms are challenged and replaced.

Piaget introduced the concept of cognitive conflict or cognitive dissonance (which will not be distinguished here). In the social constructivist account of mathematics, this has a parallel with the emergence of a formal inconsistency, or a conflict between a formal axiom system and the informal mathematical system that is its source (Lakatos,

1978a). This is analogous to cognitive conflict, which occurs when there is conflict between two schemas, due to inconsistency or conflicting outcomes. In psychology, this is resolved through the accommodation of one or both of the schemas. Likewise in mathematics, or in science, this stimulates revolutionary developments of new theories.

Overall, there is a striking analogy between theory growth and conflict in the social constructivist philosophy of mathematics and schema theory in psychology, and underlying it, between theories and schemas. Unlike the situation in the philosophy of mathematics, schema theory, as sketched above, represents the received view in psychology, lending support to a psychological parallel for social constructivism.

C. Reification and Concept Formation

The social constructivist philosophy of mathematics distinguishes two modes of concept development, vertical processes of concept formation, involving the reification of concepts into objects, and horizontal processes. These can be elaborated as part of a psychological parallel to social constructivism.

We may conjecture that psychological concept formation involves both vertical and horizontal processes. The vertical processes include the standard processes of concept formation, namely the generalization and abstraction of shared features of earlier formed concepts to form new concepts. Beyond this, we conjecture the existence of a psychological mechanism or tendency which transforms mental procedures or processes into objects. This mechanism changes a property, a construction, a process, or an incomplete collection into a mental object, a complete thing-in-itself. What is represented as a process, a verb or an adjective becomes represented as a noun. This is 'reification' or 'objectification'. Psychologically, much concept formation has this character. Even in the act of coordinating different perceptions of an external object, in sensory concept formation, we reify the set of perceptions into the concept, an enduring object-representation in a schema.

There is some parallel between this conjectured 'vertical' mechanism and Piaget's notion of reflective abstraction, the process whereby an individual's operations, both physical and mental, become represented cognitively as concepts. Thus reflective abstraction includes concept reification, although the former is a broader notion.

A number of other researchers have proposed psychological theories dealing specifically with concept reification (Skemp, 1971). Dubinsky (1988, 1989) includes 'encapsulation' as part of his explication of the notion of reflective abstraction. Encapsulation converts a subjective mathematical process into an object, by seeing it as a total entity. Sfard (1987, 1989) has been testing a theory of mathematical concept development, in which operational concepts are transformed into structural concepts, by a process of reification. Both these researchers have empirical data consistent with the hypothesis that a process of encapsulation or reification occurs in vertical concept formation. Thus there is evidence for a psychological process of vertical concept formation, parallel to the social constructivist account, and accounting for subjective belief in platonism.

D. Individualism in Subjective Knowledge

A central feature of the social constructivist theory is that subjective knowledge comprises idiosyncratic personal meanings, concepts and knowledge structures. These are subject to the constraints imposed by the external and social worlds, but this leaves room for considerable variation. A psychological version of this thesis would predict that significant variations in concepts and knowledge should occur between individuals, both within a single culture, and even more so in inter-cultural comparisons. This hypothesis seems to be confirmed empirically, although there is, of course, the methodological problem of comparing private meanings. Any evidence about individuals' personal meanings and knowledge must be based on inference and conjecture, for subjective knowledge is, by definition, unavailable for public scrutiny.

A number of different psychological approaches provide evidence of the uniqueness of individuals' subjective knowledge. First of all, there is research on errors in mathematics learning (Ashlock, 1976; Erlwanger, 1973; Ginsburg, 1977). From the patterns observed, it is clear that many errors are systematic and not random. The range of errors observed in learners suggests that they are not taught, and that learners construct their own idiosyncratic concepts and procedures. Secondly, researchers are finding that 'alternative conceptions' (i.e., idiosyncratic personal concepts) are also very widespread in science (Abimbola, 1988; Driver, 1983; Pfundt and Duit, 1988). Thirdly, researchers have tried to represent learners' cognitive structures in mathematics, using a variety of data-gathering methods. Their findings have included spontaneous (i.e., untaught) sequences of procedures in learning arithmetic (Steffe *et al.*, 1983; Bergeron *et al.*, 1986), and unpredictable growth in the links in personal concept hierarchies (Denvir and Brown, 1986).

These approaches illustrate the broad base of empirical and theoretical support for a psychological version of social constructivism. Individuals do seem to construct unique personal meanings and conceptual structures. There are, however, patterns to be found in these constructions across individuals (Bergeron *et al.*, 1986), presumably reflecting the similar mental mechanism generating subjective knowledge, and the comparable experiences and social contexts of individuals.

E. Social Negotiation as a Shaper of Thought

A central thesis of social constructivism is that the unique subjective meanings and theories constructed by individuals are developed to 'fit' the social and physical worlds. The main agency for this is interaction, and in the acquisition of language, social interaction. This results in the negotiation of meanings, that is the correction of verbal behaviour and the changing of underlying meanings to improve 'fit'. Briefly put, this is the conjectured process by means of which the partial inner representation of public knowledge is achieved.

This thesis is close to the social theory of mind of Vygotsky (1962) and his followers. Vygotsky's theory entails that for the individual, thought and language develop together, that conceptual evolution depends on language experience, and of particular relevance to social constructivism, that higher mental processes have their origin in interactive social processes (Wertsch, 1985).

> Vygotsky's point is not that there are hidden cognitive structures awaiting release through social interaction. His point is the radical one that they are *formed* through social interaction. Development is not the process of the hidden becoming public, but on the contrary, of the public and inter-subjective becoming private.
>
> (Williams, 1989, page 113)

Thus Vygotsky's social theory of mind offers a strong parallel with social constructivism, one that can also be found elsewhere in psychology, such as Mead's (1934) symbolic interactionism. A further development in this direction is the Activity Theory of Leont'ev (1978), with perceives psychological motives and functioning as inseparable from the socio-political context. Possibly less radical is the move to see knowing as bound up with its context in 'situated cognition' (Lave, 1988; Brown *et al.*, 1989), although Walkerdine (1988, 1989) proposes a fully social constructionist psychology of mathematics. Social constructionism as a movement in psychology is gaining in force, as Harre (1989) reports, and is replacing the traditional developmental or behaviourist paradigms of psychology with that of social negotiation. Harre goes so far as to propose that inner concepts such as self-identity are linguistic-related social constructions.

F. Psychological Parallels

A number of psychological parallels of social constructivism have been explored, including 'constructivism' and 'social constructionism'. Many reflect a dominant view in psychology, contrasting with the controversial position of social constructivism in the philosophy of mathematics. Thus it seems likely that a psychological version of social constructivism, enriched with appropriate empirical hypotheses, could offer a fruitful account of the psychology of mathematics.

5. Conclusion: a Global Theory of Mathematics

Social constructivism is a philosophy of mathematics, concerned with the possibility, conditions and logic of mathematical knowledge. As such, its acceptability depends on philosophical criteria. It has been shown to have more in common with some other branches of philosophy, than with traditional philosophies of mathematics, for it inescapably raises issues pertaining to empirical knowledge, and to the social and psychological domains. Despite raising such issues, no empirical assumptions concerning the actual history, sociology or psychology of mathematics have been made.

Due to the multidisciplinary nature of the issues raised, there is also the prospect of a unified social constructivist account of mathematics. The aim of this section is to propose an overall social constructivist theory of mathematics, incorporating its philosophy, history, sociology and psychology. These are distinct disciplines, with different questions, methodologies and data. What is proposed is an overarching social constructivist meta-theory of mathematics, to provide schematic explanations treating the issues and processes in each of these fields, to be developed to suit the characteristics and constraints of that field. This would result in parallel social constructivist accounts of:

1 the history of mathematics: its development at different times and in different cultures;
2 the sociology of mathematics; mathematics as a living social construction, with its own values, institutions, and relationship with society in the large;
3 the psychology of mathematics: how individuals learn, use and create mathematics.

The goal of providing such a meta-theory of mathematics is ambitious, but legitimate. Theoretical physics is currently seeking to unify its various theories into a grand theory. In the past century other great strides have been made to unify and link sciences. There have been ambitious schemes to document a shared methodology and foundations, such as the International Encyclopedia of Unified Science. The history of mathematics likewise provides many examples of theoretical unification. What is claimed here is that this is also a desirable goal for the philosophy of mathematics.

There are a number of reasons why such a project is worthwhile. First of all, as mathematics is a single discipline and social institution, it is appropriate to coordinate different perspectives of it, for the unity of mathematics should transcend the divisions between disciplines. A meta-theory which reflects this unity gains in plausibility, and reflects the characteristics of a good theory, namely agreement with the data, conceptual integration and unity, simplicity, generality and, it is to be hoped, fertility.

Secondly, beyond this general argument is the fact of the strong parallels between the social constructivist philosophy and the history, sociology and psychology of mathematics demonstrated above. These are not coincidental, but arise from genuinely interdisciplinary issues inherent in the nature of mathematics as a social institution.

Thirdly, in exploring these parallels one factor has recurred, the greater acceptability of the parallel theses in general philosophy, sociology, psychology and the history of mathematics, than in the philosophy of mathematics. In these fields, many of these theses are close to the received view or a major school of thought. In particular, social constructionist views in sociology and psychology have a great deal of support. This contrasts strongly with the position in the philosophy of mathematics, where absolutist philosophies have dominated until very recently. Thus the call for a social constructivist meta-theory of mathematics is stronger from the surrounding fields than from the traditional philosophy of mathematics.

Fourthly, one of the theses of social constructivism is that there is no absolute dichotomy between mathematical and empirical knowledge. This suggests the possibility of a greater rapprochement than hitherto, between the logical concerns of philosophy, and the empirical theories of history, sociology and psychology. An overarching social constructivist meta-theory of mathematics would offer such a rapprochement. Such a theory is therefore proposed, in the spirit of developing the self-consistent (i.e., reflexive) application of social constructivism.[7]

Notes

1 There is an interesting analogy between deductive proof and Kitcher's (1984) justification of mathematical knowledge, which might be its source. Just substitute axioms for his basis and inference as the means of deriving each stage from the next.

2 An example of this 'standard' view is that of Scheffler: 'science is a systematic public enterprise, controlled by logic and empirical fact whose purpose is to formulate the truth about the empirical world'. (Brown *et al.*, 1981, page 253)

3 A leading exponent of social constructionism as a sociology of mathematics is Sal Restivo (1984, 1985, 1988). (In addition Restivo, 1984, offers valuable insights into social constructivism as a philosophy of mathematics). David Bloor (1976, 1983) has made major contributions to both the sociology and philosophy of mathematics as a social construction.

4 Restivo (1985, page 192) also suggests, following Struik, that the separation of form from content in the objectification of mathematical knowledge is a product of the prevailing social conditions. The argument is that idealism results from, and provides a solution to problems in social outlook, during periods of social decline, such as the disintegration of the western Roman empire, and the enfeeblement of empire. Similarly Koestler (1964, page 57) suggests that Plato's idealism was a response to the decline of Greece. An interesting analogy might be drawn with the development of the rigid philosophy of logical positivism in post-Great war Austria and Germany.

5 The strength of the sociological parallels might be used to direct a charge of sociologism against social constructivism, claiming that it is a sociological *theory* of mathematical knowledge, which although avoiding overtly empirical matters, remains essentially sociological. My response is that the primary focus is on the general conditions and justification of mathematical knowledge, which is the proper concern of the philosophy of mathematics.

6 'Constuctivism' has many meanings. Below two senses of 'constructivism' are distinguished in psychology. In the philosophy of mathematics, 'constructivism' encompasses intuitionism and similar schools of thought. The psychological and philosophical senses are quite distinct (Lerman, 1989). 'Social constructivism' introduces another sense into the philosophy of mathematics. Social constructivism is also applied in the sociology of mathematics, by Restivo. 'Constructivism' also denotes a movement in the history of modern art, with proponents such as Gabo, Pevsner and Tatlin. The account of judicial reasoning of Ronald Dworkin (in his 1977 book 'Taking Rights Seriously') is termed 'constructivist', according to the Fontana Dictionary of Modern Thought.

Doubtless further 'constructivist' schools of thought exist in other disciplines. What they seem to share is the metaphor of construction: the product involved is built up by a synthetic process from previously constructed components.

7 An outcome of the social constructivist meta-theory of mathematics might be to demystify the philosophy of mathematics. For if the meta-theory is possible, then the strict demarcation of the disciplines may be seen as the reification, mystification and even the fetishization of philosophy and mathematics. The force with which the inviolability of the boundaries has been asserted (e.g. by logical positivists and empiricists) resembles a social taboo. It is surely in the interests of knowledge to offer a rational challenge to such a taboo, even if it is against the interests of the professionals who have created the mystique.

PART 2

The Philosophy of Mathematics Education

6

Aims and Ideologies of Mathematics Education

1 Epistemological and Ethical Positions

Different philosophies of mathematics have widely differing outcomes in terms of educational practice. However the link is not straightforward, and so an inquiry into the philosophies that underpin mathematics teaching and the mathematics curriculum forces us also to consider values, ideologies and the social groups that adhere to them.

Ideology

This section distinguishes a range of ideologies incorporating both epistemological and ethical views. Since the concept of 'ideology' is central, it is appropriate to clarify its meaning. Williams (1977) traces one usage to Napoleon Bonaparte, where it signified revolutionary thinking, regarded as an undesirable set of ideas threatening 'sound and sensible thinking'. This has lead to the pejorative use of 'ideology' for fanatical or impractical theories of society. Although Marx first used the term for 'false consciousness', where a thinker 'imagines false or apparent motives' (Meighan, 1986, page 174), he later used it in the sense intended here. In this more sociological sense, an ideology is an overall, value-rich philosophy or world-view, a broad inter-locking system of ideas and beliefs. Thus ideologies are understood here to be competing belief systems, combining both epistemological and moral value positions, with no pejorative meaning intended. They are not to be contrasted with the content of science and mathematics, but to underpin and permeate these fields of knowledge and to imbue the thought of the groups associated with them (Giddens, 1983). Ideology is often seen as 'the way things really are' by its adherents (Meighan, 1986), because it is the often invisible substratum for the relations of power and domination in society (Giddens, 1983; Althusser, 1971). However, the treatment of ideologies given here emphasizes epistemological, ethical and educational aspects, and social interests, power and domination will be referred to subsequently.

The aim of this chapter is to link public and personal philosophies of mathematics and education. In addition to explicitly stated philosophies we are concerned with the tacit belief-systems of individuals and groups. Such beliefs are not so easily detached from their contexts as public philosophies, being part of a whole ideological nexus. These comprise many interwoven components, including personal epistemologies, sets of values and other personal theories. Thus much more than epistemology is needed to link public philosophies to personal ideologies.

As a basis for distinguishing ideologies we adopt the Perry (1970, 1981) theory. This is a psychological theory concerning the development of individuals' epistemological and ethical positions. It is a structural theory, providing a framework into which different philosophies and sets of values can be fitted.

A. *The Perry Theory*

The Perry theory specifies a sequence of stages of development, as well as allowing for fixation at, and retreat from its levels. For simplicity we consider only three of the stages, Dualism, Multiplicity and Relativism.[1] The theory does not end at Relativism, continuing through several stages of Commitment. However these stages do not represent a radical re-structuring of beliefs, so much as the entrenchment and integration of Relativism into the whole personality. For underlying the Perry Scheme is the assumption that intellectual and ethical development begins embedded in an unquestioned set of beliefs, progresses through several levels of critical detachment, and then re-embeds itself in a commitment to a set of intellectual and ethical principles. Thus the three stages considered here suffice to distinguish structurally different ideologies.

Dualism

Simple Dualism is a bifurcated structuring of the world between good and bad, right and wrong, we and others. Dualistic views are characterized by simple dichotomies and heavy reliance on absolutes and authority as the source of truth, values and control.

Thus in terms of epistemological beliefs, Dualism implies an absolutist view of knowledge divided into truth and falsehood, dependent on authority as the arbiter. Knowledge is not justified rationally, but by reference to authority. In terms of ethical beliefs, Dualism implies that all actions are simply right or wrong.

> All problems are solved by Adherence [alignment of self with Authority]: obedience, conformity to the right and what They want. Will power and work should bring congruence of action and reward. Multiplicity not perceived. Self defined primarily by membership in the right and traditional.
>
> (Perry, 1970, end-chart)

Multiplicity

> A plurality of 'answers', points of view or evaluations, with reference to similar topics or problems. This plurality is perceived as an aggregate of

> discretes without internal structure or external relation, in the sense 'Anyone has a right to his own opinion', with the implication that no judgements among opinions can be made.
>
> (Perry, 1970, end-chart)

Multiplistic views acknowledge a plurality of 'answers', approaches or perspectives, both epistemological and ethical, but lack a basis for *rational* choice between alternatives.

Relativism

> A plurality points of view, interpretations, frames of reference, value systems and contingencies in which the structural properties of contexts and forms allow of various sorts of analysis, comparison and evaluation in Multiplicity.
>
> (Perry, 1970, end-chart)

Epistemologically, Relativism requires that knowledge, answers and choices are seen as dependent upon features of the context, and are evaluated or justified within principled or rule governed systems. Ethically, actions are judged desirable or undesirable according to the context and an appropriate system of values and principles.

A number of educational researchers have found the Perry Scheme to be a useful framework for describing intellectual and ethical development and personal beliefs. These include applications to the levels of thought of systems theory students (Salner, 1986), college and high school students of mathematics (Buerk, 1982; Stonewater *et al.*, 1988) and teachers' mathematics related belief-systems (Copes, 1982, 1988; Oprea and Stonewater, 1987; Cooney and Jones, 1988; Cooney, 1988; Ernest, 1989a). Thus the Perry theory is widely used to describe personal philosophies, particularly in mathematics.[2]

B. Personal Philosophies of Mathematics

We can relate the Perry theory to positions in the philosophy of mathematics. These are public philosophies of mathematics, explicitly stated and exposed to public debate. Here we consider personal philosophies of mathematics, which are private and implicit theories, unless thought through, stated explicitly and made public. The distinction is that between objective and subjective knowledge, made amongst others by Polanyi (1958), who argues for the importance of the role of commitment to personal knowledge, offering support for the form, if not the detail, of the Perry theory.

Applying the Perry theory to personal philosophies of mathematics, views of mathematics can be distinguished at each of the three levels. Dualistic views of mathematics regard it as concerned with facts, rules, correct procedures and simple truths determined by absolute authority. Mathematics is viewed as fixed and exact; it has a unique structure. Doing mathematics is following the rules.

In Multiplistic views of mathematics multiple answers and multiple routes to an answer are acknowledged, but regarded as equally valid, or a matter of personal preference. Not all mathematical truths, the paths to them or their applications are known, so it is possible to be creative in mathematics and its applications. However, criteria for choosing from this multiplicity are lacking.

Relativistic views of mathematics acknowledge multiple answers and approaches to mathematical problems, and that their evaluation depends on the mathematical system, or its overall context. Likewise mathematical knowledge is understood to depend on the system or frame adopted, and especially on the inner logic of mathematics, which provide principles and criteria for evaluation.

We next relate these classes of views of mathematics to distinct philosophies of mathematics, both public and private. The main distinction in the philosophy of mathematics is between absolutism and fallibilism. Absolutist schools of thought claim that mathematical knowledge is certain, but that there are rational grounds for accepting (or rejecting) it. Mathematical knowledge is established within these philosophies by means of logic applied to mathematical theories. These philosophies also acknowledge the multiplicity of approaches and possible solutions to mathematical problems, even if there is an eternal truth to be discovered by these means. Such public philosophies and belief-systems are therefore Relativistic, since knowledge is evaluated with reference to a system or framework. The same holds for published fallibilist philosophies.

However, beyond these public schools of thought, and their 'private' counterparts, are narrower personal philosophies of mathematics. The two that will be distinguished are both absolutist. The first is a Dualistic view of mathematics as a collection of true facts, and correct methods, whose truth is established by reference to authority. This perspective emphasizes absolute truth versus falsity, correctness versus incorrectness, and that there is a unique set of mathematical knowledge sanctioned by authority. This view has been termed an 'instrumental' view of mathematics (Ernest, 1989b, c, d).[3] It has been identified in empirical research on teachers' beliefs (Cooney and Jones, 1988; Ernest, 1989a; Oprea and Stonewater, 1987; and Thompson, 1984). It will be termed the 'Dualistic absolutist' view of mathematics.

A second personal philosophy of mathematics can be identified which is Multiplistic. This also views mathematics as an unquestioned set of facts, rules and methods, but does not perceive the choices and uses among this set to be absolutely determined by authority or any other source. Thus there are a plurality of 'answers', points of view or evaluations with reference to similar mathematical situations or problems, and choices can be made according to the belief-holder's preference.

Such a view can be inferred for the child Benny, in the case study of Erlwanger (1973), who sees mathematics as a mass of (inconsistent) rules, to be chosen by preference or expediency. Skovsmose (1988) implies that the unreflective use of mathematics in mathematical modelling is pragmatic, and may embody such a philosophy. Ormell (1975) reports the view of many scientists and technologists that mathematics is a collection of tools to be utilized as and when needed, each to be regarded as a 'black box' whose workings are not probed. Such views are Multiplistic, since they acknowledge a multiplicity of answers and methods in applying mathematics, but lack any principled grounds for rational choice. Selections among alternatives are made according to personal preference, or on pragmatic and expedient grounds. This is referred to as 'Multiplistic absolutism'. A number of researchers have reported teachers' mathematics-related belief systems that can be described as Multiplistic (Cooney, 1988; Oprea and Stonewater, 1987).

The level of Relativism includes subjective versions of public absolutist philosophies, as we have seen. In the terminology of Chapter 2 these consist of formal absolutist (e.g. logicism and formalism) and progressive absolutist (e.g. intuitionism) philosophies of mathematics. Fallibilist philosophies of mathematics, such as Lakatos' quasi-empiricism and social constructivism are also Relativistic, because their truths (corrigibility notwithstanding) are justified within frameworks such as informal mathematical systems or axiomatic theories. Knowledge within fallibilist philosophies is also evaluated with respect to the broader contexts of human activity and culture. These fallibilist philosophies are Relativistic because they acknowledge the multiplicity of approaches and possible solutions to mathematical problems, but require that mathematical knowledge is evaluated within a principled framework.

C. *Ethical Positions*

Individuals' ethical positions are also described by the Perry theory.

Ethical dualism

Dualism is an extreme ethical position, for it relates moral issues to an absolute authority without rational justification, and denies the legitimacy of alternative values or perspectives. Whilst minor variations within a Dualistic ethical position are possible, it describes a limited range of authoritarian outlooks.

Ethical multiplicity

A Multiplistic ethical position acknowledges that different moral perspectives on any issue exist, but lacks rational or principled grounds for choice or justification. Whilst such a position allows that individuals' preferences may be equally valid, it maintains its own set of values and interests. The lack of an absolute or principled justification for moral choice and actions, necessitates that choices are made on the grounds of whim, or the utility and expediency of their outcomes, based on pragmatic reasoning. Consequently, the set of values most compatible with this position comprises utility, pragmatic choice and expediency.

The ethical Relativist position

Just as the greatest number of personal philosophies are compatible with Relativism, so too there are different ethical outlooks which fit with Relativism. This position simply requires a consistent, principled set of values, coupled with acknowledgement of the legitimacy of alternatives. Thus to develop a theory of the aims of mathematics education, it is necessary to consider a number of sets of values, on a principled basis.

To complement the Perry theory it is appropriate to seek a psychological theory of ethics. The best known is Kohlberg's (1969, 1981) theory of moral stages. However he has been criticized for being selective or biased in his choice of ultimate moral values. His major critic Gilligan (1982) distinguishes two sets of moral values, separated and connected values, supplementing those offered by Kohlberg. Belenky *et al.* (1985, 1986) apply these sets of values to the Perry theory, as well as a synthesis, resulting in a third set. For inclusiveness I shall adopt all three sets of values, as each is consistent with Relativism.

The moral frameworks distinguished by Gilligan (1982) are briefly as follows. The separated perspective focuses on rules and principles, and objectifies areas of concern and objects of knowledge. Moral reasoning is typically based on 'blind justice', the impartial application of the rules of justice without concern for human issues. Such a perspective is said to be part of the cultural definition of masculinity.

In contrast, the connected moral perspective is concerned with human connections, with relationships, empathy and caring, with the human dimensions of situations. This view relates to the stereotypical 'feminine' role, to relate, nurture, and also to comfort and protect (socially constructed as this role may be).

These moral perspectives will be combined with Relativism, with which they are consistent. However they will not simply be regarded as sets of values. As in the proposals of Belenky *et al.* (1986), the perspectives are considered to relate to intellectual as well as ethical development. The justification for this is that the Perry theory treats both epistemological and ethical positions as forming an integrated whole.

Belenky *et al.* (1986) propose a further epistemological and ethical position, representing a synthesis of separated and connected values, together with an epistemological approach. They term this 'constructed knowing', which integrates the connected and separated 'voices'. Although an integrated epistemological and ethical position, we can isolate the ethical values involved (although Belenky *et al.* do not do this). These values combine concerns with justice and structure (separated values) with caring and human connection (connected values). The synthesis includes values concerning social justice and social structures which are liberatory and nurture the realization of individual human potential. The set of values comprises social justice, equality and human fellowship. These are largely connected values (especially fellowship, and the social aspect of social justice), but also include elements of separateness (equality, and the justice aspect of social justice).[4]

Each of the three sets of values provides a principled basis for moral reasoning. Thus each is consistent with Relativism, and can be combined with an appropriate philosophy of mathematics and epistemology to give an overall ideological perspective.

D. Combining the Distinctions

The various distinctions: epistemological frameworks from the Perry theory, personal philosophies of mathematics, and moral values are now combined to provide a model of different ideologies. There are five ideologies and before describing the individual positions, this number needs justification.

At the Dualistic level, only an absolutist view of mathematics is possible, as well as simple, Dualistic moral values.

At the Multiplistic level an absolutist view of mathematics fits best, as do the moral values of utility and expediency.

At the Relativisitic level both absolutist and fallibilist views of mathematics can be adopted consistently. Both separated and connected moral positions are consistent with the Relativistic absolutist view, so two ideological positions can be distinguished immediately, according to which set of values is adopted.

In addition, fallibilism can be combined with Relativism. Fallibilist philosophies of mathematics see mathematics as a human creation, attaching importance to human and social contexts, most fully articulated in social constructivism. The values most consistent are those of social justice, which fit well with social constructivism because of the social dimension and the link between subjective (individual) and objective (social) in both of them. This link parallels the 'constructed knowing' of Belenky *et al.*, in contrast with separated and connected values which tend to focus on one or other of the pair. Overall, the fifth ideological perspective of Relativistic fallibilism combines social constructivism with the values of social justice.

Five perspectives have been identified, although the possibility of further perspectives remains, for no claim is made for their logical exclusivity or necessity. The model of ideologies comprises the following.

Dualistic absolutism

Combining Dualism with absolutism, this view sees mathematics as certain, made up of absolute truths and depending heavily on authority. The overall perspective is characterized by two features: (1) the structuring of the world in terms of simple dichotomies, such as us and them, good and bad, right and wrong, and other simplistic dichotomies; (2) the prominence given to, and the identification with authority. Thus the values emphasize rigid distinctions, absolute rules, and paternalistic authority. Such values are consistent with an extreme version of conventional morality, identified by Kohlberg (1969) and Gilligan (1982).

Multiplistic absolutism

This combines Multiplicity with absolutism, viewing mathematics as a certain, unquestioned body of truths which can be applied or utilized in a multiplicity of ways. The overall perspective is characterized by liberality, many approaches and possibilities are acknowledged as legitimate, but it lacks a basis for choosing between alternatives except by utility, expediency and pragmatic choices. These constitute the values associated with this position, which is concerned with applications and technique, as opposed to principles or theory. Thus mathematics is applied freely, but not questioned or examined.

Relativistic absolutism

A range of perspectives fit this category, sharing the following features. Mathematics is viewed as a body of true knowledge, but this truth depends on the inner structure of mathematics (i.e. logic and proof) rather than authority. The overall intellectual and moral perspective acknowledges the existence of different points of view, interpretations, perspectives, frames of reference and values systems. Two viewpoints are distinguished, according to whether a separated or connected perspective is adopted.

Separated Relativistic absolutism. Separated moral values combined with relativistic absolutism leads to an emphasis on objectivity and rules. This ideology focuses on structure, formal systems and relationships, distinctions, criticism, analysis and argument. With reference to mathematics, this leads to an emphasis on logical inner relationships and proof, and the formal structure of mathematical theory. Because of the overall emphasis of this position on structure, rules and form, *formal absolutism* is the appropriate subjective philosophy of mathematics.

Connected Relativistic absolutism. This ideology combines an absolutist view of mathematics and contextual Relativism with connected values. On the basis of these values this viewpoint places emphasis on the knowing subject, feelings, caring, empathy, relationships and the human dimension and context. Mathematical knowledge is viewed as absolute, but emphasis is on the individual's role in knowing, and his or her confidence in understanding, mastering and coming to terms with the subject. Because of this emphasis, *progressive absolutism* is the subjective philosophy of mathematics in this position.

Relativisitic fallibilism

This position combines a fallibilist view of mathematical knowledge (social constructivism) and values concerned with social justice, within a Relativistic framework, with its acceptance of multiple intellectual and moral perspectives. Two of the central themes of this ideology are society and development. Knowledge and values are both related to society: knowledge is understood to be a social construction and the values centre on social justice. Knowledge and values are both related to development: knowledge evolves and grows, and social justice is about the development of a more just and egalitarian society. This is a highly consistent and integrated position, because human-centred principles underpin development on three levels, that of knowledge, the individual and society as a whole.

E. Appraising the Perry Theory and its Alternatives

The assumption of the Perry theory requires justification and critical appraisal. The following survey of alternatives serves to locate the Perry theory in a wider context.

Building on Piaget's work on the moral judgement of the child, Kohlberg (1969) developed a hierarchy of moral development. This has three levels: pre-conventional (ego-centric morality), conventional (moral judgements depend upon conventional norms), and post-conventional and principled (moral decisions are based upon universal principles). The last two levels offer some parallels with Dualism and Relativism, and there is a transitional level analogous to Multiplicity. Doubtless Perry was influenced by Kohlberg's theory (as he acknowledges). However, the theory of moral development, as the name implies, does not treat epistemological development. Beyond this, it has also been criticized by Gilligan (1982) for its emphasis on separated aspects of ethics (rules and justice) at the expense of connected values. This theory then does not offer a viable alternative to the Perry theory for two reasons. First, it does not treat intellectual as well as moral development. Second, it elevates one set of values above others, rather than allowing for variations in values.

Loevinger (1976) proposes a theory of 'ego development' with six stages offering some parallels with Kohlberg's theory (each of her three levels consist of two stages, as in Kohlberg's theory). Loevinger's theory has been applied to teachers' epistemological and ethical perspectives, for example, by Cummings and Murray (1989). These researchers interpreted teachers' views of the nature of knowledge (as well as the aims of education, etc.) in terms of the last three stages (conformist, conscientious, autonomous). Their approach offers some parallels to the Perry theory, but with greater discriminatory power, in terms of the range of personal beliefs. Thus the Loevinger theory evidently has potential as a means of classifying intellectual and ethical development. However, its fit with epistemology is less well articulated and more contrived than the Perry theory. Knowledge is considered in such terms as its parts, uses and sources, as opposed to its structure, basis and status. The theory is therefore less able to encompass philosophies of mathematics, and is thus less well suited to the present inquiry. Not surprisingly, in view of the theory's purpose, it offers more of a typology of ego development, than a structural analysis of theories or systems of belief.

Kitchener and King (1981) have a theory of the development of reflective judgement. This includes both the level of intellectual and ethical development, and criteria for evaluating their use in action. These details make it a valuable tool for empirical research. However other than this it offers nothing more than the Perry model, relative to present purposes. It also seems more appropriate to young adults than to lifelong development, for the highest level includes the ability to make objective judgements based on evidence. Perry's higher levels of development allow for substantial development beyond that offered by Kitchener and King's scheme. There are therefore no grounds of substance for adopting this scheme in place of the Perry theory.

Belenky *et al.* (1986) offer a developmental theory as an alternative to that of Perry. The stages of this theory are: Silence, Received knowledge, Subjective knowledge, Procedural knowledge (including the alternatives of separate knowing and connected knowing), and Constructed knowledge. These stages reflect the development of the individual as a knowledge maker, rather than the structural features of epistemological and ethical systems.

For all its many strengths, the theory suffers from two defects. First of all, it is as ethnocentric as the Perry theory, being based solely on a female sample, as opposed to Perry's almost exclusively male sample. This is admittedly an aim of the research, to balance the 'male-centredness' of the Perry theory. Nevertheless, it means that the theory proposed can only complement rather than replace the Perry theory, since its scope is only one half of humanity. The second criticism is that it is neither so broad, or so well defined and articulated a theory as that of Perry. For example, Belenky *et al.* focus on the subjective aspects of knowing, to the cost of ethics. Thus their theory does not serve to link epistemology and moral philosophy to personal beliefs as well as Perry. In particular, it does not provide the powerful connections made above between public and private philosophies of mathematics, on the one hand, and the stages of personal intellectual and ethical development, on the other. For these reasons, the theory of Belenky *et al.* will not serve as a replacement for the Perry theory here.

Belenky *et al.* offer their theory as an alternative to and criticism of the Perry theory, arguing that the Perry theory is gender-centric, being based on observations of of a sample of mainly male college students (at Harvard). Due to this bias, it is claimed that the masculine (separated) moral outlook dominates the Perry theory, and that the feminine (connected) moral outlook is omitted. I do not accept this criticism as invalidating the Perry theory. For unlike the ultimate stage in Kohlberg's (1969) theory of moral development, Perry's Relativism and succeeding positions of Commitment do not offer a unique set of moral values as the outcome of personal development. To insist on such a choice would be to give a hostage to fortune, for in different cultures different values are elevated as the highest. Kohlberg offers justice as the highest value. Gilligan argues that human connectedness should be placed no lower. One can also think of tribal peoples for whom honour is the highest value, such as some native American tribes. Doubtless other ultimate values exist. The Perry theory, by focusing on the form and structure of the ethical belief system, and the types of ethical reasoning employed by individuals, leaves open the specific set of values adopted. Thus both the connected and separated values distinguished by Gilligan are consistent with the position of Relativism.

One of the innovations of Belenky *et al.* is to relate the moral perspectives of Gilligan (1983) to levels of epistemological development. This is the course that has been followed here, within the position of Relativism. However, this also goes beyond the Perry theory, which emphasizes the form rather than the content of ideological frameworks, as we have seen. For the proposals given above also introduce specific sets of values: separated and connected values and their synthesis, social justice values. The introduction of these values complements the Perry theory, and fills out Relativism into specific ideologies. A parallel between these ideologies and the top two stages of the model of Belenky *et al.* may be noted.

Although a number of alternatives to the Perry scheme exist, they do not offer better alternatives, in the present context. There are yet further theories of intellectual or ethical development, such as Selman (1976), but none offer as useful a categorization of perspectives as that given above.

Although the Perry theory is preferred to other theories of intellectual or ethical development, two caveats are required. First of all, the adoption of the Perry theory is a working assumption. It is adopted in the spirit of a hypothetico-deductive view of knowledge. The theory provides a simple yet fruitful means of relating philosophies of mathematics to subjective belief systems. Secondly, because of its simplicity, the theory is likely to be falsified. For it posits that each individual's overall intellectual and ethical development can be located on a simple linear scale. The problem with this is that different subsets of beliefs might be located at different levels on the scale. Thus, for example, two student teachers might be at a similar stage in overall intellectual and ethical development. However, if one is a mathematics specialist and the other not, their personal philosophies of mathematics might well be identified with different Perry levels. (This hypothetical case is consistent with data in Ernest, 1989a).

Piaget's theory of cognitive development is to some extent analogous to the Perry theory. It offers a single linear scale of development, comprising a fixed number of positions. A potent criticism of Piaget's theory is that different aspects of an individual's development may be described by different positions in the developmental sequence (Brown and Desforges, 1979). Piaget acknowledged the existence of this phenomenon, terming it 'decalage', and by attempting to assimilate it into his theory. However it represents a substantial weakening, for it means that an individual's overall cognitive level can no longer be described uniquely, in terms of Piaget's theory. An analogous state of affairs can be imagined with the Perry theory, because of its simple characterization of the position or level of an individual's intellectual and ethical development. Different components of an individual's perspective might well be located at different levels. Especially when the epistemology of a single discipline, such as mathematics, is isolated from the overall intellectual and ethical position. Thus although the Perry theory is adopted as a powerful and useful tool, it is acknowledged that it may ultimately be falsified, just as Piaget's theory has been in this respect.[5]

2. Aims in Education: An Overview

A. The Nature of Educational Aims

An important feature of education is that it is an intentional activity (Oakshott, 1967; Hirst and Peters, 1970). The intentions underpinning this activity, stated in terms of their intended purposes and outcomes, constitute the aims of education. A number of different terms are used to refer to these outcomes including aims, goals, targets and objectives. Since Taba (1962) a distinction is commonly drawn in education between short-term educational goals (objectives) and 'broad aims', the longer-term and less specific goals (aims).

Hirst (1974) argues that nothing is gained by making the distinction, and prefers the term 'objectives'. Thus, for example, the index entry for 'aims' in Hirst (1974) reads 'see objectives of education'. He argues that

> this shift to a more technical term [objectives] alone indicates a growing awareness that more detailed description of the achievements we are after is desirable . . . [I]n speaking of curriculum objectives I shall have in mind as tight a description of what is to be learnt as is available.
>
> (Hirst, 1974, page 16)

Thus Hirst, in keeping with both a systems view of the curriculum and behaviourist psychology, sees aims and objectives in technical and normative terms. That is they are a means to rational curriculum design, a means of specifying what the curriculum should be. This is a widespread view throughout the literature on curriculum theory, which has been described as assuming a static society, lack of conflict, and 'the end of ideology' (Inglis, 1975, page 37).

However, a specification of the aims of education can also serve another purpose. That is the criticism and justification of educational practice, in other words, educational evaluation, whether theoretical or practical. For in its broadest sense, educational evaluation is concerned with the value of educational practices. In contrast, the technical and normative approach to aims and objectives, by focusing on particular learning outcomes, accepts much of the context and 'status quo' of education as unproblematic. The social and political contexts of education and the received view of the nature of knowledge are seen as the fixed backdrop against which curriculum planning takes place. Stenhouse acknowledges this.

> The translation of the deep structures of knowledge into behavioural objectives is one of the principal causes of the distortion of knowledge in schools noted by Young (1971a), Bernstein (1971) and Esland (1971). The filtering of knowledge through an analysis of objectives gives the school an authority and power over its students by setting arbitrary limits to speculation and by defining arbitrary solutions to unresolved problems of knowledge. This translates the teacher from the role of the student of a complex field of knowledge to the role of the master of the school's agreed version of that field.
>
> (Stenhouse, 1975, page 86)

Contra Hirst, we retain the distinction between the aims and objectives of education, and focus on the former. This allows us to avoid presupposing the unproblematic nature of the assumptions on which eduction is based. It also allows the social context and social influences on the aims of education to be considered, as opposed to being assumed as unproblematic.

Education is an intentional activity, and the statement of the underlying intentions constitutes the aims of education. However intentions do not exist in the abstract, and to assume that they do leads to a false objectification of aims. Any elucidation of aims needs to specify their ownership, for aims in education represent the intentions of individuals or groups of persons. As Sockett says: 'Intentional *human* action must stand at the centre of an account of curriculum aims and objectives.' (Sockett, 1975, page 152, emphasis added)

Beyond this, to discuss the aims of education in the abstract, without locating them socially, is to commit the error of assuming universal agreement, that all persons or groups share the same aims for education. Williams (1961), Cooper (1985) and others show that this is not the case. Different social groups have different educational aims relating to their underlying ideologies and interests.

Just as we need to consider the social context to establish the ownership of aims, so too we need to consider this context with regard to the means of attaining them. For to consider the aims of education without regard to the context and processes of their attainment is a further false objectification of aims. Others also argue that the means and ends of education are inseparable.

> Because of [the] logical type of relationship between means and ends in education it is not appropriate to think of the values of an educational process as contained purely in the various attainments which are constitutive of being an educated person. For in most cases the logical relationship of means to ends is such that the values of the product are embryonically present in the learning process.
>
> (Peters, 1975, page 241)

> The major point brought out by the notion of means as constitutive of ends is, however, that value questions are not themselves just questions of ends . . . Means may be constitutive of ends in such [teaching] activities, in that certain values are embedded in the activity, its content, and procedures: these may be attitudes which are part of what is learnt (and what is being taught) as well as part of the method of teaching.
>
> (Sockett, 1975, page 158)

> Educational aims, then, are not the end product to which educational processes are the instrumental means. They are expressions of the values in terms of which some distinctive educational character is bestowed on, or withheld from, whatever 'means' are being employed.
>
> (Carr and Kemmis, 1986, page 77)

Aims express the educational philosophies of individuals and social groups, and since education is a complex social process the means of attaining these aims must also be considered. For the values embodied in educational aims should determine, or at least constrain, the means of attaining them.

B. The Aims of Mathematics Education

The aims of mathematics education are the intentions which underlie it and the institutions through which it is effected. They represent one component of the general aims of education, combining with others to give the overall aims. Consequently the aims of mathematics education must be consistent with the general aims of education.

Many statements of the aims of mathematics education have been published. Some influential statements of aims are to be found in Whitehead (1932), Cambridge Conference (1963), Mathematical Association (1976), Her Majesty's Inspectorate (1979), Cockcroft (1982) and Her Majesty's Inspectorate (1985), from which the following is taken.

> **The aims of mathematics teaching**
> **1.1** There are important aims which should be an essential part of any general statement of intent for the teaching of mathematics. Those stated in this chapter are considered to be indispensable but it is recognized that there may be others which teachers will wish to add. These aims are intended for all pupils although the way they are implemented will vary according to their ages and abilities.
> [Pupils should have some mastery and appreciation of]
> **1.2 Mathematics as an essential element of communication**
> **1.3 Mathematics as a powerful tool**
> [they should develop]
> **1.4 Appreciation of relationships within mathematics**
> **1.5 Awareness of the fascination of mathematics**
> **1.6 Imagination, initiative and flexibility of mind in mathematics**
> [they should acquire the personal qualities of]
> **1.7 Working in a systematic way**
> **1.8 Working independently**
> **1.9 Working cooperatively**
> [and two further desired outcomes are]
> **1.10 In-depth study of mathematics**
> **1.11 Pupils' confidence in their mathematical abilities**
> (Her Majesty's Inspectorate, 1985, pages 2–6)

I wish to comment on the implicit assumptions of these aims, rather than on the aims themselves. The formulation assumes that the statement of aims is unproblematic and uncontroversial. Universal or majority acceptance of the aims is assumed, and allowance is only made for additions, in cases of omission. There is no acknowledgement that different groups have distinct aims for mathematics education. Admittedly, the set is both composite and compromise, admitting all three of 'profit, purity and pleasure' as aims, to varying degrees. However, like the other statements of aims for mathematics cited above, there is no reference to whose intentions (i.e., which social group's) are expressed in the aims.

In what follows, we therefore do not ask 'What are the aims of mathematics education?' without also asking 'Whose aims? (which social groups'?)' For educational aims need to be related to an educational and social context. This is recognized by a number of researchers, in both theoretical and empirical analyses. Morris (1981), reports the conclusion of an international meeting on the goals of mathematics education that

> every subgroup in society has a responsibility to participate in the identification of goals . . . [including] the teachers, the parents, the students, mathematicians, the employers, the employee organisations, educators and the political authorities. Involving the various groups in the process of determining goals may yield conflict.
>
> (Morris, 1981, pages 169–170)

Howson and Mellin-Olsen (1986) distinguish the aims and expectations of different social groups, including mathematics teachers, parents, employers and those in the higher levels of the education system (e.g. universities). They posit two conflicting types of socially situated aims, the S-rationale (social, or intrinsic goals) and the I-rationale (instrumental, or extrinsic goals) further elaborated elsewhere (Mellin-Olsen, 1987).

Ernest (1986, 1987) distinguishes three interest groups: educators, mathematicians and the representatives of industry and society, each with distinct aims for mathematics education.

Cooper (1985) presents a powerful theoretical case for the existence of different social groups with different interests, missions and aims for mathematics education. He demonstrates historically the existence of different interest groups concerned with mathematics education. These groups had varying aims for mathematics education, and the outcome of the struggle for ascendancy between them indicated their relative power.

Evidently the aims of mathematics education need to be related to the social groups involved, as well as to their underlying ideologies. To do this we relate the five ideologies distinguished above to five historical social interest groups, enabling us to specify their educational aims, both in general, and with regard to mathematics.

C. The Educational Aims of Social Groups: Williams' Analysis

Williams (1961) distinguishes three groups: the industrial trainers, the old humanists and the public educators, whose ideologies have influenced education both in the past and in the present. He argues for the powerful influence of these groups on the foundations of British education in the nineteenth century. He also emphasizes their continued impact on education: 'these three groups can still be distinguished, although each has in some respects changed.' (Williams, 1978, quoted in Beck, 1981, page 91).

Willams' groups are as follows. The industrial trainers represent the merchant classes and industrial managers. They have a 'petit-bourgeois' outlook, and value the utilitarian aspect of education. The aims of the industrial trainers are utilitarian, concerned with the training of a suitable workforce. Industrial trainers have had a major impact on British education, due to the

> needs of an expanding and changing economy . . . [leading both to a] protective response, the new version of 'moral rescue', very evident in the arguments for the 1870 Education Act . . . and the practical response, perhaps decisive, which led Foster in 1870 to use as his principal argument: 'upon the speedy provision of elementary education depends our industrial prosperity'. In the growth of secondary education this economic argument was even more central. [Its] great persuasiveness . . . led to the definition of education in terms of future adult work, with the parallel clause of teaching the required social character — habits of regularity, 'self-discipline', obedience and trained effort.
>
> (Williams, 1961, pages 161–162)

The old humanists represent the educated and cultured classes, such as the aristocracy and gentry. They value the old humanistic studies, and their product, the 'educated man', the cultured, well educated person. Thus their educational aim is 'liberal education', the transmission of the cultural heritage, made up of pure (as opposed to applied) knowledge in a number of traditional forms. The old humanists

> argued that man's spiritual health depended on a kind of education which was more than a training for some specialized work, a kind variously described as 'liberal', 'humane', or 'cultural'.
>
> (Williams, 1961, page 162)

The public educators represent a radical reforming tradition, concerned with democracy and social equity. Their aim is 'education for all', to empower the working classes to participate in the democratic institutions of society, and to share more fully in the prosperity of modern industrial society. Williams argues that this sector has been successful in securing the extension of education to all in modern British (and Western) society, as a right (through an alliance with the industrial trainers). Thus, the public educators can be seen as lying behind the modern comprehensive school movement. However other interest groups, especially the industrial trainers, have been successful in having a major impact on the educational aims of schooling, and on the means of achieving them.

This is a powerful historical analysis, widely accepted and quoted (Abraham and Bibby, 1988; Beck, 1981; Giroux, 1983; MacDonald, 1977; Meighan, 1986; Whitty, 1977; Young, 1971a; Young and Whitty, 1977). It has the strength of identifying educational aims with the ideologies and interests of certain recognizable social groups. The relative power of these groups is used by Williams to explain the historical ascendancy of certain educational goals over others.

Williams describes the battle waged by the old humanists against the teaching of science, technology or practical subjects (which did not include pure mathematics). Thus for example, under the sway of their influence, mathematics was taught in Victorian times using straight-edges and not graduated rulers, which were banned as 'impure'. Mathematics was taught as part of the old humanistic curriculum, but only the traditional pure mathematics, such as Euclid, and only to an élite.

Although somewhat diluted, the old humanist values remain powerful. C. P. Snow indicated this was his distinction between the 'two cultures', one humanistic and the other scientific (Mills, 1970). These represented conflicting cultural perspectives of educated persons in Britain. Science and other applied subjects are now widely accepted as part of the curriculum, partly in response to the redefinition of science as a pure theoretical subject, separating it from the more practical technology, assuaging the old humanists. However, the presence of science in the curriculum is largely a result of the modern industrial trainers' interests and power. Politicians across the whole political spectrum argue for the necessity of a skilled scientific and technologically educated workforce and population.

A Critical Review of Williams' Analysis

Although Williams' (1961) analysis is perhaps the most effective account of the overall aims and interest groups affecting British education, it is open to criticism on a number of points. In practice, as Williams admits, aims tend to be mixed, rather than pure and isolated, as described. This is, of course, true of any simplifying account of aims. Beyond this, there is the criticism that further unidentified but significant modern British interest groups exist, suggesting that the analysis can benefit from modification and revision.

The first omission is the failure to identify progressive, liberal reformers as a distinct social group with aims of their own. Admittedly, in the Victorian era, the liberal reformers and the public educators were united in a common cause, extending education to all members of society. Perhaps for this reason, Williams lists only the latter. However the two groups can be distinguished, and have increasingly distinct aims. These are the progressive, liberal reformers, termed 'progressive educators' (whose slogan might typically be 'save the child'), and the public educators, concerned with education for all to promote democratic citizenship.

In fact, Williams does distinguish between two strands of public educator, as Young (1971a) indicates. These are the democratic social reformists endorsing 'education for all', which largely correspond to the progressive educator group, and the populist/proletarian activists pressing for relevance, choice and participation in education, who have much in common with the modern, more radical public educators. Thus although the progressive educators are not distinguished by Williams as such, it could be said that they are prefigured in his analysis.

The addition of a new interest group, the progressive educators, is in keeping with the widespread distinction between two traditions in primary education: the elementary school and the progressive traditions. This can be found, for example, in Dearden (1968), Golby (1982) and Ramsden (1986). Thus, the progressive tradition in primary education is widely recognised, and is worth adding to Williams' (1961) analysis, resulting in four groups.

A comparable analysis of social groups and their educational aims is given by Cosin (1972). Cosin distinguishes four groups corresponding closely with those discussed: the rationalizing/technocratic, élitist/conservative, romantic/individualist, and egalitarian/democratic groups. The parallel between these sets of groups can be shown as in Table 6.1. The table reveals a strong analogy between the two sets of social groups. Minor differences occur in the social definitions of the old humanist group.

Table 6.1: A Comparison of Williams' (Modified) and Cosin's Groups

Williams	Cosin	Aims of Cosin's Groups
Industrial trainer	Rationalizing/ technocratic	Vocational relevance of education
Old humanist	Elitist/ conservative	Maintenance of established standards of cultural excellence through traditional methods of selection
Progressive educator	Romantic/ individualist	Development of all an individual's innate abilities
Public educator	Egalitarian/ democratic	All have an equal right to be educated

Nevertheless the analysis lends support for the modification of Williams' original analysis. A further analysis by Davis supports this conclusion.

> Educational ideologies fall into four categories: conservative, revisionist, romantic and democratic. The first, obviously, is concerned to maintain something like the status quo though conservative positions range from crude dogmatism (which characterizes many of the essays in the Black Paper) to carefully formulated versions of elite culture (T. S. Eliot, F. R. Leavis and G. H. Bantock). The revisionist arguments are couched in economic language or in a pseudo-sociological concern with the 'wastage' created by the educational system. Its emphasis is on improving the system's efficiency in terms of the job requirements of the market. Not surprisingly, successive governments — Labour and Tory — have found this the most attractive stance to adopt, and most of the official reports have incorporated its logic. The romantic attitude (which might also be called the psychological) owes much to a concern with individual development and derives from the work of Froebel, Montessori, Freud, Pestalozzi and Piaget. It was central to the establishment of the private 'progressive' schools and has had considerable influence on some forms of curriculum revision and on the primary schools. Its official monument is the Plowden Report. Finally, the democratic socialist tradition, stemming from socialist and liberal thinkers of the nineteenth century, seeks equal opportunity for all (recognizing the difficulties presented by class and patterns of socialization), and the progressive elimination of the élitist values inherent in established education. In its most recent articulate form (Williams, 1961),

> the democratic socialist approach called for a 'public education designed to express the values of an educated democracy and a common culture'.
> (Meighan, 1986, page 181)

Davis' analysis indicates the strength of the romantic/progressive tradition, further supporting identification of a separate progressive educator grouping.

The account by Davis does not fit Williams' analysis precisely, for industrial trainers are divided between the revisionist group and the conservative group, and the latter also includes the old humanists. However, what it does do, is to suggest the need to distinguish between two groups which are the intellectual descendants of the Victorian industrial trainers. Both education and the demands of employment have greatly advanced since the last century, and the descendants of the industrial trainers, whilst still valuing utilitarian preparation for work, no longer all share the same view of what this means. Some still value low-level training coupled with a moral reformist view, like the original industrial trainers. Others favour broader forms of education and competency, and lack the moral, reformist zeal. The former will be identified politically with the New Right in Britain, which favours basic skill training (Back-to-basics) as well as 'instruction in obedience' (White, 1968, quoted in Beck, 1981, page 89).

The second group 'descended' from the industrial trainers, differ from the original industrial trainers and the New Right significantly in their composition and aims for education. They are less narrow and moralistic in their outlook than the modern industrial trainers, and represent the interests of industry, commerce and the public sector employers. Thus this group is concerned with the acquisition and development of a broad range of knowledge, skills and personal qualities, notably those that prove efficacious in employment. As the level of industrialization has advanced, so too has this group expected more of education, to provide the greater skills required in employment. Indeed, the group sees social development as the outcome of the advance of industrialization and technology. Currently this group is concerned with such issues as information technology capabilities and skills, communication and problem solving skills, as well as a mastery of basics. They are an important force in modern educational policy. They are named the 'technological pragmatists' to distinguish them from the modern industrial trainers of the New Right.

Golby (1982) identifies a 'technological tradition', which describes the technological pragmatists. This tradition emphasizes utilitarian values, particularly the pursuit of science and craft, design and technology. Golby distinguishes this technological tradition from the elementary school and progressive school traditions in primary education, which correspond to other groups in the analysis. The technological pragmatist grouping also represents the rationalizing/technocratic group distinguished by Cosin (1972), which does not fit with Williams (1961) analysis. Likewise, we have seen that Davis posits a corresponding revisionist ideology. Thus there is recognition of the existence and significance of the technological pragmatist interest group in education.

When Williams wrote his account at the end of the 1950s, society and education were not yet in the grip of the 'white-hot technological revolution'. The New Right, if they existed, were marginal. Thus unlike now, the need to distinguish two utilitarian groups and ideologies was not pressing. The addition of this group constitutes a second and final modification of Williams' analysis, resulting in a total of five interest groups in education. Overall, the main criticism of Williams' (1961) model is that it needs to be refined to account for the complexity of the modern British socio-political context of education.

The ideologies of mathematics education of the groups

Below it will be shown that the five intellectual and moral positions distinguished in section 1 describe the ideologies of the five social groups. These will be identified as shown in Table 6.2.[6]

Table 6.2: The Match between Five Social Groups and Ideologies

Social Group	Ideology
Industrial trainer (New Right)	Dualist/absolutist
Technological pragmatist	Multiplistic/absolutist
Old humanist	Relativist/absolutist (separated)
Progressive educator	Relativist/absolutist (connected)
Public educator	Relativist/fallibilist

In the area of mathematics education, some empirical evidence supporting Williams' identification of aims in education can be cited. Ernest (1986, 1987) distinguishes three interest groups comprising (1) educators, (2) mathematicians, and (3) representatives of industry and society, each with their own aims for mathematics education. These aims are (1) personal development, (2) the inculcation of pure mathematics and (3) utilitarian, corresponding to the aims of (1) the public educators, combined with the progressive educators, (2) the old humanists, and (3) the industrial trainers combined with the technological pragmatists, respectively. Some empirical evidence to substantiate these attributions is provided in Ernest (1987), including the results of a large scale survey of the opinions on mathematics education of a number of different social groups concerned with education in the USA (NCTM, 1981).

Cooper (1985) provides a detailed historical study of different interest groups in Britain, concerned with mathematics education in the 1950s and 1960s. He argues that an alliance of university mathematicians, public school teachers and industrial interests succeeded in redefining school mathematics to serve their interests in the mathematical education of the élite. The mathematicians wanted the school syllabus brought closer to modern university mathematics, and the public school teachers echoed their wants. The industrialists wanted the inclusion of some modern applied topics and problem solving in the school mathematics curriculum. But above all, they wanted to be sure of a supply of mathematics graduates to serve the needs of industry. The alliance was successful in renegotiating the nature of school mathematics in line with these expressed purposes. It may be interpreted as an alliance of technological pragmatists and mathematical old humanists to defeat the aims of the public and progressive educators. For as Cooper shows, a project embodying some of the aims of these last two groups (the Midland Mathematics Experiment) failed, whereas the School Mathematics Project, associated more closely with the other groups and their interests, was successful.

D. Elements of an Ideology of Mathematics Education

A number of ideologies of mathematics education and overall intellectual and ethical frameworks have been identified and linked with social groups and their mathematical aims. Such aims, it has been argued, cannot be separated from the means of carrying them out. This raises the question: which elements in an ideology of mathematics education are needed to specify the means of attaining the aims? To answer this, a structural model of ideologies of mathematics education is proposed.

A model of ideologies of mathematics education

Meighan (1986) describes ideologies as comprising sets of beliefs operating at various levels and in various contexts with several layers of meaning. The model of educational ideologies proposed here reflects this degree of complexity. At the heart are situated the fundamental epistemological and ethical beliefs. Resting on these are a second set of beliefs concerning the aims of mathematics education and the means of attaining them. Thus the proposed model has two levels: (1) the primary level comprising the deeper elements of the ideology; and (2) the secondary level, made up of the derived elements pertaining to education.

The primary level includes the overall epistemological and ethical position, comprising an **epistemology**, a **philosophy of mathematics** and a **set of moral and other values**. However, these are very abstract elements, and ideologies must relate them to the experience of being a person and living in a society. For whether ideologies of individuals or groups are considered, the reality of being a person and relating to others, and living in a society must form a central part of consciousness, belief and world-view. Thus two further elements are included in the model of ideologies. These are a **theory of the child**, which is the special part of a theory of the person pertaining to education, and a **theory of society**. These relate to the other elements of the ideology. Epistemologies entail theories of how the knowledge of individuals' develop. That is, they entail theories of subjective knowledge as well as theories of objective knowledge. Thus epistemologies relate to theories of the person and the child. Moral values imbue and shape theories of the child, the person and society. Such theories are essential components of ideologies in general, and of *educational* ideologies in particular.

> An ideology is a system or cluster of beliefs and values held by social groups which help to bind those groups together and are used by them to further their own interests. Looked at more closely, ideologies are seen to contain beliefs and doctrines concerning man and his place in the world, the social and political structures in which he wishes to live, and his views about how best to achieve his ends and purposes.
>
> (Reynolds and Skilbeck, 1976, pages 76–77).

In their philosophies of education and its aims the classic philosophers of education first develop their theories of the child or person and society (Plato, 1941; Dewey, 1916). Likewise, modern educationists, in describing ideologies of education give a central place to theories of the child (Alexander, 1984; Esland, 1971; Phenix, 1964; Pollard, 1987; Pring, 1984), of society (Apple, 1979; Raynor, 1972; Williams, 1961; Young, 1971a), or both (Freire, 1972; Giroux, 1983; Reynolds and Skilbeck, 1976). Thus a good provenance exists for including these elements in an ideology of education.

The last component is the **aims of eduction**. Views of the nature of the child have a profound effect on the aims of schooling and the nature of education, as Skilbeck (1976) argues. Most of the authors cited include the aims of education in their treatments of educational ideologies. It represents the intentional aspect of the ideology with respect to education, drawing together elements of the underlying epistemology, system of values, theory of the child and theory of society. It is through the aims of education that the interests of the ideological group are expressed and put into practice, educationally.

The secondary level of the model comprises derived elements pertaining to mathematics education. The distinction is far from absolute, and the distinguishing characteristic is specialization to mathematics education. In ideologies of science education, for example, the secondary elements would be different, being transposed to science education.

What should the secondary elements be? First, the personal philosophy of mathematics may not be the same as the theory of school mathematics. Since mathematical knowledge is central to the whole process of mathematics education, a **theory of school mathematical knowledge** is needed, in addition to a philosophy of mathematics. Second, the specialization of the aims of education to mathematics are required. Thus the **aims of mathematics education** should be included as an element. Third, the means of attaining these aims need to be represented, as has been argued.

To attain the aims of mathematics eduction, it is necessary for mathematics to be taught, provided teaching is understood broadly enough to include liberal forms of pedagogy. Thus a **theory of the teaching of mathematics**, including the role of the teacher, is required. The teaching of mathematics has been transformed during its history, by developments in the area of resources for teaching and learning mathematics. Texts, calculating aids such as electronic calculators, and micro-computers, for example, play a central role in mathematics education. Thus it is appropriate to include a **theory of resources for mathematics education** as an element. In his general model of an educational ideology, Meighan (1986) includes such a component.

Teaching is only instrumental to learning, which represents the intended outcome of mathematics education. Thus a theory of learning mathematics, including the role of the learner, is central to an ideology of mathematics education. Theories of learning mathematics stem from both epistemological assumptions (the nature, acquisition and growth of knowledge) and moral views concerning the individual's responsibilites, and hence from theories of the child and society. Thus a **theory of learning mathematics** and the learner's role is included in the model.

The assessment of mathematics learning is of special importance, particularly with regard to its social functions. It could be subsumed under other components, but in view of its significance, it is distinguished here. Thus a **theory of the assessment of mathematics learning** is included among the secondary elements. Meighan (1986) includes it among his elements of an ideology of education, and Lawton (1984) describes it as being one of the three key components of the curriculum. This emphasis justifies its treatment as a separate element of the model.

In addition to the above elements, it is possible to distinguish elements deriving from the theories of the child and society. Implicit in a theory of the child is a theory of intelligence and ability, and its fluidity or fixity. Views vary on whether the traits of the child are inherited and fixed, or whether they are significantly influenced and formed by their environment and experiences. A theory of ability, and in particular, of mathematical ability thus derives from theories of the child, as well as from views of the social order, relating individual differences and typologies to social groupings, and from theories of the nature of mathematics and its accessibility. Concepts of ability are of especial importance in mathematics (Ruthven, 1987), and so a **theory of mathematical ability** is included in the model.

The theory of society includes within its concepts of social diversity, and of the inter-relationships between different segments. Coupled with other elements, such as views of mathematics and knowledge, this leads to personal theories concerning social diversity and its significance and accommodation in mathematics education. For this reason, a **theory of social diversity in mathematics education** is included. Theories of ability and social diversity in mathematics education are not included in models of education ideology by many other theorists. However sociologists of education have long indicated the importance of society, social relations and social diversity, as we saw above,as well as the construct of ability for education (Beck *et al.*, 1976; Meighan, 1986). In particular, Ruthven (1987) has shown the central ideological role played by teachers' concepts of mathematical ability. Increasingly, concepts of gender, race and class are being recognized as central factors in the distribution of educational opportunity in mathematics (Burton, 1986; Ernest, 1986, 1989; Ruthven, 1986, 1987). For these reasons, it is appropriate to include theories of mathematical ability and social diversity in mathematics among the secondary elements of the model.

This concludes the proposed model of educational ideology. Further elements can be included, although the model is already of sufficient, if not excessive complexity. The selection of the components is, to some extent, a matter of decision, rather than of necessity. Future developments or uses of a model of educational ideology might very well require such a decision to be revised.

The model of educational ideology for mathematics is summarized in Table 6.3. It is a tentative framework, allowing the consequences of a number of epistemological and ethical perspectives of mathematics education to be elaborated and compared. The

Table 6.3: A Model of Educational Ideology for Mathematics

Primary Elements	Epistemology
	Philosophy of Mathematics
	Set of Moral Values
	Theory of the Child
	Theory of Society
	Educational Aims
Secondary Elements	Aims of Mathematics Education
	Theory of School Mathematical Knowledge
	Theory of Learning Mathematics
	Theory of Teaching Mathematics
	Theory of Assessment of Mathematics Learning
	Theory of Resources for Mathematics Education
	Theory of Mathematical Ability
	Theory of Social Diversity in Mathematics Education

different elements of the model, such as the component theories, should not be construed as reified self-subsistent categories. They are convenient labels for aspects of a more or less integrated cluster of beliefs and values. Many of the elements distinguished are closely inter-related and interdependent, and no claim is made that they can be separated completely.

Alternative models of educational ideologies

The model can be critically evaluated by comparison with other proposals, including the following. Esland (1971) offers a model distinguishing three constitutive categories of the teacher's thought: (a) the pedagogical perspective, including assumptions about learning, assumptions about the child's intellectual status, assumptions about teaching style; (b) the subject perspective; and (c) the career perspective. These categories reflect some of the main elements proposed above. The model is generic, rather than subject-specific, so comparison requires its application to mathematics, as in Cooper (1985). On this basis, there is a partial match between Esland's model and that proposed above in terms of the elements of epistemology, aims, and theories of school knowledge, the child, ability, learning and the learner's role, teaching and assessment. Beyond these factors Esland adds a new dimension, the teacher's career perspective, which is more pragmatically related to the social and professional life of a teacher, than to the ideological framework which is the present concern. This does, however, relate to, if not treat the interests of the social groups, which their aims seek to further.

It is on this point that Cooper (1985) criticizes the model. He argues that there is insufficient treatment of the basis of the power (and its limits) of ideological groupings, nor any recognition of the possibility of conflict between ideological groups. One of the elements omitted by Esland but included in the present model, relates to this point. This is theory of society. The inclusion of this element, and other factors, means that Cooper's criticism does not apply to the model proposed in this chapter. For the social location of ideological perspectives, and the discussion of their relative power and conflicts in, for example, the development of the National Curriculum (Chapter 10), explicitly treat interests, power and conflict.

Hammersley (1977) proposes an ideological model of teacher perspectives, distinguishing five components: view of knowledge, view of learning, pupil action, teacher's role and teaching techniques. These are broken down further, revealing that assessment is included in the last component. Overall, in terms of its elements, this model represents a proper subset of that proposed in this chapter. As an educational ideology, it is open to the criticism directed by Cooper at Esland's model. However, it is intended to describe teacher perspectives, rather than group ideologies, and much of its strengths lie in its identification of the constructs in each of its components.

Meighan (1986) offers a more refined model of educational ideology, which includes eight components, each of which is itself a personal theory, in the sense used above. Its components are the following.

1 A theory of knowledge, its content and structure.
2 A theory of learning and the learner's role.
3 A theory of teaching and the teacher's role.
4 A theory of resources appropriate for learning.
5 A theory of organization of learning situations.
6 A theory of assessment that learning has taken place.
7 A theory of aims, objectives and outcomes.
8 A theory of the location of learning.

Most of these components have analogues in the model proposed above. Once again it is a generic, rather than subject specific model. It contains two components not included in the model proposed above: theories of the organization of learning and of its location. These introduce aspects of the social context of schooling, if not the full social and political milieu. However Meighan makes it clear that the theory of aims in the ideology includes social aims. For he distinguishes this component as part of an ideology of legitimation, which serves to buttress a social group's interests. In contrast, the other components are said to be part of an ideology of implementation, that is, they are the means of implementing the aims.

A criticism that can be directed at Meighan's model is that whilst he treats well the secondary elements of an ideology of education, he does not treat the core beliefs on which these are based. Most of his components are part of an 'ideology of implementation', and do not represent the core of an epistemological and ethical belief-system.

Ernest (1989c, d) offers an analysis of the beliefs of the mathematics teacher including four components: views of the nature of mathematics, its teaching and its learning, and educational principles, which includes both educational values and views regarding social issues. Of the models surveyed, it is the only one directly applied to mathematics education. It is open to Cooper's criticism, that it does not recognize the relationship between aims, power and the interests of social groups.

In an overview of the forces underpinning educational policy, Lawton (1984) distinguishes three socially situated categories of educational ideology: beliefs, values and tastes (or choices). These general categories subsume all the elements considered, but are too general to be of much use.

This brief survey provides a partial justification of the model proposed. For virtually all of its elements are included and justified in one of the models surveyed. One criticism can be directed at several of the models. This is the lack of any theoretical basis for their components. In contrast, the present model rests on a systematic basis, namely the epistemological and moral dimensions of an underlying ideology, as well as the literature, and thus has a theoretical foundation.

Notes

1 Capitalization (e.g. Dualism) denotes technical usage of the Perry terms.

2 A Perry Network exists in the USA (chair: William S. Moore, previously Larry Copes).

3 The term 'instrumental' used here corresponds to the unquestioning use of mathematics (Skemp, 1976) or mathematics texts (Goffree, 1985) as a tool presumed to be correct. In philosophy, instrumentalism denotes a different view of theories as predictive tools not assumed to be true (Lakatos, 1978b; also Dewey and Glasersfeld).

4 This does not accurately represent the 'constructed knowing' of Belenky *et al.*, (1986), which is concerned with the relationship between the knower and known, including the knower's personal engagement and involvement with the known.

5 Chapter 11 criticizes the view that an individual can be identified with a fixed level in an hierarchy, such as of mathematical ability.

6 The match of the five ideologies with the five social groups results from design, not coincidence. I originally identified the aims of three groups (Ernest, 1986, 1987) and three personal philosophies of mathematics (Ernest, 1989b, c, d). Thanks to Neil Bibby, I made the connection with Williams (1961). The model was subsequently grounded more deeply in philosophy and sociology. Over a period of time I saw that two further groups needed to be distinguished, and identified appropriate personal philosophies and values for them. Thus the ideologies and groups have evolved together. Naturally the model remains imperfect and conjectural, to be judged by its fruitfulness, and ultimately replaced.

7

Groups with Utilitarian Ideologies

1. Overview of the Ideologies and Groups

A. Five Ideologies of Mathematics Education: an Overview

In this section we provide a brief overview and comparison of the groups and their ideologies. Although necessarily superficial, it serves an orientating function, an advance organiser (Ausubel, 1968). The overview, in Table 7.1, uses the elements of the model of educational ideologies (Table 6.3) for its categories. It differs in that two elements are omitted for brevity, and the political ideology (and name) of the social interest group is added, reflecting the social location, aspirations and interests of the group.

A number of patterns can be seen in Table 7.1. First, all the secondary elements cohere with and derive from the philosophy of mathematics, the set of moral values and the theory of society. These primary elements imbue all the aspects of mathematics education within an ideological cluster, illustrating a central thesis of the book, that ideologies have a powerful, almost determining impact on mathematical pedagogy.

Further patterns can be discerned, including the social reproduction implicit in the first four ideologies. The teaching of mathematics through these groups serves in differing ways to reproduce the existing stratification of society, serving the group interests. The theme of 'purity' is shared by the third and fourth ideologies, concerning the purity of subject matter or with pure creativity and personal development. It also relates to the first ideology, which is concerned with moral purity. Finally, the theme of 'social relevance' is shared by the first two and the last ideologies. However, this cleaves into the reproductive-utilitarian bent of the first two, and the social engagement for change, of the last ideology. These themes will be further developed, subsequently.

Table 7.1: Overview of the Five Educational Ideologies

Social group	Industrial trainer	Technological pragmatist	Old humanist	Progressive educator	Public educator
Political Ideology	Radical right, 'New Right'	meritocratic, conservative	conservative/liberal	liberal	Democratic socialist
View of Mathematics	Set of Truths, and Rules	Unquestioned body of useful knowledge	Body of structured pure knowledge	Process view: Personalized maths	Social constructivism
Moral values	Authoritarian 'Victorian' values, Choice, Effort, Self-help, Work, Moral Weakness, Us-good, Them-bad	Utilitarian, Pragmatism, Expediency, 'wealth creation', Technological development	'Blind' Justice, Objectivity, Rule-centred Structure, Hierarchy, Paternalistic 'Classical' view	Person-centred, Caring, Empathy, Human values, Nurturing, Maternalistic, 'Romantic' view	Social Justice, Liberty, Equality, Fraternity, Social awareness, Engagement and Citizenship
Theory of Society	Rigid Hierarchy Market-place	Meritocratic Hierarchy	Elitist, Class stratified	Soft Hierarchy Welfare State	Inequitable hierarchy needing reform
Theory of the Child	Elementary School Tradition: Child 'fallen angel' and 'empty vessel'	Child 'empty vessel' and 'blunt tool' Future worker or manager	Dilute Elementary School view Character building Culture tames	Child-centred, Progressive view, Child: 'growing flower' and 'innocent savage'	Social Conditions view: 'clay moulded by environment' and 'sleeping giant'
Theory of Ability	Fixed and inherited Realized by effort	Inherited ability	Inherited cast of mind	Varies, but needs cherishing	Cultural product: Not fixed

Mathematical aims	'Back-to-Basics': numeracy and social training in obedience	Useful maths to appropriate level and Certification (industry-centred)	Transmit body of mathematical knowledge (Maths-centred)	Creativity, Self-realization through mathematics (Child-centred)	Critical awareness and democratic citizenship via mathematics
Theory of Learning	Hard work, effort, practice, rote	Skill acquisition, practical experience	Uderstanding and application	Activity, Play, Exploration	Questioning, Decision making, Negotiation
Theory of Teaching Mathematics	Authoritarian Tranmission, Drill, no 'frills'	Skill instructor Motivate through work-relevance	Explain, Motivate Pass on structure	Facilitate personal exploration Prevent Failure	Discussion, Conflict Questioning of content and pedagogy
Theory of Resources	Chalk and Talk Only Anti-calculator	Hands-on and Microcomputers	Visual aids to motivate	Rich environment to explore	Socially relevant Authentic
Theory of Assessment in Maths	External testing of simple basics	Avoid cheating External tests and certification Skill profiling	External examinations based on hierarchy	Teacher led internal assessment Avoid failure	Various modes. Use of social issues and content
Theory of Social Diversity	Differentiated schooling by Class Crypto-racist, Monoculturalist	Vary curriculum by future occupations	Vary curriculum by ability only (maths neutral)	Humanize neutral maths for all: Use local culture	Accommodation of social and cultural diversity a necessity

B. Limitations of the Account

The account suffers from a number of limitations, which need to be clarified. First of all, a great deal of simplification is entailed. Doubtless more interest groups than those listed exist; they need not be stable over time, neither in social group definition terms, nor in aims, ideology and mission terms; within a single grouping there will not be a single fixed ideological position, rather a family of overlapping ideologies; group members may subscribe to composites including components of several of the ideologies; and the ideological positions themselves are simplified and to some extent arbitrary in the choice of elements included.

Such complexity is routinely recognized by sociologists. For example, the professions have been characterized as:

> loose amalgamations of segments pursuing different objectives in different manners, and more-or-less delicately held together under a common name at a particular period of history.
>
> (Bucher and Strauss, 1961, page 326).

Similarly, Crane (1972) describes science in terms of hundreds of research areas constantly being formed and progressing through stages of growth, before tapering off. The groups described here can similarly be expected to be in a state of flux and change, for they are held together even more loosely by shared interests, underpinning a larger family of overlapping ideologies.

This complexity means that over-simplification inevitably occurs. Consequently, the account is tentative, and always open to revision. Of course, this is the theme of the treatment of the philosophy of mathematics, above. Even the most certain of knowledge is tentative and fallible. It applies across the whole realm of human knowledge, even more so to these tentative proposals.

Secondly, the ideologies need not be bounded by nationality. However, the social interest groups pertain to British history, and so whatever parallels might exist elsewhere, this account is directed at the British context.

2. The Industrial Trainers

A. The New Right as Industrial Trainers

I wish to establish that Dualistic absolutism describes the ideology of New Right in Britain, and that this group represents the modern industrial trainers (Williams, 1961). The 'New Right' is the radical conservative group, which include Margaret Thatcher and like-minded members of the Conservative Governments of Britain of the 1980s. It is also most vocally represented by the associates of a number of right-wing think-tanks and pressure groups (Gordon, 1989).[1]

Dualistic absolutism characterizes the views of the New Right concerning knowledge, moral values and social relations. According to this perspective knowledge, including mathematics, is cut and dried, true or false, and established by authority. Moral values likewise are organized in dualities such as right and wrong, good and bad, and are established with reference to authority. In our society authoritarian paternalism sanctions the most traditional values and moral perspectives in the culture. Most notable of these are Victorian values and the Protestant work ethic, which place a premium on work, industry, thrift, discipline, duty, self-denial and self-help (Himmelfarb, 1987). Thus to the Dualist moralist, these are all morally desirable and good. In contrast play, ease, self-gratification, permissiveness and dependence are all bad.

These values are also associated with a view of humanity and social relations. From the valuing of authority and some of the traditional Judeo-Christian values comes a hierarchical view of humankind — good at the top (close to God) and base at the bottom.[2] This was the accepted model of the world in medieval times, the 'Great Chain of Being', and its influence persists (Lovejoy, 1953). Such a view sees persons as fundamentally unequal, and identifies social position with moral value. It also sees children as bad, or at least naturally inclined to error, with all the moral connotations thereof, particularly lower-class children. Another consequence of this perspective is to trust only those who agree with these views (one of us),[3] and to distrust and reject those who disagree (them).

This view is conservative, in that it accepts social stratification and class as fixed, representing the proper order of things. It is self-serving, for it seeks to enhance trade and commerce, and to reward those with the virtues it admires, namely the petit-bourgeoisie.

In support of this identification, it should first be noted how closely these assertions fit the definition of Dualism (Perry, 1970). It is not claimed that Dualist absolutist thinkers must logically be followers of the New Right. The connection is contingent, and depends on the British cultural and historical context. In the USA Dualist absolutist thought is likely to be found among fundamentalist Christians and radical-right groups such as the Heritage Foundation, for example, although the puritan ethic and business values are widespread in American education (Carier, 1976). What is claimed is the converse, that New Right thinking in Britain can be categorized as Dualist and absolutist.

The key representative of New Right thinking is the British Prime Minister of the 1980s, Margaret Thatcher, and a case study of her ideology follows. As the prime mover behind social and educational policy, her ideological position is of key importance. Thus Thatcher's personal ideology, being the engine of her policy, has undoubtedly had a major impact on government legislation and policy, and she is thus the most important representative of the New Right. In addition, the right wing think-tanks and pressure groups are currently very influential precisely because the views they publish resonate with those of Thatcher. In fact it was she, together with Keith Joseph, who in 1974 founded possibly the most important of these groups, the Centre for Policy Studies (1987, 1988). This was established as a radical alternative to the conservative party research department, which was dominated by moderates loyal to Edward Heath (Gordon, 1989). Prior to 1979, these think-tanks had no impact on policy, whatever the complexion of government. Overall, their rise to influence reflects a major ideological shift in the power base, for the industrial trainer perspective was marginal in influence in modern Britain prior to the 1980s.

Despite Thatcher's undoubted importance in policy making, to justify the identification of one of the main intellectual and moral frameworks with a social group on the basis of a single case study would be unsound. Thatcher's importance is not personal, it stems from the fact that she epitomizes a social group that has newly found power. So corroborating evidence is provided, both historical and contemporary, to show the broader base for this position. Hence it will be seen that Thatcher's ideology, the epitome of the New Right outlook, fits well within a modern industrial trainer grouping.

B. Case Study: The Ideology of Margaret Thatcher

Thatcher was greatly influenced by her father, a grocery-shop keeper, with values typical of the petit-bourgeoisie. She adopted his Victorian values of hard work, self-help, rigorous budgeting, the immorality of extravagance, duty instead of pleasure.[4] These values were also reinforced by Methodism (she still reads improving works by moral theologians[5]) and said of Wesley 'He inculcated the work ethic, and duty. You worked hard, you got on by the results of your own efforts.'[6] She has a dualistic view of the world, a belief in right versus wrong, good versus evil, coupled with a certain knowledge of her own absolute and incontestable correctness.[7] The us-right, them-wrong view is applied to academics and educationists, who are regarded as incorrect, by questioning her dogma[8] and untrustworthy, through being self-serving. The imposition of 'equalisation' instead of acknowledging ability differences denies 'educational opportunities for those prepared to work'.[9] She also developed a strong market-forces, anti-collectivist and anti-interventionist view, from an early age[10], stating that 'There is no such thing as society, there are individual men and women, and there are families.'[11] Thatcher 'felt an uncommon need to link politics to a broad, articulated philosophy of life.'[12] She believes that the values of free society come from religion, and that the key value is freedom of will and personal choice, reinforcing the market-forces, anti-collectivist ideology. She said the most important book she had ever read 'A Time for Greatness' by H. Agar, which has the theme that a 'moral regeneration of the west [is] needed . . . we must fight inner weakness'.[13] Elsewhere she has emphasized that she bases on Christianity her own, and hence the only correct 'view of the universe; a proper attitude to work; and principles to shape economic and social life'.[14]

The central elements in Thatcher's ideology are Victorian values for the personal (the virtues of work, self-help and moral striving), which have social stratification as an unquestioned backdrop, and the related market-place metaphor for the social (industry, welfare and education). The pre-eminence of this social model in the policies of the government stems from the adherence to it of Thatcher and her early advisers Joseph and Hoskyns. There is also an anti-intellectualism stemming from a dualistic certainty of belief, which sees argument as the attempt to dominate, coupled with a distrust of professionals, who challenge the dogma because they are morally flawed and self-seeking.[15]

This brief case study shows that a central figure of the New Right has a Dualistic-absolutist ideology. (For simplicity, it is assumed that this ideology is shared by all members of the New Right[16]).

C. *The Origins of the Industrial Trainer Ideology*

The industrial trainer perspective is the petit-bourgeois ideology of the merchant classes (Williams, 1961). This describes Thatcher and the New Right, who represent the small merchants and shopkeepers, or their children, rather than the professions, and have their narrow petit-bourgeois ideology. Historically, both moral and epistemological components can be identified in this ideology. Thus, this group not only values a narrowly utilitarian education, but expects of it 'teaching the required social character — habits of regularity, "self-discipline", obedience and trained effort.' (Williams, 1961, page 162).

The roots of this thinking can be traced through the Judeo-Christian tradition. Thus we find in the Bible that:

> Foolishness is bound in the heart of a child, but the rod of correction shall drive it far from him.
>
> (Proverbs 22:15)
>
> The rod and reproof give wisdom, but a child left to himself bringeth his mother to shame.
>
> (Proverbs 9:152)

These moral values were reaffirmed in the puritan revolt, Calvinism and Wesley's Methodism.[17] Thus John Wesley's

> mother was also a Tory with an intense and deeply held personal religion . . . Her avowed intention was 'to break the child's will', to impart Christian habits, especially industry
>
> (Plumb, 1950, page 91)
>
> Wesley considered play unworthy of a Christian child . . . and idle minutes he regarded as the greatest danger to a child's soul. For the sake of its everlasting soul, it ought to be at work.
>
> (Plumb, 1950, page 96)

This moral outlook is one source of an ideological model of childhood, implicit in the 'Victorian values' of the elementary school tradition. This sees children as 'fallen angels' (harking back to original sin), who are sinful by nature and must be kept busy ('the devil finds work for idle hands'). Implicit in this is the identification of work and industry with virtue, and ease or play with evil.

The second source is the view of learning and knowledge, according to which children are 'empty buckets' who must be trained and fed the right facts by the teacher. For left to their own devices their minds will fill up with inappropriate and disorganized material.[18] These views have their origins in Aristotle, according to whom:

> Each should receive the minimal training necessary to perform his political function ... children are regarded as *empty buckets* of differing capacities, which should be filled up appropriately by agents of the state; if they are left about, they will fill with rubbish.
>
> (Ramsden, 1986, page 4)

John Locke in 1693 held that the mind began as an empty page, a 'tabula rasa', waiting to be filled up with characters. This model has long persisted. In the Victorian era, the 'empty bucket' view is exemplified in the following piece of social commentary, albeit from a work of fiction, the well-known words of Gradgrind.

> Now what I want is facts. Teach these boys and girls nothing but facts. Facts alone are wanted in life. Plant nothing else, and root out everything else.
>
> (Dickens, 1854, page 1)

In this tradition, useful knowledge is considered to be a set of discrete facts and skills: true facts and correct skills.[19]

One further factor which strongly influenced the elementary school tradition is a stratified view of society, social relations and social roles. Much of this can be traced back in Western thought, if not to Aristotle himself, at least to his influence on the medieval notion of the 'Great Chain of Being'. According to this idea, all creatures, including all of humankind, are arranged in a single hierarchy of excellence, with God at the top (Lovejoy, 1953). In the human part of the scale, the King is highest, followed by the aristocracy, next the gentry, all the way down to the lowliest serf. Lovejoy shows how such a hierarchical notion of 'place' in society has persisted as part of the received world-view, and is reflected in such writings as the poetry of Herbert, Pope and Thomson.

This stratified view is reflected in the education offered to different social classes, preparing them for their role or 'place' in society. Thus, as Crabbe wrote in 1810: 'To every class we have a school assigned, Rule for all ranks and food for every mind.' (Howson, 1982, page 101). Such views were widespread, for example Andrew Bell, who played a significant part in spreading the monitorial system, and who campaigned for elementary schools, argues:

> It is not proposed that the children of the poor be educated in an expensive manner, or all of them taught to write and cipher. [Such] schemes ... confound that distinction of ranks and classes of society on which the general welfare hinges ... there is a risque of elevating by an indiscriminate education the minds of those doomed to the drudgery of

> daily labour, above their condition, and thereby rendering them discontented and unhappy in their lot.
>
> (Howson, 1982, page 103)

Overall, the elementary school tradition combined Dualistic conceptions of both morality and knowledge with a social purpose, to 'gentle the masses', the preparation of the children of the working classes for a docile acceptance of their place in society and a life of industry, toil or servitude (Glass, 1971; Lawton and Prescott, 1976; Williams, 1961). Thus the industrial trainer view encompasses far more than utilitarian aims for education. It also incorporates a wide reaching Dualistic moral and intellectual perspective, with maintenance of the existing social divisions and order as a part. Because of this Dualism, the perspective can be identified with the authoritarian 'custodial pupil control ideology'.[20]

D. The Ideology of the Industrial Trainers/New Right

The ideology and 'Victorian values' of the elementary school tradition describe the educational and moral outlook of the New Right in education, as the following quotations show.

> [The Elementary School] 'was based on the authority of the teacher, whose task it was . . . to ensure that the most important facts in those lessons were remembered . . . he had a valuable store of information to impart to the untutored children in his care.'
>
> (Froome, 1970, page 15)[21]
>
> The merit of 'Victorian arithmetic' was that it was . . . based on the hard facts of number, and it dealt in measureable concrete quantities like gallons of beer, tons of coal and yards of linoleum . . . Victorian arithmetic wins hands down on the new maths . . . standards of skill in computation and manipulation of number are far more important than dabbling in a new mathematics which seems too theoretical
>
> (Froome, 1970, page 109)
>
> 1. Children are not naturally good. They need firm, tactful discipline from parents and teachers with clear standards. Too much freedom for children breeds selfishness, vandalism and personal unhappiness.
> 5. The best way to help children in deprived areas is to teach them to be literate and numerate.
> 8. Examinations are essential for schools . . . Without such checks, standards decline.
> 10. You can have equality or equality of opportunity; you cannot have both. Equality will mean the holding back (or the new deprivation) of the brighter children.
>
> (Cox and Boyson, 1975, page 1)

> There are certain forms of knowledge, understanding and skill which it is necessary for children to acquire . . .
>
> 1. A sure grasp of the English language, including its grammar and spelling.
> 2. An ability to read . . .
> 3. Basic mathematical skills, both numerical and graphical, together with an ability to apply those skills in day-to-day situations.
>
> Subjects like sociology, peace studies, world studies and political education have no place in the school curriculum.
>
> (Campaign for Real Education, 1987)

The ideology of the New Right, including Thatcher, can now be spelled out explicitly. Its key features are the dualistic view of knowledge, the moral vision regarding work as virtuous, and the authoritarian-hierarchical views of the child and society.

Moral values

The morality consists of 'Victorian values' and the 'protestant work ethic' (Himmelfarb, 1987). Key principles are freedom, individualism, inequality, and competition in the 'market-place'. However the fallibility of human nature means that strict regulation is needed (Lawton, 1988).

Theory of society

Society is stratified into social classes, reflecting differences in virtue and ability. All individuals have their place in society, which they can keep, or slightly improve, if they fulfil their social duties and responsibilities, and practice 'virtuous' living. However, competition between individuals in the market-place ensures that those who slacken off, or fall below the standards of their position, find a new, lower place in society.

In addition, this ideology includes a powerful theory of British society and culture, as follows. The British nation has a unique and characteristic Christian cultural heritage, of which we should be proud. Any attempt to weaken and dilute it with foreign religions and multicultural elements should be opposed as a threat to our national identity. Any immigrants or foreigners who choose to live here must agree to embrace our culture and values, or go back to where they came from. Anti-racists, who object to these doctrines, are nothing but extremist, politically-motivated agitators (Brown, 1985; Palmer, 1986). Evidently this perspective is monoculturalist, crypto-racist and xenophobic.

Epistemology and philosophy of mathematics

Knowledge stems from Authority, be it the Bible, or from experts. True knowledge is certain, and above question. Mathematics, like the rest of knowledge, is a body of true facts, skills and theories. The facts and skills must be learned correctly but the theories are complex and should be reserved for the more able, the future élite.

Theory of the child

The child, like all of humanity, is tainted by original sin, and slips easily into play, sloth and evil unless checked and disciplined. Strict authority is needed as a guide, and 'one must be cruel to be kind'. Competition is needed to bring out the best in individuals, for only through competition will they be motivated to excel.

Educational aims

These vary according to the social location of the students. The aims for the masses are the mastery of basic skills (Letwin, 1988), and training in obedience and servitude, in preparation for a life of work, as befits their station. For the higher social strata, mastery of a more extended range of knowledge, as well as training in leadership, serve as preparation for future occupations and life-roles.

E. The Industrial Trainer Ideology of Mathematics Education

Theory of school mathematics

Mathematics is a 'clear body of knowledge and techniques' (Lawlor, 1988, page 9), made up of facts and skills (as well as 'complicated and sophisticated concepts more appropriate to academic research'). The skills include 'understanding simple mathematics' and the facts include '$2+2=4$' (Letwin, 1988). School mathematics is clearly demarcated from other areas of knowledge, and must be kept free from the taint of cross-curricular links and social values (Lawlor, 1988, page 7). Social issues have no place in mathematics (Campaign for Real Education, 1987), which is completely neutral, and concerns only such objective contents as number and computation.

> children who needed to count and multiply were learning anti-racist mathematics — whatever that might be. Children who needed to be able to express themselves in clear English were being taught political slogans.
>
> (Margaret Thatcher)[22]

Thus the intrusion of social issues such as multiculturalism, ethnicity, anti-sexism, anti-racism, world-studies, environmental issues, peace and armaments, is rejected outright. Not only are they perceived as irrelevant to mathematics, they undermine British culture (Palmer, 1986). The consideration of any critical social issue in mathematics teaching is to be strongly opposed (Campaign for Real Education, 1987;

Lawlor, 1988). Mathematics ia a value-free tool, and so to include such issues usually represents a sinister attempt to undermine its neutrality.

Aims of mathematics education

The aims of mathematics education are the acquisition of functional numeracy and obedience. The unquestioning mastery of basics must precede all else (Letwin, 1988). Thus the aim is primarily

> to ensure that children leave school literate, numerate and with a modicum of scientific knowledge, it should not extend beyond these three core subjects, nor attempt to do more than set minimum standards in basic knowledge and technique.
>
> (Lawlor, 1988, page 5)

Theory of learning mathematics

Learning, like success in life for the masses, depends on individual application, self-denial and effort. Learning is represented by the metaphor of 'work' or hard labour. In addition, learning is isolated and individualistic.

> There is no reason to imagine that pupils learn from talking . . . The acquisition of knowledge require[s] effort and concentration. Unless children are trained to concentrate and make the effort to master knowledge they will suffer.
>
> (Lawlor, 1988, pages 18–19)

What is important is paper and pencil work, and drill and rote learning (Lawlor, 1988, page 15). It is wrong to say that learning must take place without effort and in the guise of games, puzzles and activities (Lawlor, 1988, page 7). It is inappropriate to make subject matter relevant to the interests of the child (Lawlor, 1988, page 18). All that is needed is hard work, practice and application. Competition is the best motivator.

Theory of teaching mathematics

The industrial trainer theory of teaching is authoritarian, it involves strict discipline, and the transmission of knowledge as a stream of facts, to be learned and applied. Teaching is a matter of passing on a body of knowledge (Lawlor, 1988, page 17). The moral values provide a view of schooling as consisting of hard work, effort and self-discipline. Hence the view of teaching is that of rote learning, memorization, the practice of skills, hard application in school 'work' at the subject (i.e. mathematics). Mathematics is *not* 'having fun' (Prais, 1987a). 'Teaching is a hard grind, and no attempt to transform it into happy informality can ever be successful.' (Froome, 1979, page 76).

As this quotation illustrates, there is also a strong rejection of progressive education (Letwin, 1988). Child-centredness, children's choices, investigational mathematics and calculator use are all castigated as leading to permissiveness, moral slackness, laziness, and evasion of the hard work necessary (Froome, 1970; Prais, 1987). 'Proper' teaching is needed **not**

> salesmanship; enthusiastic teaching; surveys of people's opinions; attractive resource materials; investigative activities; games, puzzles, television material.
>
> (Lawlor, 1988, pages 13–14)

Theory of resources for learning mathematics

As we have seen above, the theory of resources for learning mathematics is largely negative. 'It is the quality of teachers which matters, rather than . . . their equipment.' (Cox and Boyson, 1975, page 1). Learning is based on paper and pencil work, not on the irrelevant distractions of attractive resource materials, games, puzzles or television.

> Keeping pace with the development of the new maths, educational suppliers have released onto the market dozens of practical aids . . . there is a very real danger of teachers placing too great emphasis on their use because they are fashionable and 'trendy' . . . 'discovery' can be made from books as well as from things.
>
> (Froome, 1970, page 106)

In particular, we should restrict the use of the calculator (Lawlor, 1988, page 17). We 'need to recognise the risks which . . . calculators in the classroom offer' (DES, 1988, page 100). Their use prevents the development of computational skills, for it offers an easy way out of the hard work of computation (Prais, reported in Gow, 1988). Instead 'much drill and rote learning' is needed (Prais, 1987, page 5).

Theory of assessment of mathematics learning

The industrial trainer perspective is authoritarian, with humanity arranged hierarchically. It is the responsibility of each stratum in the hierarchy to control and to check up on the levels below. Thus tests are needed to check up on the acquisition of mathematical knowledge and skills by pupils, and to ascertain that the moral obligations of schoolwork are fulfilled. Consequently errors in mathematics are castigated as the failure of self application, or as moral dereliction, even. Discussion and cooperation are rejected because they risk the temptation of cheating, i.e. getting the answers without the hard work, succumbing to the temptations of laziness. Competition is morally necessary, it leads to the survival of the fittest, that is, the rewarding, through prosperity and success, of the virtuous (Cox and Dyson, 1969a).

Clear and simple targets are needed, and children's ability to reproduce knowledge and apply it correctly must be tested. Tests provide external standards. If children are protected from failure, the tests are a sham. Passing tests correctly is the goal (Lawlor, 1988; Prais, 1987, 1987a).

Theory of ability in mathematics

Children are born with different abilities in mathematics, fixed by heredity, so streaming and selection are necessary to allow children to progress at different rates. Indeed, a hierarchy of schools of graduated quality is necessary, to accommodate the different types and abilities of children. To penalize the 'better' children (who it might be inferred are in some sense more worthy) by 'equalling down' would be unnatural and morally incorrect (Cox and Dyson, 1969). The inferior children may better themselves if they try hard enough to overcome their inheritance, through moral self-help. Selective schools for 'better' children, and public schools for the future élite, mean that all levels in society are catered for.

> Stupid children, after all, can't help being stupid, and it is no credit to clever children that they were born with brains.
>
> (Sparrow, 1970, page 65)

Theory of social diversity in mathematics

As we have seen, social issues and the interests of social groups have no place in mathematics, which is completely neutral. Anti-racism, anti-sexism and even multiculturalism, are all rejected outright. Not only are they irrelevant to mathematics, they represent political propaganda aimed at undermining British culture and values (Palmer, 1986). The intrusion of social issues into mathematics is the work of Marxist agitators and propagandists. 'Schools are for schooling, not social engineering' (Cox and Boyson, 1975, page 1). Social diversity therefore does not concern mathematics, except that pupils need to be stratified by ability, and that economic differences allow social 'betters' to buy better (i.e. private) education. In summary, this view is agressively reproductive of the existing social order and inequalities, as well as being monoculturalist and crypto-racist.

F. Review of the Industrial Trainer Aims

The overt mathematical aims of this position concern the acquisition of functional numeracy, but the aims extend beyond this to social control and the reproduction of the social hierarchy. In support, the means of teaching are such as to discourage critical thinking and originality, and to foster obedience and docility. For the masses, the aim is to prepare children for their future place at the lower levels of society. For the minority in selective education, to which entry is negotiated through 'ability', wealth or class, the aim is higher qualification, to serve the needs of industry, trade, commerce and the professions at the upper levels (Young, 1971a).

Criticism of the industrial trainer aims

The industrial trainer aims for mathematics education can be criticized on epistemological, principled and pragmatic grounds. First of all, there is a fundamental epistemological weakness. The view of mathematics is Dualistic, incompatible with any of the public philosophies of mathematics. It ignores the rational theoretical basis of mathematics, and is rejected by all reasoned philosophical viewpoints. Mathematics is not simply a body of true facts, techniques, or even theory, whose veracity is determined by authority. Mathematics is above all else, a rational discipline in which claims are established from axioms by proof. Despite philosophical conflicts over its nature, no philosophy of mathematics denies this. Thus the Dualistic epistemology of the industrial trainers is uniquely and profoundly anti-rational and anti-intellectual.

Secondly, the industrial trainer aims are based on an extreme and largely rejected set of principles and moral values. Thus, equality of opportunity for all is a principle adopted by the majority of individuals and governments, and enshrined by British law. The 'protestant work ethic' and other moral values, such as 'original sin', are an extremist and inappropriate foundation for educational policy. Their basis is one that is rejected by the weight of Western intellectual thought. A number of other notions, such as the fixed, inherited nature of intelligence and mathematical ability, are widely questioned. Beyond this, the extreme Jingoistic values of monoculturalism, cryptoracism and xenophobia are morally repugnant to the majority of modern liberal thinkers.[23]

Thirdly, there are pragmatic grounds for rejection. The industrial trainer aims for mathematics do not serve the needs of modern industrial society. Functional numeracy combined with low life-expectations fail to meet the needs for an ever increasingly skilled work-force, as expressed by employers. Basic mathematical skills are insufficient, and docility and lack of initiative are counter-productive in many, if not most, sectors of employment.

Overall, this is a damning criticism of the industrial trainer aims of the New Right. They are based on insupportable and extremist assumptions, and ironically, are counter-productive in terms of training for industry.

3. The Technological Pragmatists

A. The Background to the Technological Pragmatist Group

In the Victorian era, the industrial trainers' overt aim for mathematics education was to provide a training in basic skills and numeracy for all, thereby meeting the needs of industry and commerce. The backdrop was a social context which lacked public or any schooling for most children. Thus their goal was to extend education, for utilitarian ends. Education has moved on since then, and in Britain the schooling of all students well into their 'teens', has been compulsory for some decades. The vast majority of children study a course of mathematics, and not just arithmetic or numeracy, during most of their schooling.

Thus the educational aims of the New Right, including mathematics, are in some ways the opposite of those of the Victorian industrial trainers. They are regressive, aiming to restrict the content and form of education, reinstating basic skills and authoritarian teaching and rote learning.

The other modern group descended from the traditional industrial trainers, the technological pragmatists, promote a modernized version of the original utilitarian aims. As their name signifies, their values are pragmatic (as opposed to the Dualism of the New Right), and they are concerned with the furtherance of industrial interests through technological development.

B. The Technological Pragmatist Ideology: Multiplistic Absolutism

Epistemologically, the technological pragmatist perspective views knowledge as unproblematic and given, something which like a tool, can be applied in practical applications. In particular mathematics is seen as fixed and absolute, but applicable. Thus, the philosophy of mathematics is an unquestioning, 'black box' absolutism.

Ethically, the position is a pragmatic, not based on ethical principles, but on utility or expediency. It is Multiplistic, admitting a plurality of points of view, which cannot be distinguished by means of principles or reasoned judgement. So moral judgement is based on utility and expediency, and choice is determined with reference to personal or sectorial interests. There is a belief, based on the utilitarian values, that scientific and technological developments through furthering industrial production also further social progress. Just as knowledge is accepted unquestioningly, so too the existing economic and social order, and its further development is accepted, which utility is seen to further. Thus the ideology values utility and wealth, accepts knowledge and the social *status quo* without question, and regards scientific and technological progress as the means to social development and the fulfilment of its values.

The distinction between the ethics of this position, and the utilitarianism of J. S. Mill (1893) should be drawn. Mill's ethical system has as its base principle the promotion of pleasure or happiness for the greatest number of persons (and the avoidance of their opposites). Thus utilitarianism has an end (pleasure or happiness) which is valued for all (or the greatest number) irrespective of who is making the evaluation, or their allegiances. Fundamentally, utilitarianism is democratic, in that equality of persons is assumed by the greatest happiness principle. This contrasts with the technological pragmatist perspective, which values means (the creation of wealth) as opposed to ends, and favours the needs and wants of industry and commerce over other sectors. Thus it lacks both proper ethical ends and the principles of democracy as fundamental values. For this reason the ideology of technological pragmatism is not regarded as being based on as rational, principled or defensible a moral basis as that of utilitarianism (for all the latter's weaknesses, Langford, 1987), and should be sharply distinguished from it.

The technological pragmatist grouping is rather diffuse, being the largest of the five ideological groups. Outside education it includes the majority of politicians, industrialists, technologists and bureaucrats. In education it includes the utilitarian minded, including applied mathematicians, scientists and technologists. Because of its diffuseness and pragmatic basis, explicit statements of the underlying ideology are rare, although the educational aims are widely acknowledged (Dale, 1985; Golby, 1982; Holt, 1987; Pollard *et al.*, 1988; Raffe, 1985; Watts, 1985).

The main features of the ideology are an unquestioning acceptance of existing structures and models (epistemological, social and human) coupled with an action-orientated world-view, treating intellectual and ethical matters in terms of practical outcomes.

Epistemology

The view of pure knowledge is one of unquestioning acceptance. In contrast, applied knowledge is seen to reside in the knowledge, skill and experience of the professionals and skilled practitioners who apply it. Such knowledge is Multiplistic, and experts will have differing opinions and disagreements about the best approaches to and uses of knowledge. Many equally valid methods and points of view are acknowledged, and choice between them is made on pragmatic grounds of utility, expediency and self or group interest.

Philosophy of mathematics

This exemplifies the overall epistemological viewpoint. Pure mathematics is accepted unquestioningly, and so the philosophy of mathematics is Absolutist. But there is no best method of application, rather it depends on the knowledge and skill of the professional experts who apply it. Choices between approaches are made not on the basis of principles, but on pragmatic utilitarian grounds.

Set of moral values

As we have seen, the values of this position consist of utility, expediency, pragmatism and self or group interest. These are perceived to be best served in modern society through industrial production and wealth creation. Scientific and technological progress are also valued, for they serve these ends, as well as social development.

Theory of society

Industrial and technical growth are understood to be the engine of social development and progress, so science, technology and industry are located at the heart of society. Existing social and political structures are accepted as an underlying reality. Thus a hierarchical model of society is accepted, with experts, technocrats and bureaucrats in an elevated position. However, the social hierarchy is not seen as rigid: social mobility is possible. Society is seen as meritocratic (or technocratic), and those who acquire the necessary scientific and technological knowledge and skills are rewarded by increased wealth, status and power.

Theory of the child

The theory of the child comes about from an unquestioned acceptance of the received elementary school view, without the moral zeal of the industrial trainers. Thus the child is seen as an empty vessel needing to be filled up with facts and skills. There is also a valuing of experience, as the source of skills, as well as their future deployment in industry. So the child is also seen as a 'blunt tool', sharpened through training, for use in the world of work.

C. The Complexity of the Technological Pragmatist View

The technological pragmatist perspective can be identified both historically and in the present. Its mark is the espousal of utilitarian and industrial aims in education, without the Dualistic moral overtones of the industrial trainers. One problem with identifying this group is that it is, in reality, a shifting alliance between representatives of many sectors, including industrialists, other employers, members of the government, bureaucrats, scientists, mathematicians, technologists, teachers and educationists. No one of these groups is itself monolithic, only some of its members will subscribe to these aims. Furthermore, these aims are themselves a composite, and include at least three interwoven elements. Education is for:

1. The acquisition of the knowledge and skills necessary to serve the **immediate** needs of industry, commerce and employment;
2. The acquisition of scientific, mathematical and technological knowledge and skills necessary to serve the **future** technological needs of industry and society;
3. The certification of potential employees, by means of examinations and tests, to facilitate the processes of selection for employment.

According to the breadth of vision, industrial pragmatists may lay greater stress on aim 1 or 2. It is also well documented that industrial pragmatists, especially bureaucrats, often hold the instrumental aim 3, which values the certification over the content of learning (Weber, 1964).

A further source of variation is the fact that technological pragmatist groups do not hold the same utilitarian aim for all children in schools. The aim varies for children of different ages, thus secondary schooling may be expected to be more related to future occupations than primary schooling. However, the greatest variation arises for different sectors of the school population. These are expected to have different occupations. Typically there is a tri-partite division of children and classification of career outcomes.

The lowest level, representing the majority, are expected to be workers in one occupation or another, requiring basic education plus vocational training. The second tier represents the second largest grouping who will hold responsible decision-making positions such as middle managers, civil servants, teachers and professionals. This group requires both extensive education and certification. The third and smallest tier comprises the future élite, who will be the most powerful and wealthy. Their education involves an élite route (typically, a public school such as Eton followed by Oxford University) and it is the form and value (social cachet) attached to this education, rather than its content, which prepares the students for their careers.

However, a further complexity is that this picture varies with time, with the educational demands on the middle and bottom tier increasing over the course of this century (both in terms of increased content requirements and increased certification), and the highest tier becoming a less certain entry route to the ruling élite. Technological pragmatist aims typically concern only the bottom two tiers, where the content and certification of education has the greatest bearing on future occupations.

To avoid the complexity of both the groupings and the variant aims, a set of simplifying assumptions will be made. Two loose groupings will be considered. First, those societal groups largely outside of education but concerned with its utilitarian outcomes. This will include industrialists, other employers, members of the government, bureaucrats, scientists and technologists. This grouping represents the reforming pressures on the overall education system to meet the utilitarian aims, and who have a major influence on the allocation of educational resources. Secondly, those within the education system, including educational administrators and reformers, scientists, mathematicians, technologists, teachers and educationists. This group represents the reformers working inside the education system, who affect any changes in the school mathematics curriculum. Clearly these two categories are not disjointed, for the government department of education can be said to belong to both, since it allocates resources and increasingly controls the content of the curriculum.

D. The History of the Technological Pragmatist View: Social Pressures

In the later nineteenth century, the technological pragmatist view is a major part of the industrial trainer outlook. Following the Great Exhibition of 1851, it was felt by many that British industrial pre-eminence was threatened by rapid advances in industrialization in continental Europe.

> Fears raised by continental competition in the mid nineteenth century led eventually to a wider definition of elementary education. Similarly, internation rivalry in the space programmes in the 1950s was a stimulus to the proliferation of new mathematics and science projects in Britain and the USA.
>
> (Gordon, 1978, page 126)

The former, coupled with 'the needs of an expanding and changing economy' (Williams, 1961, page 161), led to the 1870 Education Act. In introducing this to parliament W. E. Forster said:

> Upon the speedy provision of elementary education depends our industrial prosperity. It is of no use giving technical education to our artisans without elementary education . . . if we leave our workfolk any longer unskilled . . . they will become over-matched in the competition of the world.
>
> (Dawson and Wall, 1969, page 30)

The 1870 Education Act extended elementary education to all, with the technological pragmatist aims at the fore. These aims were equally endorsed by the 1895 Bryce Report on Secondary Education:

> Secondary education . . . is a process of intellectual training and personal discipline conducted with special regard to the profession or trade to be followed. All secondary schools . . . insofar as they qualify men for doing something in life, partake more or less in the character of institutes that educate craftsmen.
>
> (Dawson and Wall, 1969, pag 42)

The pressures on the education system to satisfy the technological pragmatist aims have been intensified as a consequence of the country waging war.

> Passing on to the effects of war, we observe that, from the time of Napoleon, educational reform has occurred in post-war eras. (See, for example, some of the major Education Acts of the present century, 1902 and 1944.) War has also had a startling effect in displaying shortcomings of knowledge in industrial and technical skills. The South African War which required an army ten times the size of the Boers to succeed, exposed the poor state of science teaching in schools.
>
> (Gordon, 1978, pages 125–126)

The technological pragmatist aims have had a marked impact on educational provision, although the inclusion of scientific and technical elements have been slower to arrive.

> Indeed, while technical education never really arrived until after the *Second* World War, before the First, the 'demand for general education

> preparatory to vocational training on the job . . . was of great consequence to the expansion of formal education at all levels — elementary, secondary and higher — and at all levels the requirements were essentially the same' . . . the provision of appropriate academic competence, the development of general intelligence and attitudes efficient and adaptable to employment.
>
> (Davies, 1976, page 58, quoting Wardle)

A further source of impetus for the technological pragmatist view has been the growth of bureaucracy.

> A second influence on the curriculum was the rise of *bureaucratic procedures* over the past hundred years . . . the formal organisation of institutions . . . increasingly characterized by impersonal rules and an authority structure with rational procedures. A good example of this is the growth of the examination system.
>
> (Gordon, 1978, page 124)

According to Weber (1964) and others, the bureaucratization of modern industrial and post-industrial societies has led to certification via external examinations as a major end for schooling (Macdonald, 1977). Certification serves as a means of selection for employment (including indirect access, via further and higher education) and thus satisfies the utilitarian aim for education. In this way, certification serves as a mediating factor between the educational outcomes of schooling, and the real demands of employment, without necessarily reflecting either (Dore, 1976; Oxenham, 1984).

A growing component in the technological pragmatist view of education, is what Benton terms the ideology of technical and technological development.

> According to this conception . . . technical development is the main cause of all forms of social and cultural developments, so that, in general, the progress of mankind comes to be seen as assured, given that all irrational obstacles to technological development are removed. This ideology of progress, technological determinism . . . in the early and middle sixties . . . proliferated to an enormous scale . . . The ideology which gained dominance within the [Labour] party at this time . . . was . . . a form of technological determinism. Technological progress had, it was argued, transformed the manpower requirements of the economy. An increasing proportion of highly skilled and technically educated personnel was required. For the educational system one clear implication was drawn: there must be an overall increase in the provision of education — and particularly science education . . . [This] was now presented as a demand imposed on the education system by economic and technical necessity. Since technology was perceived as the main agent of social progress, a *general* subordination of the educational system to the requirements of economic development was called for . . . The distribution of resources . . . within education should be

> governed by the levels and proportions of the different skills, competences and knowledges required and expected to be required by the economy.
> (Benton, 1977, pages 126–127)

As Benton shows, this ideology was subscribed to in Britain by the political left (as well as the moderate right), where it was combined with egalitarian principles.

> [I]n the age of scientific revolution . . . the improvement of our living standards and our survival as a free democracy depend largely on the quality of our scientific, technological and technical education.
> (Labour Party, quoted in Benton, 1977, page 127)

This ideology is now regarded as uncontroversial.

> Technical teachers who would strenuously deny any taint of liberalism are teaching to your unacademic apprentices principles of mathematics and science that twenty years ago were regarded as sophisticated academic work. They cannot help themselves because technical progress demands it.
> (Robinson, quoted in Benton, 1977, page 127)

In modern times, the technological pragmatist aims, including the ideology of technological development, have underpinned external pressures on the school curriculum, and in particular, on the mathematics curriculum.

In his Ruskin College speech of 1976, Prime Minister James Callaghan focused and gave a further impetus to these forces:

> I am concerned . . . to find complaints from industry that new recruits from schools sometimes do not have the basic tools to do the job that is required . . . There seems to be a need for a more technological bias in science teaching that will lead towards practical applications in industry . . . Then there is concern about the standards of numeracy of school leavers. Is there not a case for a professional review of the mathematics needed by industry at different levels?
> (Callaghan, 1976)

Reflecting this technological pragmatist critique of the outcomes of schooling, a committee of inquiry into mathematics teaching was set up, chaired by Sir W. Cockcroft:

> To consider the teaching of mathematics in primary and secondary schools . . . with particular regard to the mathematics required in further and higher education, employment and adult life generally, and to make recommendations.
> (Cockcroft, 1982, page ix)

Despite its reputation as an endorsement of progressive mathematics education, the utilitarian emphasis is clear in the committee's report (Cockcroft, 1982), which

> fall[s] into three main parts. Part 1 presents an overview of the mathematical needs of school leavers in an industrial society.
>
> (Department of Education and Science, 1983, page 2)

The current reform of the content of British education, the National Curriculum, explicitly states its technological pragmatist aims.

> The aim is to equip every pupil with the knowledge, skills, understanding and aptitudes to meet the responsibilities of adult life and employment.
>
> (Department of Education and Science, 1987a, page 35)

In keeping with the technological pragmatist aims the reforms give an elevated place to science and technology, for in the National Curriculum

> The core subjects are English, mathematics and science. The other foundation subjects are technology (including design) . . . The foundation subjects . . . will cover fully the acquisition of certain key cross-curricular competences: literacy, numeracy and information technology skills.
>
> (Department of Education and Science, 1989a, pages 7–8)

This represents a major shift towards technological pragmatism, compared with the traditional school curriculum, for science and information technology are elevated to essential basic skill status, alongside literacy and numeracy. In addition, the traditionally low status subject of Craft, Design and Technology comes first in the list of compulsory foundation subjects.

Thus a technological pragmatist tradition can be traced from Victorian times to the present in Britain. During this time it has gained in momentum, as the role of technology in society has grown. This tradition emphasizes the vocational training aspect of education, the certification of achievement as an aid to selection for employment, and the needs and value of technological development, which is to be served and aided by education. It elevates the pursuit of technological studies, including applied mathematics and science, close to the status of being an end in itself (Golby, 1982).

E. Technological Pragmatist Aims in Mathematics Education

In addition to the general social pressures to meet the technological pragmatist aims, there have also been internal pressures for the reform of the mathematics curriculum to meet these aims. In the Victorian era, the mathematics curriculum for the few in selective secondary schools was pure mathematics, whilst for the masses in elementary schools it was basic arithmetic. At the turn of the century, John Perry, an engineer and former science teacher, was influential in pressing for the reform of the mathematics curriculum (Griffiths and Howson, 1974; Howson, 1982; Ministry of Education, 1958; National Council of Teachers of Mathematics, 1970). He argued for utilitarian aims and against the old humanist aims which dominated mathematics teaching in selective schools. For example, in 1901 he addressed the British Association for the Advancement of Science, in Glasgow, as follows.

> The study of Mathematics began because it was useful, continues because it is useful and is valuable to the world because of the usefulness of its results, while the mathematicians, who determine what the teacher shall do, hold that the subject should be studied for its own sake.
>
> (Griffiths and Howson, 1974, page 16)

Just over a decade later, the International Commission on the Teaching of Mathematics reported from its surveys that: 'Utilitarian aims were becoming increasingly more important'. (National Council of Teachers of Mathematics, 1970, page 183).

Despite these pressures the school mathematics curriculum still consisted largely of arithmetic for the masses, and pure mathematics for those in selective education (Cooper, 1985). The Board of Education of 1934 argued that the narrow utilitarian goal of basic arithmetic for the masses should be liberalized, and that utility was better served by broader goals:

> Purely utilitarian requirements . . . are substantially less than the content of the ordinary school course up to 14 years of age, but cultural and civic requirements are much less restricted. In arithmetic what is needed by the ordinary man is an intelligent power of dealing with numbers whenever and wherever he meets them.
>
> (Ministry of Education, 1958, page 30)

A quarter of a century later the next official report on mathematics in secondary education was published. The first words of its forward set the tone as follows:

> Our standard of living and our position in the world depend upon our ability to remain in the forefront of scientific advance, both pure and applied. Mathematics is the basis of science . . . its [mathematics] ready application cannot be cultivated too early.
>
> (Ministry of Education, 1958, page iii)

This statement represents a shift towards the full technological pragmatist position, in which mathematical and scientific education are seen to drive technological and industrial development, and hence social progress and prosperity.

Cooper (1985) documents how at this time applied mathematicians and industrial employers pressed for a move towards 'problem-orientated applied mathematics' (Cooper, 1985, page 100) in schools. This pressure (as well as the resources that came with it) was marshalled in support of the School Mathematics Project, which came to dominate and redefine the school mathematics curriculum of Britain (excluding Scotland) in the late 1960s and 1970s. Although this project clearly shows the influence of the Modern Mathematics Movement, it also represents a major shift towards the utilitarian aims in mathematics education.

> The main intention was not to prepare students for university, but to introduce them to modern applications of mathematics in a technological society. Thus SMP, which was funded by industry, laid emphasis on a wide variety of applications such as computer mathematics, statistics, probability, operations research (linear programming, critical path analysis, transport flow).
>
> (Howson, Keitel and Kilpatrick, 1981, page 173)

Many continue to press the utilitarian aims of mathematics teaching and link them with the impact of technology, such as H. Pollak, head of research at Bell Laboratories, and a long-time proponent of these views.

> Mathematics education has at least four major purposes: The mathematics needed for everyday life, the mathematics needed for intelligent citizenship, the mathematics needed for your vocation or profession, and mathematics as a part of human culture . . . the four kinds of mathematics [needs] . . . continue to change . . . Because the technology changes, because the applications of mathematics change
>
> (Pollak, 1988, pages 33–34)

Such an emphasis is widely endorsed by the computer lobby in mathematics education, with additional emphasis on the role of information technology.

> It is important that the rapid changes in our society in technology, in methods of communication . . . are reflected in changes in mathematics education . . . One of the aims of recent initiatives taken by the Government is to make the educational system more relevant to the future social and economic needs of this country; in particular . . . to support the information based sectors of commerce and the technology industries and to prepare young people for life in an increasingly 'technological' society. Mathematics has an important contribution to make in meeting these aims.
>
> (Ball *et al.*, 1987, pages 7–8)

Another group within the mathematics education community pressing for the utilitarian technological pragmatist aims are the applied mathematics and modelling lobby. This is reflected in the report of an international group on applications and modelling in the mathematics curriculum.

> The ultimate reason for teaching mathematics to all students, at all educational levels, is that mathematics is useful in practical and scientific enterprises in society.
>
> (Carss, 1986, page 199)

Within Britain, a number of influential lobbyists have been pressing the technological pragmatist aims, in particular Lighthill (1973), Burkhardt (1979, 1981) and Burghes:

> So with what should we fill up the mathematics curriculum? One

> suggestion is very simple: don't fill it up at all! Why do we continue to teach mathematics . . . ? . . . Could we accept the thesis that children do not have to succeed at *academic* mathematics to be worthwhile citizens? What we must aim for is to make sure that they are numerate . . . Another suggestion is to teach mathematics through applications and contextual situations . . . I have already outlined the theme of motivation through relevance.
>
> (Burghes, 1989, pages 86–87)

This is a very clear statement of the technological pragmatist position, accepting mathematics as given, but questioning its educational value unless its immediate utility can be demonstrated. This confirms Skovsmose's (1988) characterization of atheoretical proponents of mathematical applications and modelling as pragmatic and expedient. However not all supporters of a modelling approach to mathematics share this perspective. For example Ormell (1980, 1985) offers a model of mathematics as many-layered and embedded in the world, with applicable mathematics at the outside shading into pure mathematics driven by 'intrinsic' values at its centre. This view is much closer to social constructivism than to the view subscribed to by technological pragmatists.

This and the previous section have indicated some of the specific lobbies within the mathematics education community (and without) promoting the technological pragmatist aims. They also provide evidence supporting the identification of the ideology of the group, particularly the primary elements of epistemology, values and the theory of society. We turn next to the secondary elements.

F. The Technological Pragmatist Ideology of Mathematics Education

Mathematical aims

The aims of this group for the teaching of mathematics are utilitarian; students should be taught mathematics at the appropriate level to prepare them for the demands of adult employment. This aim has three subsidiary components: (1) to equip students with the mathematical knowledge and skills needed in employment, (2) to certify students' mathematical attainments to aid selection for employment, and (3) to further technology by thorough technological training, such as in computer awareness and information technology skills.

Theory of school mathematical knowledge

School mathematics is seen to have two parts. First, there are the pure mathematical skills, procedures, facts and knowledge. These are the dry bones of the subject, which are simply tools to be mastered. Second, there are the applications and uses of mathematics. This is the vital, living part of mathematics, which justifies and motivates the study of the subject. School mathematics is an outward facing subject, looking to applications of increasing complexity. Intrinsic values, creativity and pattern are not deemed significant.

Theory of mathematical ability

The technological pragmatist view of mathematical ability is that it is inherited, but that it requires teaching to realize its potential. Thus, for example, according to Burghes:

> Although children have particular talents for various topics, a good teacher can make all the difference between a pupil losing interest (which is exceedingly difficult to recover in mathematics), or working conscientiously, enjoying the subject . . . all my experience shows that good maths teachers are born with talent, and I am not yet convinced that training does any good.
>
> (Burghes, 1989, page 85)

There is also an indication of these views in the National Curriculum.

> Basing objectives for attainment on 10 levels . . . means that . . . differences in ability . . . can be accommodated.
>
> (Department of Education and Science, 1989a, section 8.2)

This illustrates the technological pragmatist view of mathematical ability as fixed.

Theory of learning mathematics

The view of learning associated with this perspective is analogous to apprenticeship: the acquisition of knowledge and skills largely through practical experience.

> to become proficient at modelling you must fully experience it — it is no good just watching somebody else do it, or repeat what somebody else has done — you must experience it yourself. I would liken it to the activity of swimming. You can watch others swim, you can practise exercises, but to swim, you must be in the water doing it yourself.
>
> (Berry *et al.*, 1984, page xiii)

Theory of resources for mathematics education

Resources play a significant role in the view of pedagogy. The teacher uses them to illustrate and motivate teaching. Learning is practical, so students must have access to resources for experimental and practical learning. Information technology skills are important, so students must have hands on experience of computers, interactive video, and similar resources. Underpinning all this is the experiential view of learning requiring that students have practical experiences.

Theory of teaching mathematics
The view of teaching mathematics associated with the technological pragmatist perspective is that of skill instruction and motivation through work relevance. The focus is on 'the art of *teaching* the art of applying mathematics' (Lighthill, 1973, page 98). In higher education, there is an emphasis on applied problem solving.

> In teaching modelling at least three different activities should be involved:
> (1) Practice in tackling complete problems.
> (2) Exercises on specific modelling skills.
> (3) Development of awareness of effective strategies.
>
> (Burkhardt, 1979, page 241)

Theory of assessing mathematics learning
The view of assessment is that of external tests to provide certification of attainment and skills. For the less academic, the emphasis is on the provision of records of achievement and skill profiles (Hodkinson, 1989). However, for modern technological pragmatists

> job specific vocational preparation [in skills] alone is not enough . . . technological change is likely to alter the structure of many jobs: therefore vocational preparation should provide young people with a range of skills that have utility across a range of vocational areas . . . The term *generic skills* is used to describe such skills.
>
> (Further Education Unit, 1982, page 3)

The emphasis in the skill profiles are most frequently on a broad range of generic skills, including numeracy, graphicacy, literacy, communication, practical skills, problem solving, decision making and responsibility (Further Education Unit, 1982), to which may be added computeracy.

Theory of social diversity in mathematics education
With regard to social diversity and education, the technological pragmatist focus is on the utilitarian needs of employment and further education. Social diversity is seen in terms of future occupation, and culture, race and gender are immaterial except insofar as they pertain to employment. Mathematics is seen as neutral, unless it is applied, in which case it is related to industry and technology, not cultures.

With regard to the relations between social divisions, education and employment, technological pragmatism is essentially reproductive of the *status quo*. However, technological progress means that skills levels are expected to rise to serve the increased technological demands of employment. Consequently, there have been complaints about skill standards (Callaghan, 1976; Institute of Mathematics and its Applications, 1984), as the mathematical demands of employment have risen faster than school-leavers' skills (Bell *et al.*, 1983; Cockcroft, 1982).

G. A Critique of the Technological Pragmatist Aims

The technological pragmatist perspective can be criticized in general terms, and on the basis of its aims for mathematics education. However, it should first be stated that a strength of this position is that it relates education to the goals and needs of society, or at least, to an industrialist's perception of these goals and needs. It must be right that education is seen and evaluated as part of a broader social context, provided a proper and sufficiently broad basis for this evaluation is adopted. This very point leads to the first group of criticisms of the perspective. Namely that it lacks a proper epistemological, moral or social foundation.

The technological pragmatist view takes knowledge, including mathematics, as given, and is unconcerned with the growth and development of mathematics as a discipline. The focus is instead on immediate applications of mathematics, and their short term payoffs. This view neglects the fact that the most fundamental developments in mathematics (and other areas of knowledge), those which have ultimately the greatest practical impact, are often initially purely theoretical.

Similarly, the nature of society is accepted without question, with technological and industrial development seen as positive forces for social progress. A central belief of the technological pragmatism is that technological development is the engine of industrial growth and social progress, and as such, is a virtue and an end in itself. There is no critical scrutiny of these relationships, nor any recognition that the issues are problematic, both for society and the environment (Dickson, 1974; Ellul, 1980; Gorz, 1989; Marcuse, 1964). Indeed, the view is technocentric, that is, tending to give centrality to the objects of technology, in preference to people, cultures and goals (Papert, 1980, 1988). Much of the myopia of this position arises from its narrow and extrinsic set of values (in addition to its unquestioning epistemology). Technological pragmatism rests on utility and expediency, and so there is no principled moral basis. Thus there is no concern with the good of all individuals, or with the good of society as a whole (Marcuse, 1964). Likewise, no moral basis for education is offered, beyond purely instrumental values.

With regard to the aims of mathematics education, the concern with needs of industry alone may be narrow and counterproductive. For as some of the industrial lobby realize already, general knowledge and transferable skills suit industry better than narrow vocational skills. Education rather than training better serves society, by potentiating its individual members and enabling them to better adapt to new demands and responsibilities (Abbs, 1987). Unless creative and aesthetic aims are included alongside the utilitarian aims of mathematics education, the teaching of the subject will be stilted, and fail to contribute fully to the education of the whole person (Isaacson, 1989). Beyond this, the emphasis on certification as the outcome of education means that the utilitarian knowledge and skills of a previous epoch survive unquestioned in the curriculum, serving neither the needs of individuals nor those of society and employment, as Benjamin (1971) so aptly illustrates. Finally, the overvaluing of technology means that attention and resources are diverted from the human interactions of education, to technological wonders which, like programmed learning machines, may prove to be educational *cul de sac*.

Overall, the technological pragmatist position rests on an inadequate epistemological and moral basis. It does not offer an adequate basis for the teaching of mathematics. Furthermore, in terms of its own instrumental values it can be counter-productive. For the means it adopts may fail to be the most efficient for meeting its own ends. The technological pragmatist perspective rightly recognizes that the social function of mathematics and the role of computers are potentially of tremendous importance for mathematics education, as well as for society. But whilst these elements need recognition, technological pragmatism fails to incorporate them in a sufficiently broad or well-founded perspective.

Notes

1 Examples of such groups and New Right members actively concerned with education include: Centre for Policy Studies (S. Lawlor, O. Letwin); Hillgate Group (C. Cox, J. Marks); Campaign For Real Education (N. Seaton, R. Boyson).
2 This pun illustrates how our positional language is laden with these values: high, elevated or lofty mean good; low, fallen and base mean bad.
3 A biography of Margaret Thatcher is entitled 'One of us', indicating both the importance attached to this last distinction by Young (1989), and his attribution of a Dualistic perspective to Thatcher.
4 Young (1989), pages 5–6.
5 Young (1989), page 409.
6 Young (1989), page 420.
7 Young (1989), page 216.
8 This point is made by Warnock (1989).
9 Young (1989), page 69.
10 Young (1989), page 82.
11 *The Guardian*, 6 September 1989, page 25.
12 Young (1989), page 405.
13 Young (1989), page 405.
14 Quotation from Thatcher's address to the Assembly of the Church of Scotland, reported (and analysed by Jonathan Raban) in *The Observer*, 28 May 1989, page 33. (See Raban, 1989).
15 Young (1989), page 414.
16 Lawton offers a more refined analysis of the educational philosophies of the New Right, deriving from their political ideologies (Lawton, 1988). He divides them into 'privatisers', who want all of education privatised and left to unregulated market forces, and 'minimalists', who want a cheap minimal state education system retained. Both groups strongly agree on a market-place view of education, but differ on how far to take it. What I describe as the New Right comprises 'minimalists', since a back-to-basics state curriculum is urged by them. Margaret Thatcher has sympathy with the privatisers (stating, for example, that she expected 80 per cent of schools to opt out of state control), but wishes to regulate education because of her authoritarian moral vision, and so settles for minimalism.
17 Many Methodists are of course not Dualistic-absolutists, as letters in *The Observer*, 4 June 1989, page 34, testify, responding to Raban (note 14).
18 The ideological models of childhood are from Ramsden (1986). A similar set is in Gammage (1976).
19 The persistence of this view in the popular conception of knowledge is borne out by two observations. First, 'education' sections in remaindered book stores (e.g. County Books, Exeter and

elsewhere) consist largely of books of facts, e.g. '2000 Science Facts', '1000 Facts of History', and dictionaries. Second, media quiz games deal almost exclusively with 'facts', similarly. For example, the TV quiz 'Mastermind' delves in great detail into academic and popular specialisms, but tests participants solely on their memory of isolated facts. Both examples suggest that knowledge is perceived to be an unstructured collection of facts.

20 Quoted in Abraham and Bibby (1988), page 9.

21 S. Froome is a prominent radical right contributor to the Black Papers on Education (Cox and Dyson, 1969, 1969a, 1970; Cox and Boyson, 1975).

22 From Thatcher's address to the Conservative Party Conference, October 1987 (paraphrased in *The Guardian*, 3 November 1987; quoted in *Mathematics Teaching* No. 125, 1988, page 4.)

23 These are the values of individuals characterized as having an 'authoritarian personality' by Adorno *et al.* (1950).

8

Groups with Purist Ideologies

1. The Old Humanists

The old humanist group considers pure knowledge to be worthwhile in its own right. In particular, the mathematical old humanists regard mathematics as intrinsically valuable, a central element of culture. Mathematics is a supreme achievement of humankind, 'queen of the sciences', a perfect, crystalline body of absolute truth. It is the product of an élite, a small band of genius. Within mathematics rigour, logical proof, structure, abstraction, simplicity, elegance are valued. Based on these values the aim for mathematics education is the communication of mathematics for its own sake. The ideology of this group is separated Relativistic absolutism.

A. The Separated Relativistic Absolutist Ideology

Set of moral values

The separated perspective focuses on rules and principles, and objectifies areas of concern and objects of knowledge. Moral reasoning is based on truth, fairness and 'blind justice', the impartial application of the rules of justice to all, without concern for individual human issues and concerns. This perspective corresponds to The Post-conventional and Principled Level in Kohlberg's theory of moral development.

> Moral decisions are generated from rights, values or principles that are (or could be) agreeable to all individuals composing or creating a society designed to have fair and beneficial practices.
>
> (Kohlberg, 1981, page 411)

These values can consist of prior rights 'The right is upholding the basic rights, values and legal contracts of a society, even when they conflict with the concrete rules and laws of the group'. At a higher stage the set of values 'assumes guidance by universal ethical principles that all humanity should follow'. (Kohlberg, 1981, pages 411–412)

According to Gilligan (1982) these values are part of the cultural definition of masculinity. It gives pride of place to absolute, rational standards and pure reason, and leads to the rejection of human and connected elements of judgement as detracting from fairness and universality.

> The morality of rights is predicated on equality and centred on the understanding of fairness, while the ethic of responsibility relies on the concept of equity, the recognition of differences in need. While the ethic of rights is a manifestation of equal respect, balancing the claims of other and self, the ethic of responsibility rests on an understanding that gives rise to compassion and care.
>
> (Gilligan, 1982, pages 164–165)

The separated values of this ideology also lead to the rejection of expediency and utilitarian judgements, as counter to the principles of fairness and justice. Because of the centrality of reason, rationality and logic, there is a purist aesthetic, valuing simplicity, clarity, purity and objectivity in moral reasoning, indeed in all reasoning. This valuing of purity is a central characteristic of the position.

The location of the values at one pole of the separated-connected dichotomy is analogous to one of the poles of a number of other dichotomies, such as masculine versus feminine, subject-centredness versus child-centredness in teaching, and classical versus romantic traditions in the arts and education (Jenkins, 1975). Each of the poles corresponding to separation values rules, structure, form and objectivity, over feelings, expression and subjectivity.

Epistemology

As the overall perspective is Relativistic, a plurality of points of view, interpretations and frames of reference are acknowledged, and their structural features provide the grounds for analysis, comparison and evaluation. On the basis of the values the emphasis is on reason, logic and rationality as the means of establishing, comparing and justifying knowledge. The valuing of logic, rigour and purity leads to an internalist view of bodies of knowledge as a consistent, self-subsistent, richly interconnected structures, which are pure, neutral and value-free. Thus knowledge is seen to be objective and independent of human and social values and concerns.

Philosophy of mathematics

The absolutism of this ideology means that mathematics is seen as a body of pure objective knowledge, based on reason and logic, not authority. Thus it is a logically structured body of knowledge, leading to the view of mathematics as hierarchical. It is also a system of rigour, purity and beauty, and hence neutral and value-free, although it has its own internal aesthetic. Applied mathematics, in contrast, is seen as an inferior, mere technique, the earthly shadow of an eternal, celestial body of truth. The roots of this view originate with Plato, who saw mathematical knowledge in absolute, transcendental terms as pure, true and good (Brent, 1978).

Theory of society

This position is essentially conservative and hierarchical in its theory of society, although politically it may be liberal. Above all, it values the knowledge and cultural tradition of the West, for its own sake, and strives to conserve it. In particular it is concerned with the pure élitist culture of the educated middle and upper classes. Thus the position aims to conserve existing cultural traditions and the associated social structures. Underpinning these aims is the unquestioned assumption of a hierarchical, stratified society, the structure inherited from the past. This is seen to separate 'gentle' cultured persons from 'common folk'. The cultured élite are those fitted to rule society, for the masses do not have the same finely balanced judgement. Society is seen primarily as the means of conserving and creating high culture, which provides the measure of its level of civilization.

Theory of the child

Such a view sees persons as determined by their inherited 'essences' or character. Being conservative it includes, in diluted form, the ideological model of childhood of the elementary school tradition. Children are 'fallen angels' and 'empty vessels'. However, according to this ideology, those of a 'gentle' disposition can be tamed through character building and educated by an exposure to traditional culture. For these experiences will inculcate the appropriate spiritual, moral and aesthetic values and tastes.

Educational aims

The central educational goal of this position is the transmission of pure knowledge and high culture and its associated values. Thus the aim of education is to produce the liberally educated person, with an appreciation of culture for its own sake, and the discriminatory powers and tastes that accompany it. Only a minority will achieve this, those fit to govern and lead society. Thus the educational aims are élitist, for they can only be achieved by a minority. The rest of the population may fall short of these goals, but they will be the better for aiming at them.

B. The Old Humanists as Separated Relativistic Absolutists

Williams (1961) identifies the 'old humanists', who value humanistic studies for their own sake. Such studies have included mathematics since ancient times. Plato, for example, advocated the study of pure disciplines with the 'power of turning the soul's eye from the material world to objects of pure thought . . . yielding *a priori* certain knowledge of immutable and eternal objects and truths.' (Plato, 1941, page 230). Only pure subjects (including mathematics) were considered appropriate for study, practical knowledge and 'the manual crafts . . . were all rather degrading'. (Plato, 1941, page 232).

Around 500 AD Boethius influenced the content of a 'liberal education', which he determined was to include the trivium of grammar, logic and rhetoric, as well as the mathematical quadrivium. This curriculum survived as the course of study at the new universities of Oxford and Cambridge around the end of the fourteenth century. Such learning was cultivated for its own sake. However, it also provided access to status and power, as the graduates of these universities achieved highest offices of church and state (Howson, 1982).

During the Renaissance, as the dominance of Artistotle and the churchmen over the world of learning waned, the influence of Plato and other novel ideas grew. Consequently,

> The humanists, scholars like Erasmus (c. 1466–1536), believed in the powers of the human intellect and in the value of the study of works of great men . . . 'It was the substitution of humanism for divinity, of this world for the next, as the object of living, and therefore of education, that differentiated the humanists from their predecessors . . . The humanist's progress consisted in the adoption of the dogma *The noblest study of mankind is man*'
>
> (Leach, quoted in Howson, 1982, page 9)

This quotation indicates the origins of the name 'old humanists', although the central ideas of the tradition were already almost 2000 years old.

Around the middle of the last century, the Junior Censor of Christ Church, Oxford, indicated the continued hold of the tradition, in his description of the initial studies of undergraduates.

> A couple of plays of Euripides, a little Virgil, two books of Euclid, or the like, form the occupation of a large part of our men during their first university year.
>
> (Ministry of Education, 1958, page 2)

Thus the old humanist tradition endured until modern times, represented by the cultured and educated middle classes and traditional academic élite. This group values 'liberal education' for its own sake, for its contribution to the educated or cultured person; and rejects or holds in low esteem practical or technical knowledge (Williams, 1961). According to Hirst and Peters this group reached full flower in the last century.

> [T]he notion of 'educated' as characterising the all round development of a person morally, intellectually and spiritually only emerged in the nineteenth century . . . though previously to the nineteenth century there had been the ideal of the cultivated person . . . the term 'educated man' was not the usual one for drawing attention to this ideal. Nowadays . . . the concept of an educated man as an ideal has very much taken root . . . [But] for us education is no longer compatible with any narrowly conceived end.
>
> Hirst and Peters, 1970, page 24)

> [A] teacher might teach a subject such as science with purely vocational or economic ends in view . . . equipping people for vocations or serving a national need for trained manpower, without much thought of the development of the individuals concerned, as individuals . . . But teaching people with these limited ends in view should be distinguished from educating people.
>
> (Hirst and Peters, 1970, page 28)

The central element of this ideology is that education and knowledge are a good, an end in themselves, and not a means to a baser, utilitarian end. Thus, according to Cardinal Newman, a prominent old humanist of the last century:

> Though the useful is not always good, the good is always useful . . . Knowledge is capable of being its own end. Such is the constitution of the human mind that any kind of knowledge, if it be really such, is its own reward . . . Knowledge is not a mere extrinsic or accidental advantage, which is ours today and another's tomorrow . . . which we can borrow for the occasion, carry about in our hand and take into the market; it is an acquired illumination, it is a habit, a personal possession, and inward endowment.
>
> (Brent, 1978, page 61)

Young identifies old humanism as part of the 'Liberal/conservative' ideology, originally deriving from the 'Aristocracy/gentry' grouping with educational policies which are 'Non-vocational — the 'educated' man, an emphasis on character' Young (1971a, page 29).

Similar analyses are made by others. Raynor (1972) describes the aristocratic ideology of education, which sees education as the means of preparing the young person for his social role as a gentleman and as a leader. Cosin (1972) describes the élitist/conservative perspective which is concerned to maintain established standards of cultural excellence through traditional methods of selection.

The mainstay of the old humanist view of culture and character building in education is the English public school tradition. This has maintained traditional subjects and élitist views of knowledge as pure and unrelated to the immediate needs of life, as well as always catering for the children of the aristocracy and the gentry (Meighan, 1986).

> Public School[s provided] the door of entrance into the 'governing class' . . . In many respects the Public Schools were a success . . . But the subjects which they taught were too much confined to the Classics to meet all the requirements of the new age, though they formed the basis for a high development of literary culture at Oxford and Cambridge, in Tennyson's England at large. In the microcosm of 'public school' life . . . character gained more than originality.
>
> (Trevelyan, 1944, page 520)

There are many modern proponents of the old humanist viewpoint, who stress the value of the traditional disciplines and culture far above practical or technical knowledge.

> Liberal understanding should precede technical know-how; in this way the accidental events of everyday life will be encountered by a mind already prepared to meet such accidents by reference to philosophic principle and contextual significance.
>
> (Bantock, 1975, page 15)

> [T]he ideal of university education [is a] belief in culture, civilization and disinterested criticism . . . its function is to civilize, to refine, self consciously to 'make' itself a culture . . . [This is threatened by the] pressure to perform — to answer to social needs, technological need, industrial need, economic need.
>
> (Cox and Dyson, 1969, page 60)

The humanistic knowledge embodied in high culture is seen to be intrinscially valuable, and indeed, to justify the class system.

> [C]lass itself possesses a function, that of maintaining that part of the total culture of the society which pertains to that class . . . in a healthy society this maintenance of a particular level of culture is to the benefit, not merely of the class which maintains it, but of the society as a whole.
>
> (Eliot, 1948, page 35)

Such statements are typical of the stereotypical university 'dons' of Oxford and Cambridge, who eschew practical matters in favour of pure reason and culture. They make up the humanistic strand (as opposed to the technological strand) of the 'two cultures distinguished by C.P. Snow (Mills, 1970).

The old humanist views also fit well with the 'classical' view of the curriculum, with its emphasis on structure and rationality, and above all on excellence and culture:

> The notion of excellence, historically, leans back towards the Greek 'upbringing'. It emphasizes quality and reputation and its standards are consistent and objective . . . Within the classic tradition this ultimately means an emphasis on the high culture . . . [In] the classic tradition, that culture depends upon a settled conceptual apparatus, stable 'expectations' and a community of outlook.
>
> (Jenkins, 1975, pages 18–19)

> The old humanists 'argued that man's spiritual health depended on a kind of education which was more than a training for some specialized work, a kind variously described as "liberal", "humane", or "cultural".'
>
> (Williams, 1961, page 162)

Williams describes the battle waged by the old humanists against the teaching of science, technology or practical subjects. This did not include pure mathematics, which was considered to develop the capacity for pure thought, the basis of rationality.

C. The Mathematical Old Humanists

Pure knowledge, including mathematics, is claimed to originate in societies which separated the work of the hand and brain (Restivo, 1985). In societies such as Ancient Greece, intellectual work was dissociated from manual work, becoming the province of the more powerful social class, the élite, closely associated with the rulers of society.

Through the millennia, the study of pure mathematics has been associated with high culture and the liberal education of the élite. Plato's Academy bore a sign over the portal denying entry to any who had not studied geometry. The Roman Boethius made sure of the place given to mathematics in a liberal education. He adjoined the mathematical quadrivium of arithmetic, geometry, music and astronomy to the trivium at the core of the liberal curriculum. Beyond the curriculum of his day (c. 480–524), Boethius influenced British education for the following millennium, through his textbooks (Howson, 1982).

Although its fortunes varied, pure mathematics was a central part of the public school curriculum throughout the Victorian era, largely represented by Euclid's Elements. It was valued for its contribution to the development of thinking, as the Royal Commission of 1861 reported:

> mathematics at least have established a title of respect as an instrument of mental discipline; they are recognized and honoured at the universities.
>
> (Ministry of Education, 1958, pages 2–3)

The teaching of geometry was criticized by the Taunton Commission of 1868, but there was no threat to the purity of the syllabus in the nineteenth century (Howson, 1982). Indeed, only in the twentieth century did elements of applied mathematics begin to be included in the selective school curriculum, leading to the Applied Mathematics 'A' level course in the 1950s (Cooper, 1985).

The fact that the pressures of the industrial trainers and technological pragmatists for applied knowledge were resisted for so long is an indication of the strength of the old humanist lobby, the mathematical old humanists in particular. The majority of academic mathematicians, certainly in the first half of this centry, prized the purity of mathematics and disparaged utilitarian or applied mathematics. Thus none of the traditional philosophies of logicism, formalism or intuitionism even refer to the applications of mathematics, as we saw in Chapter 1. Mathematics is identified with pure mathematics, and its applications are not considered to be the concern of either 'real' mathematicians, or of the philosophy of mathematics. In discussing the nature of mathematics, neither Frege, Russell, Hilbert, Bernays, Brouwer nor Heyting refer to anything other than pure mathematical knowledge.

The values of purity are so pervasive, that they become invisible. Hardy epitomizes these values:

> [I]f a chess problem is, in the crude sense, 'useless', then that is equally true of most of the best mathematics; that very little of mathematics is useful practically, and that little is comparatively dull.
>
> (Newman, 1956, page 2029)

> I have never done anything 'useful'. No discovery of mine has made, or is likely to make . . . the least difference to the world.
>
> (Newman, 1956, page 2038)

The distinguished mathematician Halmos (1985) illustrates these values with the title of a paper: 'Applied Mathematics is Bad Mathematics', in which he distinguishes the pure aesthetic of the 'knowers' from the practical concerns of the 'doer'. The following echoes and amplifies this sentiment.

> The reputed superiority of mind over matter finds mathematical expression in the claim that mathematics is at once the noblest and purest form of thought, that it derives from pure mind . . . and there is a pervasive unspoken sentiment that there is something ugly about applications.
>
> (Davis and Hersh, 1980, page 85)

The mathematical old humanists celebrate the intrinsic beauty of pure mathematics. Many mathematicians have affirmed the elegance, beauty, harmony, balance and depth of mathematical results (Davis and Hersh, 1980).

> Many pure mathematicians regard their speciality as an art, and one of their terms of highest praise for another's work is 'beautiful'.
>
> (Halmos, 1981, page 15)

> Beauty is the final test: there is no permanent place in the world for ugly mathematics.
>
> (Hardy, quoted in Steen, 1981, page 36)

Like the other old humanists, the mathematical variety often hold élitist views as to who can contribute to high culture. Thus according to the mathematician Adler:

> Each generation has its few great mathematicians, and mathematics would not even notice the absence of the others . . . There is never any doubt about who is and who is not a creative mathematician, so all that is required is to keep track of the activities of these few men
>
> (Adler, quoted in Davis and Hersh, 1980, page 61)

The influence of the mathematical old humanists, and their values, has been evident in a number of official reports on education, such as the Spens Report of 1938:

> No school subject, except perhaps classics, has suffered more than mathematics from secondary rather than primary aims, and to emphasise extraneous rather than intrinsic values. As taught in the past, it has been informed too little by general ideas . . . It is sometimes utilitarian, even crudely so, but it ignores considerable truths in which actual mathematics subserves important activities and adventures of civilized man . . . We believe that school mathematics will be put on a sound footing only when teachers agree that it should be taught as art and music and physical science

> should be taught, because it is one of the main lines which the creative spirit of man has followed in its development.
>
> (Ministry of Education, 1958, page 17)

Cooper (1985) shows that university mathematicians allied with élite public school teachers were successful during the early 1960s in bringing the school mathematics curriculum closer to modern university mathematics (although technological pragmatists were also successful in introducing more applications into its contents).

The old humanist view of mathematics, with variations, has a tradition two and a half millennia old. It identifies mathematical talent with pure intelligence, regarded as an inherited character. The view is that of the mathematics establishment, and is centred on the structure of mathematical knowledge, and on the values of mathematicians, so it is 'mathematics-centred'. It was epitomized by pure mathematicians at universities, sitting on the Examinations Boards and by the Mathematical Association, in Britain.

D. The Mathematical Aims and Ideology of the Old Humanists

Aims of mathematics education

The aims of this group for mathematics education are those of the old humanists applied to mathematics: a concern with the transmission of mathematical knowledge, culture and values. The goal is to transmit pure mathematics *per se*, with an emphasis on the structure, conceptual level and rigour of the subject. The aim is to teach mathematics for its intrinsic value, as a central part of the human heritage, culture and intellectual achievement. This entails getting students to appreciate and value the beauty and aesthetic dimension of pure mathematics, through immersion in its study.[1] A very important subsidiary aim is the education of future pure mathematicians, which introduces the element of élitism.

Theory of school mathematical knowledge

School mathematics is understood to be, like the discipline itself, a pure, hierarchically structured self-subsistent body of objective knowledge. Higher up the hierarchy, mathematics becomes increasingly pure, rigorous and abstract. Students are encouraged to climb up this hierarchy as far as possible, according to their 'mathematical ability'. As they ascend, they get closer to 'real' mathematics, the subject taught and studied at university level.

This theory is implicit in many mathematics textbooks and schemes, albeit combined with other less purist perspectives. Thus unique hierarchical structures are found in many books and workcard schemes, such as the School Mathematics Project books and schemes of the 1960s and 1970s.

Theory of learning mathematics

The theory of learning concerns the reception and understanding of a large, logically structured body of mathematical knowledge, and the modes of thought associated with it. The successful learner internalizes the pure conceptual structure of mathematics: a hierarchical network of concepts and propositions interconnected by logical links, mathematical relationships and fundamental ideas, mirroring the organization of mathematics. Properly learned, mathematical knowledge allows the learner to solve mathematical problems and puzzles. Students are expected to come up with different approaches and methods, in applying this knowledge, according to their talents and ingenuity.

Theory of teaching mathematics

The teacher's role, according to this perspective, is that of lecturer and explainer, communicating the structure of mathematics meaningfully. The teacher should inspire through an exciting delivery, should enrich the mathematics course with additional problems and activities, adapting the structured textbook approach. At best, various approaches, demonstrations and activities are employed to motivate and facilitate learning and understanding. Teaching entails a benign master-pupil relationship; the master, possessor of the knowledge, transmits it to the pupil, as effectively as possible. Thus, according to Hardy

> In mathematics there is one thing only of prime importance, that a teacher should make an honest attempt to understand the subject he teaches as well as he can, and should expound the truth to his pupils to the limits of their patience and capacity.
>
> (Ministry of Education, 1958, page iii)

Overall, the ethos is that of 'teaching mathematics', as opposed to 'teaching children'; the traditional secondary school rather than the modern primary school ethos.

Theory of resources in mathematics education

The 'purist' ideology leads to a restricted view of the resources appropriate for school mathematics. Text books and traditional aids to pure mathematical construction are admitted, such as straight-edge and compasses. Electronic calculators and computers may also be used as tools in mathematics, but only for older students who have demonstrated mastery of the basic concepts. Models, visual aids and resources may be used by the teacher to motivate or to facilitate understanding. However, the 'hands-on' exploration of resources by students is practical work, inappropriate to pure mathematics, and is thus reserved for low attainers, who are not studying 'real' mathematics, anyway.

Theory of mathematical ability

According to this view, mathematical talent and genius are inherited, and mathematical ability can be identified with pure intelligence. There is a hierarchical distribution of mathematical ability, from the mathematical genius at the top, to the mathematically incapable, at the bottom. Teaching merely helps students to realize their inherited potential, and the 'mathematical mind' will shine through. Educational provision is needed for the mathematically gifted, to enable them to fully realize this talent. Since children vary greatly in mathematical ability, they need to be streamed in school for mathematics. It is an élitist theory of mathematical ability, seeing it as hierarchical and stratified, and valuing those at the top most.

Theory of assessment of mathematics learning

According to this theory, formative assessment of mathematics learning may involve a range of methods, but summative assessement requires external examinations. These should be based on a hierarchical view of mathematical subject matter, and at a number of levels, corresponding to mathematical 'ability'. However difficult, the excellence of the mathematically talented will shine through, and any move to make examinations more accessible or less trying for students, must represent a dilution of standards. Competition in examinations provides a means of identifying the best mathematicians.

Theory of social diversity in mathematics education

Mathematics is viewed as pure and unrelated to social issues, so no room is allowed for the accommodation of social diversity. Mathematics is objective, and attempts to humanize it for educational purposes, however well intentioned, compromise its essential nature and purity (Ernest, 1986, 1988b). For those members of society who cannot cope with it because of differences in ability or background, a smaller dose is required, perhaps offering access only to the bottom rungs of the hierarchical ladder of mathematics.

E. A Critique of the Mathematical Old Humanist View

A strength of this perspective is its emphasis on the organization and structure of mathematics as a theoretical discipline, with central unifying concepts. Related to this is the appreciation of the beauty and aesthetics of mathematics. The focus on the intrinsic value of mathematics means that these important aspects are not neglected, as they are in utilitarian perspectives.

A critique of the ideology

The underlying ideology of the old humanists is open to a number of criticisms. First of all, there is the purist-absolutist view of mathematics which denies the connection between pure mathematics and its applications. To view mathematics as a pure entity, divorced from the base shadow of its applications is to subscribe to a dangerous, unsustainable myth. Many advances in pure mathematics, such as Newton's contributions to the calculus, cannot be separated from scientific problems and stimuli. Currently developments in computing are profoundly shaping the development of pure mathematics (Steen, 1988). The schizophrenic existence of mathematics as both queen and servant of science can no more be separated, than can waves and particles in quantum theory. Pure and applied mathematics must be regarded as two facets of the same thing.

Secondly, the 'ivory tower' academicism, and associated élitism of this position is morally unsound. It denies that mathematics has any involvement in, or responsibility for broader, social issues. However mathematics is but one part of knowledge, just as mathematicians are but one segment of society, and there is no moral basis for denying the responsibility of the part to the greater whole, in either case. To base a perspective on the notion that mathematics is intrinsically good, in some sense, and then to deny responsibility if its outcomes should be bad, through a negative impact on the lives of others in society or school, is morally irresponsible and incorrect. No area of knowledge or life has the right to this royal prerogative, in a democratic society all are accountable.

Thirdly, there is the unjustified assumption of a fixed view of human ability, related to a stratified and élitist view of society and human nature. Whatever part inheritance plays in determining human and mathematical ability, it is widely accepted that environmental influences have a major impact on its realization (Beck *et al.*, 1976).

Negative educational outcomes

These weaknesses have grave consequences for education. First of all, there are problems stemming from the 'top down' view of the mathematics curriculum. This sees the primary function of 'A' level mathematics to prepare students for university mathematics, the function of GCSE mathematics is to prepare students for 'A' level mathematics, and so on down the years of schooling. The absurd consequence is a mathematics education for all serving the needs of the few, the tiny minority of less that 1 per cent who study pure mathematics at university. One major outcome of the reforms of the 1960s was just this (Howson and Wilson, 1986). Syllabuses for all constructed as 'watered down' versions of higher status academic preparation syllabuses. Thus the opportunity to develop more appropriate free-standing syllabuses was not taken. Consequently, many students study a mathematics curriculum not designed with their needs at the fore, whatever they might be (Cockcroft, 1982).

This 'top down' view of the curriculum extends more widely than just to mathematics, and can be transposed to any academic school subject. It is explained by the old humanists' elevation of pure knowledge above applied knowledge and practical skills in education. The result is inappropriate curricula for the majority of students, curricula not designed with either their needs or interests in mind. The only rational explanation for this is that education is serving the interests of the old humanists, at the expense of the interests of society as a whole. For a purist curriculum, with its associated values, provides a source of new recruits to the old humanist grouping. It also educates those who fail to gain entry to the group to respect its values and thus help to secure its status and power.

The mathematics curriculum, in particular, prepares a tiny minority of students to be mathematicians whilst teaching the rest to stand in awe of the subject. To let one group distort the aims of education like this, to serve its own interests, is wrong and anti-educational. It results in more persons being disadvantaged than are advantaged, meaning that on utilitarian grounds alone, the system is insupportable.

A second consequence of this perspective, is that mathematics is presented to learners as objective, external, cold, hard and remote (Ernest, 1986, 1988b). This has a powerful negative effect on attitudes and affective responses to mathematics (Buerk, 1982). In particular, this separate conception of mathematics is considered to be a contributing factor to females' negative attitudes to mathematics, and their subsequent underparticipation (Open University, 1986). Mathematics viewed as unrelated to the world, human activity and culture alienates students, irrespective of gender. The emphasis on mathematical structure and logic, of this perspective, can be stultifying. Polya quotes Hadamard:

> The object of mathematical rigour is to sanction and legitimate the conquests of intuition, and there never was any other object for it.
>
> (Howson, 1973, page 78)

An emphasis away from the human, process aspects of mathematics in favour of the objective and the formal reduces personal involvement in the learning of mathematics. Most recent authoritative reports on mathematics have stressed the importance of active participation in its learning in humanized contexts, especially problem solving, applications and investigational work (National Council of Teachers of Mathematics, 1980, 1989; Cockcroft, 1982; Her Majesty's Inspectorate, 1985). In contrast, an over-emphasis on rigour, structure and formalism leads to incomprehension and failure.

Thirdly, the assumption that mathematical ability is fixed by inheritance, is damaging for those not labelled as mathematically gifted. The labelling of individuals in terms of others' perceptions of their abilities, is known to be self-fulfilling (Meighan, 1986). The result is to reduce the level of attainment of those labelled as low-ability, damaging mathematical achievement (Ruthven, 1987). It also contributes to the gender problem in mathematics, where the stereotyping of mathematical ability as a masculine trait is regarded as a major causal factor.

2. The Progressive Educators

A. The Connected Relativistic Absolutist Ideology

Set of moral values

The moral values of this position are the connected values:

> the ethic of responsibility relies on the concept of equity, the recognition of differences in need . . . [it] rests on an understanding that gives rise to compassion and care.
>
> (Gilligan, 1982, pages 164–165)

These values are concerned with human relationships and the connections between persons, with empathy, caring and the human dimensions of situations. They are stereotypically attributed to the 'feminine' role (socially constructed as it is): to relate, nurture, comfort and protect.

The connected perspective parallels the romantic tradition in art and education, valuing expression, style, diversity, experience and subcultures (Jenkins, 1975). Expression represents a particularly individualistic and person-centred value; likewise experience as the source of 'personal knowledge'.

Connected values can be identified with the leading component of a number of other pairs, including child-centred versus subject-centred, content versus structure, progressive versus traditional, intuitive versus rational, and Dionysian versus Apollonian. The parts corresponding to connectedness suggest expression, creativity, feelings, subjectivity and dynamic growth in place of rules, structure, logic, objectivity and static form. These are some of the connotations of connected values, if not logically entailed by them.

Epistemology

Epistemologically, this position is rationalist, but also includes elements of empiricism (Blenkin and Kelly, 1981). It views knowledge as innate, re-created by individuals as part of their process of development and unfolding. The mind contains within it the seeds or forms of knowledge which develop in the process of maturation and, in response to experience. This epistemology has its roots in the thought of Plato, Descartes, Kant, and the rationalist tradition. Recently, it can be discerned in the work of Piaget and Chomsky, who see logico-mathematical and linguistic knowledge, respectively, as innate.

There is also an empiricist strand to the epistemology, as in British empiricism, Dewey and the progressive tradition in education (Dearden, 1968; Blenkin and Kelly, 1981). Experience is an essential stimulus for the unfolding of the child's innate knowledge. The knowledge of the child develops through interaction with the world.

Both empiricism and rationalism accept the existence of objective truth. However, this ideology believes that although our knowledge is perfectible, progressing steadily towards absolute objective truth, it has not yet achieved it (as in Popper, 1979).

Philosophy of mathematics

The philosophy of mathematics is absolutist, viewing mathematical truth as absolute and certain. But it is progressive absolutism, because great value is attached to the role of the individual in coming to know this truth. Humankind is seen to be progressing, and drawing nearer to the perfect truths of mathematics. On the basis of the connected values, mathematics is perceived in humanistic and personal terms, and mathematics as a language, its creative and human side, and subjective knowledge are valued and emphasized. But this is coupled with absolutism. Thus the view of mathematics is progressive absolutist, the absolutism coloured by the humanistic, connected values.

Theory of the child

The theory of the child regards children as having full rights as individuals, and needing nurturing, protection and enriching experiences to allow them to develop to their full potential. The child is seen as an 'innocent savage' and a 'growing flower' (Ramsden, 1986). The innocent savage is born good, an individual whose needs and rights are paramount, who learns and grows through experience of the physical and social world. The connected values are the source of the protective ethos, and for the fostering of creativity and personal experience. As a growing flower a child is born with all it needs for full mental and physical growth, and given the proper nurturing environment and experiences will autonomously develop to his or her full potential. Rationalism locates the seeds of this growth within the child. When exposed to the appropriate experiences (empiricism), this allows for the full potentiation and realization of the human being, in terms of overall development and knowledge.

Theory of society

The focus of the ideology is on the individual rather than on the social matrix, except as a context for individual development. Ideally, society is seen as a supportive and nurturing environment, but the reality of social ills requires caring responses to individuals. Due to this individualism, the structural features of society are down-played. The theory of society does not encourage any questioning of the social structures which sustain inequalities and deny opportunities. The connected values lead to a commitment to the amelioration of the conditions and suffering of individuals in society. Thus the theory of society is progressive and liberal, concerned to improve conditions, but without any questioning of the social *status quo*.

Educational aims

According to this perspective, the goal of humanity is the self-development and personal fulfulment of each individual in 'becoming a person' (Rogers, 1961). The aim of education is to promote the self-realization of individuals by encouraging their growth through creativity, self-expression and wide-ranging experience, enabling them to reach full flower. These are progressive, child-centred aims, deriving from the connected values and the epistemology. The aims are purist, because they concern the development of the child for its own sake, as something of intrinsic value.

B. The Progressive Tradition as Connected Relativistic Absolutism

Origins of the progressive tradition

The theory of childhood which sees the child as an 'innocent savage' and a 'growing flower', is part of the progressive tradition of thought. The roots of this tradition lie in Plato's epistemology. Plato argues that all are born with latent knowledge, and learning is becoming aware, or recollecting, that which lies latent and forgotten in each of us.

Rousseau takes this as his starting point, arguing that the child has the inner potential for learning, and will develop according to his or her own plan. But in contrast to the elementary school tradition, with its particular Judeo-Christian roots, Rousseau denies that children are tainted with original sin. Thus the opening lines of Emile are:

> God makes all things good; man meddles with them and they become evil.
>
> (Rousseau, 1762, page 5)

Rousseau argues that from this pristine state we develop, following an inner pattern, subject to education.

> Plants are fashioned by cultivation, man by education . . . All that we lack at birth, all that we need when we come to man's estate, is the gift of education. This education comes to us from nature, from man, or from things. The inner growth of our organs and faculties is the education of nature, the use we learn to make of this growth is the education of men, what we gain by our experience of our surroundings is the education of things.
>
> (Rousseau, 1762, page 6)

Rousseau's emphasis is on the inner potential of the child, the child's needs, and the role of practical activities, play and experience in education. Because of its pristine, uncultivated state, the view of childhood is that of the 'noble savage', in Dryden's phrase. Through his emphasis on the child's unsullied nature, its rights, experiences, and knowledge-getting, Rousseau offers the first child-centred vision for education.

This vision is echoed in Pestalozzi and in Froebel, who adds the metaphor of the 'growing flower' to that of the 'noble savage'. As a growing flower a child is born with all it needs for full mental and physical growth, and given the proper environment, a kindergarten or child-garden, will grow to his or her full potential. Such an environment should encourage and be based on spontaneity, enjoyment, play, practical and experiential learning and group activities (Ramsden, 1986).

The vision of childhood embodied in the writings and the practices of Rousseau, Pestalozzi and Froebel provides the basis for the progressive tradition in education. Following Rousseau, Pestalozzi emphasized the need for concrete experience in education (Blenkin and Kelly, 1981). Froebel argued that 'play is the child's work', essential in the unfolding of the child's potential (Dearden, 1968). A twentieth century contributor to this tradition is the philosopher and educator John Dewey. Dewey (1916) argued for the experimental method in education: children ought to put received knowledge to the test, and should learn by activity, problem solving and the 'project method'. He emphasized the importance of knowing and the process of coming to know in education, as he did in philosophy.

Dewey's philosophy of mathematics, that of the Pragmatist Movement, is more empiricist, or even fallibilist, than absolutist. However, he never worked out fully the consequences of his views for the philosophy of mathematics. Thus, he had little impact on the absolutist philosophy of mathematics of the progressive tradition, except perhaps to shift it towards progressive absolutism. However his pragmatist view of knowledge was readily assimilated to the epistemology described above, with its focus on the empirical roots of knowledge (empiricism), as well as on the structured unfolding of knowledge in the mind of the individual, in accord with some absolute pattern (rationalism).

Another contributor to the progressive tradition in education in the early years of this century is Maria Montessori. The 'noble savage' and 'growing flower' view of childhood are explicitly contained in her ideas.

> Her theories view . . . the child's nature as essentially good and education as a process of unfolding what has been given the child at birth . . . liberty as an essential ingredient for this unfolding, and [the need for] . . . sense experiences in this process of development.
>
> (Lillard, 1973, page 10)

Psychological theories

Another strand in the progressive tradition is that of psychological theories of development, from Herbart and Freud to Piaget and Bruner. Piaget's theory of intellectual development focuses on two central aspects of the progressive view of childhood. First, on the centrality of children's experience, especially physical interaction with the world. Second, on the unfolding logic of children's thought, which differs from that of the adult. These are intertwined, as the child represents its experiences and actions mentally, and transforms these representations by means of the developing sequence of logical operations.

Piaget is a constructivist, proposing that children create their knowledge of the world. However, he also believes that in the creation and unfolding of their knowledge, children are constrained by absolute conceptual structures, especially those of mathematics and logic. Thus Piaget accepts an absolutist view of knowledge, especially mathematics. He also provides psychological support for the progressive views of children's minds as unfolding during development and the importance of experience. The notion of children's development reaching stages of 'readiness' also gained theoretical support from Piaget's work.

The progressive tradition in education

One strand of this thought is the private 'progressive' school tradition, including A.S. Neill's (1968) Summerhill and Dartington Hall (Meighan, 1985). However the main impact of the progressive tradition has been on primary education in Britain. This was greatly assisted by two official reports on education, the Hadow Report (1931) and the Plowden Report (1967), (Blenkin and Kelly, 1981). The Hadow Report provided a very influential articulation of the progressive tradition in education and rejection of the elementary school tradition of the industrial trainers. Building on the view of childhood described above, the report emphasized the process of education as active coming to know rather than the passive reception of knowledge.

> [T]he curriculum is to be thought of in terms of activity and experience rather than of knowledge to be acquired and facts to be stored. Its aims should be to develop in a child the fundamental human powers
>
> (Hadow Report 1931, page 93)

The progressive educators' 'liberal romanticism' (Richards, 1984) received its fullest expression, as well as its strongest official endorsement, in the Plowden Report (1967).

> At the heart of the educational process lies the child. No advances in policy, no acquisitions of new equipment have their desired effect unless they are in harmony with the nature of the child, unless they are fundamentally acceptable to him.
>
> (Plowden Report, 1967, quoted in Pollard, 1987, page 1)

> A school is not merely a teaching shop, it [transmits] values and attitudes. It is a community in which children live first and foremost as children and not as future adults ... The school sets out deliberately to devise the right environment for children, to allow them to be themselves and to develop in the way and at the pace appropriate to them ... It lays special stress on individual discovery, on first hand experience and on opportunities for creative work. It insists that knowledge does not fall into neatly separate compartments and that work and play are not opposite but complementary.
>
> (Plowden Report, 1967, quoted in Richards, 1984, page 71)

Thus as its critic Peters (1969) summarized, the Plowden Report represented the liberal/progressive view that the child has a 'nature' which 'develops' given the appropriate environment, that self direction (including both autonomy and discovery) are vital to this development, that knowledge is not compartmentalized and the curriculum should be integrated to reflect this, and the teacher must be a guide, an arranger of the environment rather than an instructor, enabling the child to proceed from discovery to discovery when they are 'ready' with minimum intervention.

Protectiveness

A further factor in the progressive tradition is the view that the child needs to be protected and shielded from the harshness of everyday life. This is found in Rousseau:

> Tender, anxious mother, I appeal to you. You can remove this tree from the highway and shield it from the crushing force . . . From the outset raise a wall around your child's soul
>
> (Rousseau, 1762, pages 5–6)

This factor also underpins modern progressive primary practice. Kanter observed that the provision of an appropriate set of experiences in a nursery school, including play and personal exploration 'involved dealing with the children and structuring the environment in such a way as to limit the seven experiences seen as "anxiety producing"' (Dale *et al.*, 1976, page 168). The factors limited to protect the children are uncertainty, strangeness, mystery, coercion, accountability, unpleasantness and peer conflict. Limiting these factors to protect and shield the child leads to the routinization of play, the discouragement of personal responsibility and the avoidance of dissonance and conflict.

Explicit statements of progressive ideology

A clear statement of the progressive ideology of education, articulated as a set of 'assumptions about children, learning and knowledge' emphasizing curiosity, active learning, sequential development and subjective knowledge, is due to Barth.

> Children are innately curious and display exploratory behaviour quite independent of adult intervention.
>
> Active exploration in a rich environment, offering a wide array of manipulative materials, facilitates children's learning.
>
> Play is not distinguished from work as the predominant mode of learning in early childhood.
>
> Children will be likely to learn if they are given considerable advice in the selection of the materials they wish to work with and in the selection of the questions they wish to pursue with respect to those materials.
>
> Children pass through similar stages of intellectual development, each in his own way, and at his own rate and in his own time.

> Intellectual growth and development take place through a series of concrete experiences followed by abstractions.
>
> Knowledge is a function of one's personal integration of experience and therefore does not fall into neatly separate categories or disciplines. There is no minimum body of knowledge which it is essential for everyone to know.
>
> (Alexander, 1984, page 10)

Similarly, Richards provides an explicit statement of the progressive educator ideology:

> *liberal romanticism* — which starts from, and constantly refers back to the individual child when developing educational principles. Compared with other ideologies it advocates a more equal partnership of teacher and taught, with teachers learning 'alongside' children, and it offers children a relatively high degree of choice in the type, content and duration of activities.
>
> (Richards, 1984, page 62)

Criticism of the elementary school tradition

The progressive educators strongly, and often emotively reject the elementary school tradition of the industrial trainers. This is characterized as: 'mass teaching, chanting of tables, mechanical reading around the class . . . rigid timetables and the silencing of bells . . . the same textbook for every child'. (Kirby, 1981, page 11). Both the strict regimentation and denial of individuality, and the lack of respect for each child becoming a person at their own rate, are anathema to the progressive educators. Alexander analyzes their writings, and finds that

> the language lulls and cradles, suggesting an affinity with a romantic conception of the natural order, a regression to childhood innocence and security, a pot-pourri of firelight and warmth against the cold night. 'At the heart of the educational process lies the child': Plowden's use of the words 'heart' (rather than the more neutral 'centre'), and 'lies', with the image of the child recumbent, dormant and maternally cradled, seems hardly accidental . . . [Whereas] the vocabulary used to characterise non-child-centredness is . . . harsh, suggestive of restriction, repression, violence even: 'nourish', 'sensitive' versus 'crude', 'inert', 'restriction', 'impose'.
>
> (Alexander, 1984, page 18)

The ideology of the progressive educators has had widespread support this century, especially in primary school education. Consequently most discussions of the primary school make reference to this ideology, although a number of different terms are used including 'liberal romanticism' (Richards, 1984), the 'progressive tradition' (Golby, 1982), the 'developmental tradition' (Pollard, 1987) and 'open schooling' (Silberman, 1973). However, the widespread endorsement of the progressive educator ideology is not simply an indication of its ascendancy. It rather represents a need for legitimation during a period of criticism. For since the publication of the Plowden Report (1967) the progressive educator ideology has been attacked by industrial trainers, technological pragmatists and old humanists (Cox and Dyson, 1969, 1969a, 1970; Cox and Boyson, 1975; Callaghan, 1976; Peters, 1969).

C. The Progressive Tradition in Mathematics Education

The progressive educator ideology in mathematics education is largely a matter of the past hundred years. Three inter-related strands of this tradition in mathematics can be distinguished.

1. The provision of an appropriately structured environment and experiences for the learning of mathematics;
2. The fostering of active and autonomous inquiry in mathematics, by the child;
3. A concern with the child's feelings, motivation and attitudes and the shielding from negative aspects.

Since the turn of the century progressive educators have been trying to provide appropriately structured environments and experiences for children. Considerable ingenuity has been involved in developing structural mathematical apparatus for topics such as number and algebra. Montessori developed a range of materials for learning mathematics such as the 'golden bead' representations of the decimal number system (Williams, 1971). Stern, Cuisenaire and Gattegno developed further materials and approaches, including structured 'number rod' embodiments. Dienes (1960) developed a range of different structural materials and games for learning mathematics, including embodiments of number, logic, sets and algebra. The modern orthodoxy of primary mathematics education is that children learn best through experiences of a range of manipulative learning aids (Williams, 1971).

In 1953 the Association for Teaching Aids in Mathematics was founded with an emphais on aids and materials in the learning of mathematics (Cooper, 1985). This was to become the Association of Teachers of Mathematics, an organization representative of the progressive movement in mathematics education.

In 1956 the Mathematical Association brought out a report on the teaching of mathematics in primary schools embodying many of the precepts of progressive education, including a chapter on the use of 'material aids to teaching' mathematics.[2] The report begins:

> No one who teaches young children today can avoid the challenge of the words 'Activity and Experience'. Modern educationists are united in believing that . . . children . . . must be *doing* if they are also to think
>
> (Mathematical Association, 1956, page 1, emphasis added)

The report explicitly subscribed to the progressive educator aims.

> mathematical specialists and... teachers engaged in the work of Primary Schools [are] concerned with children's mathematical development as only one facet of their whole growth... because it is the wholeness of a child's growth which they (and we) consider all important... it is essential that Mathematics... should be seen to be an integral and worthy part of a young child's education.
>
> (Mathematical Association, 1956, page vii)

In the 1960s the progressive tradition in mathematics became widespread and an influential statement by one of its leading proponents Edith Biggs (1965), sold 165,000 copies in three years. During this period the progressive orthodoxy developed, increasing the emphasis on discovery, problem solving and children's attitudes to mathematics. An influential statement of this philosophy came from the Nuffield Mathematics Teaching Project (1965). This emphasized activity through both content and title ('I do and I understand'), devoted a chapter to discovery learning, and asserted the importance of attitudes to mathematics.

> [T]he most vital factor has been left until now — does the child *enjoy* and *succeed* at his work? It is recognized that attitudes towards mathematics are largely formed in the primary school and most probably in the first few years. In order to prevent the continuation of certain [negative] attitudes... care must be taken to prevent the possibility of their early establishment.
>
> (Nuffield Mathematics Teaching Project, 1965, page 5)

The Association of Teachers of Mathematics (1966) indicated the importance it ascribed to children's mathematical activity and problem solving with a report presented at the International Congress of Mathematicians, Moscow 1966 ('The Development of Mathematical Activity in Children and the Place of the Problem in this Development').

An influential base for the progressive philosophy of mathematics education lay in the teacher training colleges. The Mathematics Section of the Association of Teachers in Colleges and Departments of Education (1966) published an influential statement on the education of mathematics teachers. This stressed the importance of creative mathematical activity among students and children and introduced the term 'mathematical investigation' to describe open-ended problem posing and exploration in mathematics, as well as providing a selection of 'starting points'.[3]

At the end of the 1960s the Mathematical Association published a further report on primary mathematics, which began by endorsing the progressive ideology of the earlier report.

> children, developing at their own individual rates, learn through their active response to the experiences that come to them; through constructive play, experiment and discussion children become aware of the relationships and develop mental structures which are mathematical in form and are in fact the only sound basis of mathematical techniques... the aim of teaching

> mathematics in the primary school is 'the laying of this foundation of mathematical thinking' about . . . objects and activities.
>
> (Mathematical Association, 1970, page 3)

The Cockcroft report (1982) endorsed the progressive tradition in mathematics education with its emphasis on problem solving, practical and investigational work, discussion and on learners' attitudes to mathematics. This was further endorsed by Her Majesty's Inspectorate (1985), and the National Criteria in Mathematics for the General Certificate of Secondary Education (Department of Education and Science, 1985) followed this with an emphasis on problem solving and investigational work and projects, as part of the national assessment of 16 year olds in mathematics. Finally, the interim report of Mathematics Working Group of the National Curriculum (Department of Education and Science, 1987) represented a paradigmatic statement of the progressive perspective emphasizing first and foremost, learners' attitudes to mathematics, second children's processes of mathematization, and only last, the significance of mathematical content.

The supporters of the progressive tradition in mathematics education include mathematics educators, advisors and teacher educators, as well as progressive teachers. The tradition had grown in strength this century, embracing both primary and secondary education in Britain. This can be partly attributed to progressive educators forming an alliance with progressive technological pragmatists, on the basis that practical, confident problem solvers serve the needs of employment (Hodkinson, 1989). Such an alliance can be inferred from Cockcroft (1982), in which both progressive and utilitarian aims for mathematics education are endorsed.

Overall, this perspective is epitomized in British mathematics education by the Association of Teachers of Mathematics. However, this identification is not so clear cut, for there has long been a radical wing to the association, and some leading members have public educator sympathies.[4] Furthermore, the Mathematical Association increasingly displays progressive educator sympathies.[5]

The progressive ideology worldwide

The progressive educator ideology in mathematics education is represented worldwide, in such places as continental Europe, North America and Australia. Thus in the USA the National Council of Teachers of Mathematics published an influential 'agenda for action' which stated as its first recommendation:

> Problem solving must be the focus of school mathematics in the 1980s . . . Fundamental to the development of problem solving activity is an open mind, an attitude of curiosity and exploration . . . Mathematics teachers should create classroom environments in which problem solving can flourish . . . [it] is essentially a creative activity.
>
> (National Council of Teachers of Mathematics, 1980, pages 2–4)

More recently, the National Council of Teachers of Mathematics (1989) has published a statement of intended 'standards in school mathematics' embodying much of the progressive educator ideology, as one of its authors comments.

> A spirit of investigation and exploration should permeate instruction . . . teachers need to provide a caring environment . . . Students should be actively involved in the learning process, investigating and exploring individually and in groups . . . Teachers should be facilitators of learning, not merely dispensers of knowledge.
>
> Cooney (1988, page 355)

D. The Educational Ideology of Progressive Mathematics Educators

Theory of school mathematical knowledge

The emphasis of this ideology, according to Marsh, is on the 'Experience, not curriculum . . . Child, not curriculum' (Alexander, 1984, page 16). Mathematics is a vehicle for developing the whole child, so the curricular emphasis is on mathematics as a language, and on the creative and human side of mathematical experience. The processes of mathematical problem solving and investigating, such as generalizing, conjecturing, abstracting, symbolizing, structuring and justifying, figure more prominently than the specification of mathematical content. Mathematics is just a part of the whole curriculum, so child-led applications of 'mathematics across the curriculum' are also valued as part of school mathematics.

Aims of mathematics education

The mathematical aim of the progressive educator is to contribute to the overall development of the growing human being, to develop the child's creativity and self realization through the experience of learning mathematics. This involves two things. First, the development of the child as an autonomous inquirer and knower in mathematics. Second, the fostering of the child's confidence, positive attitudes and self-esteem with regard to mathematics, and shielding the child from negative experiences which might undermine these attitudes.

Theory of mathematical ability

The progressive educator theory of mathematical ability is individualistic. Its central assumption is that there are innate, inherited differences in mathematical ability leading to differential individual rates of development or unfolding. These, in turn, lead to different levels of 'readiness' for further mathematical development. However, each individual's mathematical ability needs an appropriate set of experiences to be fully realized, otherwise the child's growth may be stultified. Two contradictory forces are at work, deriving from rationalist and empiricist epistemology. There is the pull of inherited ability and the innate stages of thinking, as well as the impact of experiences and the environment.

Theory of learning mathematics

The most elaborated of the theories of the progressive educator is the theory of learning mathematics. This involves the students' active responses to environment, autonomous inquiry by the child, seeking out relationships and creating artefacts and knowledge. Learning involves investigation, discovery, play, discussion and cooperative work. The environment in which learning takes place must be rich and challenging, but must be secure, fostering self confidence, positive attitudes and good feelings. Thus the learning of mathematics is first and foremost active, with the child learning through play, activity, investigations, projects, discussion, exploration and discovery. A second key feature is that of self-expression, with the child's own methods of solution and recording encouraged. Children's own mathematical ideas and projects are particulary valued.

Theory of teaching mathematics

The teaching of mathematics, according to this perspective, consists of encouragement, facilitation, and the arrangement of carefully structured environments and situations for exploration. Ideally, it will involve the use of teacher or school constructed mathematics curriculum, offering a 'circus' of different mathematical activities around the classroom, and employing integrated multi-disciplinary projects. The role of the teacher is seen to be that of manager of the learning environment and learning resources, facilitator of learning, with non-intrusive guidance and shielding from conflict, threat and sources of negative feelings.

Theory of resources in mathematics education

The theory of resources for learning mathematics plays a central part, for learning is understood to involve activity. Thus the classroom must be a rich environment, structural apparatus and other equipment are provided to facilitate concept formation and external representations of mathematical ideas. Resources for creating, expressing and making are needed, as is the environment beyond the confines of the classroom, to link mathematics with the whole of the child's experience. Furthermore, the child's access to resources needs to be self determining.

Theory of assessment of mathematics learning

The theory of assessment is that of informal or criterion-based teacher assessment of positive achievements, with the avoidance of failure and the labelling of children's creations as 'incorrect'. External assessment is not valued for fear that it will blight the development of the child. The correction of errors in children's work is avoided, or refomulated in some way (ticks but no crosses, or 'see me', to avoid outright marking wrong). The child is to be protected from conflict and pain.

This is the source of a contradiction. The absolutist view means that mathematical tasks are understood to have a right or wrong answer, but the protectiveness of the ideology means that children's 'errors' cannot be explictly corrected or indicated as wrong, for fear of hurt and emotional damage. Instead euphemisms for incorrectness or failure are employed (such as 'see me'), which take on the socially defined meaning that something is 'wrong'.

Theory of social diversity in mathematics education

The connected values require that cultural and racial diversity be catered for, to bring the mathematics into the cultural sphere of each child. The cross-curricular approaches favoured by this position also facilitate multi-cultural references in mathematics. Thus this position acknowledges children's different cultural origins and tries to utilize aspects of these cultural features in the teaching of mathematics. Great value is attached to meeting the needs of each and every child, and to give them emotional support and succour, to build up their self-esteem, and to avoid conflict for them. Consequently, only the positive or neutral aspects of multicultural issues are utilized. Racism, and other areas of conflict are denied or minimized to protect the sensibilities of children. Overall, the theory of social diversity is individualistic, striving to accommodate cultural and linguistic differences and to meet the diverse needs of individuals, as perceived. However, the harsh realities of social conflict or racism are denied, from a desire to protect.

E. A Critique of the Aims of the Progressive Educators

The overarching strength of this perspective and its aims are that it attends to the nature, interests and needs of the learner (as they are perceived). The aim is to enhance the learner in terms of self-esteem and as a confident epistemological agent in mathematics. This is a very great strength. The purpose of education must primarily be to benefit and enhance the learner, and any intended benefits for society or human culture derive from this primary aim.

In addition, these aims value creativity in mathematics, without concern for utility. This is a vital but neglected aspect of mathematics education (Isaacson, 1989, 1990).

Theory of school mathematical knowledge

There is a contradiction in this view, arising from the tension between an absolutist view of mathematics and a child-centred theory of school mathematics and schooling. When child-centredness is opposed to mathematics-centredness, the result is a focus on the experience of the child as opposed to engagement with mathematics. This may be at the expense of mathematics, failing to develop mathematical concepts and structures to a sufficient depth to give children confidence in their use as 'tools for thought' (Mellin-Olsen, 1987). Further, if learning experiences are not compartmentalized into subject areas, the learner may not develop a sense of mathematics, and the unique characteristics of its knowledge and methods of inquiry.

Theory of teaching mathematics

The theory of teaching is inadequate, de-emphasizing the role of the teacher. The teacher has at least three essential roles, which the progressive educator perspective fails to recognize adequately. First, the teacher mediates between the corpus of mathematical knowledge and the learner, involving the selection and representation of mathematical knowledge (Peters, 1969). This is essential in constructing the learning environment and in planning learning experiences. Secondly, the teacher must monitor children's learning and intervene in their sense making, by communicating feedback and managerial instructions to the children, challenging children to reconsider their responses, regulating interaction. Thirdly, the teacher provides a role model for the children through her behaviour and social interactions. In each of these ways the teacher is central to the process of education, and insufficient acknowledgement is given to this.

Over-protectiveness

The third criticism is that the progressive educator perspective is overprotective, shielding children from the dissonance and conflict needed to provoke intellectual growth. Such overprotectiveness may mean that children's 'errors' are not explicitly corrected, for fear of hurt and emotional damage. Instead euphemisms for incorrectness or failure are employed (such as 'see me'), which children understand very well to denote error, adding a layer of deceit to the meaning. Awareness of error is essential for learning, and dissonance and cognitive conflict are likewise necessary for cognitive growth in learning mathematics. Beyond this, dealing with interpersonal conflict and with controversial issues are essential life skills for citizenship in modern society. However, this perspective tries to sustain an artificial harmony through the denial of conflict in the classroom and the world beyond. By shielding the child from such experiences the progressive educator view is obstructing the child's cognitive, emotional and social growth.

Theory of society

The fourth criticism concerns the inadequacy of the theory of society. The ideology is naively apolitical, ignoring the social matrix and inequalities that surround education, and instead focuses exclusively on the individual. Social progress, according to this view, has individual solutions dependent on self-realization and self-esteem. Thus, there is little or no recognition of the social or political causes of the life-conditions of individuals, nor of the social realities to be faced by children in adulthood, let alone preparation for them.

Theories of childhood and of learning mathematics

The fifth criticism is that theories of childhood and the nature of children's learning are over-romanticised, unrealistic and based on unchallenged assumptions and theories. Children are neither 'innocent savages' nor 'growing flowers'. Such metaphors are inadequate, and in particular, the importance of the social dimension in psychology is increasingly recognized (Mead, 1934; Vygotsky, 1962; Donaldson, 1978), as it is in educational theory (Meighan, 1986). Language is a social acquisition, and with language comes thought and a view of the world (Sapir, 1949; Vygotsky, 1962). Thus this progressive view of childhood is inadequate, for children's nature and development cannot be isolated from the impact of social interchange.

The over-romanticism of the progressive educator ideology extends further, leading to a disparity between rhetoric and practice in education. Research shows that practical activity in learning mathematics is far less common than the progressive educator orthodoxy might lead us to suppose. For example, over half of a representative sample of 11 year olds were found to use number apparatus *never or less than once a term* in 1982 (Assessment of Performance Unit, 1985). Desforges and Cockburn (1987) found that the practice of primary mathematics in the schools they studied was largely routine and mechanical, and did not succeed in stimulating autonomy and higher order thinking. Mathematics learning for most children consists of working through a commercially published mathematics text or scheme routinely, on an individual basis (Assessment of Performance Unit, 1985).[6]

Notes

1 As Cooper (1985) suggests, there is an element of self interest behind these aims (as well as a commitment to the mathematical old humanist values). For unless an appropriate number of students are recruited into the study of pure mathematics, the strength and resource basis of the group is threatened.

2 A key influence, unrepresentative of the association, was that of the progressive educator Caleb

Gattegno (Howson, 1982). During the completion of the report Gattegno was founding the rival association ATAM (Cooper, 1985).

3 Although mathematical investigation is implicit in earlier publications, including Polya (1945), I can find no earlier use of the term.

4 Examples of politicized statements are Stanfield-Potworowski (1988) and Lingard (1984).

5 The Mathematical Association's publications by G. Hatch and M. Bird, for example, epitomize the child-centredness of the progressive educator, being process and activity led, and focusing non-judgementally on children's creative productions.

6 Two tensions in the progressive educator ideology arise from the great disparity between (1) the romanticized ideology and rhetoric of progressive education and its practice, and (2) the safe haven of the classroom and social and economic reality, condemning many students to lives of drudgery and poverty. When these contradictions surface, progressive educators may become disillusioned and pessimistic. My thesis is that this can partly account for the 'deschooling' movement in education, represented by Holt (1964, 1972), Goodman (1962), Illich (1970) and Lister (1974). They emphasize the regimentation and denial of individuality of children in schools, anathema to the progressive educator ideology. They propose the abandonment or dramatic re-structuring of schooling, to properly cater for the individuality and interests of students, that is to fulfil the progressive educator ideals. Whilst the deschooling view contains some insights, it is both too pessimistic and over-deterministic. Many of the outcomes of schooling are valuable, and social conditions can be changed by political and educational action.

Of course, the 'deschooling' movement is more complex than the simplistic sketch given here (Meighan, 1986).

9

The Social Change Ideology of the Public Educators

1. The Public Educators

A. The Ideology of Relativistic Fallibilism

The primary elements of this ideology are as follows.

Philosophy of mathematics

The philosophy of mathematics of this ideology is social constructivism. As we have seen, this entails a view of mathematical knowledge as corrigible and quasi-empirical; the dissolution of strong subject boundaries; and the admission of social values and a socio-historical view of the subject, with mathematics seen as culture-bound and value-laden. It is a conceptual change view of knowledge (Confrey, 1981).

Epistemology

The overall epistemology of this position is fallibilist, and conceptual-change orientated (Toulmin, 1972; Pearce and Maynard, 1973), consistent with the philosophy of mathematics. Thus the epistemology recognizes that all knowledge is culture-bound, value-laden, interconnected and based on human activity and enquiry. Both the genesis and the justification of knowledge are understood to be social, being located in human agreement. In view of the social and political awareness of this ideology, it is a critical epistemological perspective, which sees that knowledge, ethics, and social, political and economic issues are all strongly inter-related. In particular, knowledge is seen to be the key to action and power, and not separated from reality.

Set of moral values

The moral values of this position are those of social justice, a synthesis of the separated and connected values. From the separated perspective comes a valuing of justice, rights, and a recognition of the importance of social, economic and political structures. From the connected perspective comes a respect for each individual's rights, feelings and sense-making, and a concern that all might live in society as in an ideal extended family. Underpinning these concerns are the principles of egalitarianism and the desire for caring social justice, which are based on three fundamental values: equality, freedom and fraternity (or fellowship). There are also two derived values: democratic participation (equality plus freedom) and humanitarianism (equality plus fraternity) (Lawton, 1988).

These values can be loosely identified with the political left. They can be traced at least as far back as the American and French revolutions. Thus the American Declaration of Independence begins with the assertion of equality and liberty as universal human rights.

> We hold these truths to be self evident, that all men are created equal, that they are endowed by their creator with certain inalienable rights; that among these are life, liberty, and the pursuit of happiness.
>
> (Ridgeway, 1948, page 576)

The triad was completed by the French revolutionaries who asserted the rights to 'liberté, egalité and fraternité', thus adding fraternity to liberty and equality.

Theory of the child

The theory of childhood is that of individuals who are born equal, with equal rights and in general, equal gifts and potential. These individuals develop within a social matrix and are profoundly influenced by the surrounding culture and social structures, especially class. Children are 'clay to be moulded' by the powerful impact of social forces and cultures. However this over-emphasizes the malleability of individuals at the expense of their inner forces of development. For children and other persons are seen as active and enquiring makers of meaning and knowledge. Language and social interaction play a central role in the acquisition and creation of knowledge in childhood. The psychological theories which describe this position are those of Vygotsky (1962) and Leont'ev (1978), among others, namely that psychological development, language and social activity are all essentially inter-related. This is the 'social constructionist' view that the child's knowledge and meanings are internal constructions resulting from social interactions and the 'negotiation of meaning' (Pollard, 1987).

Theory of society

The theory sees society as divided and structured by relations of power, culture, status and the distribution of wealth, and acknowledges social inequalities in terms of rights, life chances, and freedom for the pursuit of happiness. This view sees the masses as disempowered, without the knowledge to assert their rights as citizens in a democratic society, and without the skills to win a good place in the employment market, with the remuneration it brings. The theory of society is also dynamic, for it sees that social development and change are needed to achieve social justice for all. It is concerned with the difference between social reality and the social ideals, and is a view committed to change in order to achieve its social values.

This perspective also sees the masses as a 'sleeping giant' which can be awakened by education to assert its just rights. Unless individuals have their consciousness raised to question the *status quo*, the forces of the 'hidden curriculum' in schooling and society will tend to reproduce their economic and cultural class identity (Giroux, 1983).

Educational aims

The goal of this position is the fulfilment of the individual's potential within the context of society. Thus the aim is the empowerment and liberation of the individual through education to play an active role in making his or her own destiny, and to initiate and participate in social growth and change. Three interwoven constituent aims can be distinguished.

1. The full empowerment of an individual through education, providing 'tools for thought' enabling that person to take control of their life, and to participate fully and critically in a democratic society.
2. The diffusion of education for all, throughout society, in keeping with the egalitarian principles of social justice.
3. Education for social change — the move towards a more just society (and world) in terms of the distribution of wealth, power and opportunity (by means of 1 and 2).

Overall, this ideology is socially orientated, with its epistemology based on social construction, and its ethics based on social justice. Since it is relativistic, in all domains it acknowledges the validity of alternative perspectives.

B. The Public Educators as Relativistic Fallibilists

The Relativistic Fallibilist ideology is that of the public educators, who represent a radical reforming tradition, concerned with democracy and social equity (Williams, 1961). Their aim is 'education for all', to empower the working classes, and others, to participate in the democratic institutions of society, and to share in the prosperity of modern industrial society. For education, this aim means to develop the faculties of independent critical thought, enabling students to question received knowledge with confidence, whatever the authority of its source, and to accept only that which can be rationally justified. Two outcomes of this aim are that received knowledge is no longer accepted as abolute, and that 'high' culture is no longer valued more than popular or the 'people's' culture. This extends to the distinction between practical or culturally embedded knowledge and academic knowledge. Whilst the latter is valued for its theoretical structures, this is not at the expense of the former which is valued as being part of people's culture and conditions of life.

The origins of the public educator ideology

The roots of the public educator and progressive educator traditions are intertwined. Thus the provision of elementary education for all in the 1870 Education Reform Act, represented a victory for both groups (in alliance with the industrial trainers). However, not all shared the public educator aim for this Act to politically empower the masses. Rather, amongst other things, it was hoped that it would moderate their exercise of power, following the enfranchisement of most urban workers in 1867. In the contemporary words of Robert Lowe:

> From the moment you entrust the masses with power, their education becomes an imperative necessity . . . You have placed the government of this country in the hands of the masses and you must therefore give them an education.
>
> (Dawson and Wall, 1969, page 28)

There were moves to bring universal education to the masses independently of the progressive educator tradition. In the late eighteenth century thinkers like Malthus and Bentham argued that a state education for all was necessary to remedy the ignorance and conditions of the poor (Dawson and Wall, 1969).

An early Victorian movement the 'science of common things', related science education to the everyday lives and experiences of the people. The 'engineer of this reform' was Henry Moseley, whose aims have much in common with the public educator perspective. He argued that 'to give . . . a child the mechanical power to read without teaching it to comprehend the language of books' (Layton, 1973, page 86) would not be empowering. In considering the components of a suitable curriculum , he argued that

> arithmetic, if looked upon as the logic of the people and developed with relevance to the intellectual culture of the working class child, was . . . an essential ingredient; but no branch of secular instruction was likely to be more effective in elevating the character of the labouring man than a knowledge of those principles of science which had an application to his welfare and future occupations. Armed in this way, the child had a resource of immense value for his future struggle with the material elements of existence. He would be equipped to avoid the degradation of mindless labour.
>
> (Layton, 1973, page 87)

This view of education as a means of enabling working people to have greater power over their lives and material conditions represents an early example of the public educator perspective.

Although Moseley was at first successful in securing funding for practical scientific equipment and resources for student experimentation in schools, science did not become a basic part of the elementary school tradition. Instead 'object lessons' became commonplace, in which the teacher exhibited a common object, such as a piece of coal, or its representation, such as a picture of a horse, and then elicited from pupils its description, definition and properties (Layton, 1973). This was a far cry from the 'science of common things' and the public educator curriculum.

Williams describes a further source of the public educators. This was a group, drawn from the working classes, who had an impact through adult education by introducing the elements of the 'the students' choice of subject, the relation of disciplines to actual contemporary living, and the parity of general discussion with expert instruction' Williams (1961, page 165).

An early twentieth century proponent of the public educator position is Dewey. He espoused three inter-related sets of beliefs pertaining to this view. These were, first, through Pragmatism, the view that all knowledge is tentative and fallible. In this respect, Dewey was far ahead of his time, for the 'perfectibility of knowledge' was the orthodoxy of his time. Second, Dewey believed in education for democracy, and in particular, in the importance of critical

> reflective thought [which] is the active, careful and persistent examination of any belief, or purported form of knowledge, in the light of the grounds that support it and the further conclusions towards which it tends.
>
> (Stenhouse, 1975, page 89)

Third, Dewey argues that the gap between the child's interests and experiences, and the different subjects of the curriculum, must be bridged. The child's experience and culture should provide the foundation for school learning which

> takes the child out of his familiar physical environment, hardly more than a square mile or so in area, into the wide world — yes, and even to the bounds of the solar system. His little span of personal memory and tradition is overlaid with the long centuries of the history of all peoples.
>
> (Golby, Greenwald and West, 1977, page 151)

Dewey believes that education should begin with children's interests and culture, and that it should then build outwards, towards the enquiring disciplines of the curriculum from this foundation.

Thus Dewey was a proponent of the public educator aims. Although simultaneously a key contributor to the progressive tradition in education, he was also a critic of it in its over romanticized form (Silberman, 1973), and was deeply committed to the values of the public educator ideology.

Some powerful statements of the public educator ideology have come from post-colonial countries outside Britain, concerned with social development. One example is the Tanzanian programme of 'Education for Self-Reliance' initiated by Julius Nyerere, with the following aims:

> to prepare people for their responsibilities as free workers and citizens in a free and democratic society, albeit a largely rural society. They have to be able to think for themselves, to make judgements on all the issues affecting them; they have to be able to interpret the decisions made through the democratic institutions of our society . . . The education provided must therefore encourage the development in each citizen of . . . an enquiring mind, an ability to learn from what others do.
>
> (Nyerere, quoted in Lister, 1974, page 97)

Paulo Freire has developed a comprehensive public educator ideology, with the following tenets. All knowledge is tentative, and cannot be divorced from the subjective knowing of individuals.

> [W]orld and consciousness are not statically opposed to each other, they are related to each other dialectically . . . the truth of one is to be gained through the other; truth is not given, it conquers itself and remakes itself. It is at once discovery and invention.
>
> (Freire, quoted in Lister, 1974, page 19)

According to Freire the aim of education is to achieve critical consciousness or 'conscientization' which is

> a permanent critical approach to reality in order to discover it and discover the myths that deceive us and help to maintain the oppressing dehumanizing structures.
>
> (Freire, quoted in Dale *et al.*, 1976, page 225)

This critical awareness is achieved through 'problem posing' education in which students actively choose the issues and objects of study, are co-enquirers with the teacher, and are free to question both the curriculum and pedagogy of schooling (Freire, 1972). This is opposed to 'banking' education, in which students are the passive and powerless recipients of knowledge. Freire developed his educational ideology (evidently under the influence of Marxism) through teaching literacy to peasants in Brazil with the aim of empowering them to engage with the social structures of society and to take control over their own lives.

An increasing number of other authors, world wide, have argued for elements of a public educator curriculum, including critical reflection on received knowledge and the nature of society (Postman and Weingartner, 1969; Giroux, 1983) and increased democracy and student control over the form and content of schooling (School of Barbiana, 1970; Hansen and Jensen, 1971).

In Britain, Williams (1961) proposed a public educator curriculum to give students mastery of the languages of English and mathematics; to introduce students to the culture — including popular culture — of the society around them, and practise in the critical reading of newspapers, magazines, propaganda and advertisements; to prepare them to participate in the democratic institutions of society; to engage in the methods of enquiry of the sciences and to understand the history and social effects of the sciences. In short 'a public education designed to express and create the values of an educated democracy and common culture.' (Williams, 1961, page 176).

Although many such projects never got beyond the planning stage, the Humanities Curriculum Project was successfully implemented with some of the public educator aims explicitly in mind.

> The pedagogical . . . aim of the project is to develop an understanding of social situations and human acts and of the controversial value issues that they raise.
>
> (Stenhouse, 1976, page 93)

This project used controversy and contained conflict (argument) as part of its methodology to raise students' critical awareness. The teacher was cast in the role of neutral chairperson, to avoid partisan teaching and the charge of indoctrination. A problem arose in the treatment of racism, where it was felt that neutrality was inadmissible. This lay behind the adoption of a deliberately committed

> Aim to educate for the elimination of racial tensions and ill-feeling within our society — which is and will be multi-racial — by undermining prejudice, by developing respect for varied traditions, and by encouraging mutual understanding, reasonableness and justice.
>
> (Stenhouse, 1976, page 131)

Although this project includes elements of the public educator aims, it does not fully address the social change and political aims. Elsewhere, educators have proposed public educator curricula addressing the full range of aims, for example, as 'urban education' (Hall, 1974; Raynor, 1974; Raynor and Harden, 1973; Zeldin, 1974). A clear statement of the aims and principles of one such project is provided by Zimmer.

> 1 There would be no more class teaching. Everything would be done through projects.
> 2 The projects should fulfil the needs of a working class which aims to achieve self-determination.
> 3 The principle of self-determination should also apply within the school, and in the choice of projects.
> 4 The school should not live in a world of its own, but should move back into society in those areas where change is needed.
> 5 The children should be given every chance of self-fulfilment. They should be happy, and their needs should be satisfied, as far as this is possible within a school context.
> 6 The children should not be cut off from society — otherwise they might apply their demand for self-realisation only to their own, limited environment. They should argue for their interests in the light of the interests of society as a whole, and they should negotiate and achieve their interests in a democratic way.
>
> (Lister, 1974, pages 125–126)

Thus Zimmer proposes that the life situation of the learner is the starting point of educational planning; knowledge acquisition is part of the projects; and social change is the ultimate aim of the curriculum. He suggests that the curriculum should be based on projects to help the pupil's self-development and self-reliance, with topics such as 'conflict at the factory' and 'the social welfare office'.

> Both the factory and the welfare office projects offer possibilities for parallel and follow-up projects. In the first one could study mathematics and its contribution in the production process, not regarding it as a matter of transmitting mathematical skills in isolation from their possible application in everyday life. People should learn how to analyse how values which lie outside mathematics can be transformed into mathematical symbols, rules, and processes. Conversely, people should be able to recognise the nature and the values of the things that lie behind the formal mathematical symbols. People need to be able to do this particularly in situations where the technological processes and the mathematical activities associated with them give an impression of objective rationality, while the interests which lie behind them remain hidden.
>
> (Lister, 1974, page 127)

Zimmer is thus able to see how that most recalcitrant of subjects, mathematics, has a full part to play in the achievement of the public educator aims.

Proposals for a public educator curriculum continue to this day. Jones (1989), for example, proposes a charter for education which constitutes a full and powerful statement of the position.

Overall, Williams (1961) argues that the public educators have been successful in securing the extension of education to all in modern British (and Western) society, as a right. This was achieved by expedient alliances with industrial trainers and others, resulting, most notably, in the expansionist Education Acts of 1870 and 1944. The public educator goal of 'education for all' in terms of universal free schooling has thus been achieved.

However, the public educators have not been successful in changing the content and transactional style of schooling to reflect their educational aims. Thus even the most successful of the projects described above, the Humanities Curriculum Project, was a short lived experiment. This means that equality of educational opportunity has not been achieved in Britain. A number of social groups, including female, ethnic minority and working class students, are served less well by the educational system, in terms of life chances, than male, white and middle class students (Meighan, 1986).

C. The Public Educator Group in Mathematics Education

The emergence of a public educator group specifically in mathematics education is relatively recent, due to the late emergence of fallibilist and social constructivist philosophies of mathematics. A milestone was reached in 1963–4, with the first publication of the bulk of Lakatos (1976), in journal articles. Only after this has a fallibilist view of mathematics gained some legitimacy and currency. Until this happened the full public educator ideology *with regard to mathematics* was not possible. However, from the outset, the educational implications of fallibilist philosophies of mathematics (and science) have been apparent to their originators (Lakatos, 1976; Kuhn, 1970) and elaborators (Hersh, 1979; Davis and Hersh, 1980; Agassi, 1980, 1982; Tymoczko, 1985).

An early explicit statement of a fallibilist view of mathematics in education, although rather subjectivist in flavour, is due to the Association of Teachers of Mathematics.

> Mathematics is made by men and has all the fallibility and uncertainty that this implies. It does not exist outside the human mind, and it takes its qualities from the minds of men who created it. Because mathematics is made by men and exists only in their minds, it must be made or re-made in the mind of each person who learns it. In this sense mathematics can only be learnt by being created.
>
> (Wheeler, 1967, page 2)

More recently, public educators have come to acknowledge that mathematics is a social construction which is culturally embedded and culture-bound. This applies to informal mathematical practices termed 'ethnomathematics' by D'Ambrosio (1985), representing both the everyday uses and sources of mathematics.

> Mathematics . . . is therefore conceived of as a cultural *product*, which has developed as a result of various activities . . . Counting . . . Locating . . . Measuring . . . Designing . . . Playing . . . Explaining . . . Mathematics as cultural knowledge, derives from humans engaging in these six universal activities in a sustained and conscious manner.
>
> (Bishop, 1988, 182–183)

This cultural embeddedness also applies to formal and academic mathematics and its applications, which is part of 'the social institution of mathematics' (Abraham and Bibby, 1988). Mathematical knowledge itself is explicitly acknowledged to be a social construction (Restivo, 1988; Abraham and Bibby, 1989).

In addition to views of the nature of mathematics the public educator group has views of the nature of mathematics education and its relationship with society. First concerning the aims of mathematics education

> It is of democratic importance, to the individual as well as to society at large, that any citizen is provided with instruments for understanding th[e] role of mathematics [in society]. Anyone not in possession of such instruments becomes a 'victim' of societal processes in which mathematics is a component. So, *the purpose of mathematics education should be* to enable students to realize, understand, judge, utilize and sometimes also perform the application of mathematics in society, in particular to situations which

> are of significance to their private, social and professional lives.
>
> (Niss, 1983, page 248)

In order to empower learners and give them greater control over their lives (Frankenstein and Powell, 1988), mathematics teaching should encourage student autonomy and student choice of problem areas for study.

> It is not only that there is value in having students actually move in both directions — from situations to posing and from posing to de-posing — but it is also worth designing a curriculum which exhibits the difficulties people had in making such moves on their own in the history of [the] discipline [of mathematics].
>
> (Brown, 1984, page 16)

> Mathematics education should involve some individual and group generation of mathematical problems . . . But we do not believe that this is sufficient . . . in addition we wish to include that part of a public educator perspective which emphasises *democratic citizenship* . . . [which] means not only having the skills to generate one's own mathematical problems, but also having some understanding of how and why other pervasive mathematical problems are generated and maintained along with their most important consequences for democracy and citizenship.
>
> (Abraham and Bibby, 1988, page 4)

Thus mathematics education should lead to both personal and social engagement on behalf of the learner (Mellin-Olsen, 1987). This involves critical thinking and 'conscientization' through mathematics, which is

> the crucial process by which the relationships between mathematics and society (especially the social institution of mathematics) are related to the personal development/situation of the pupils or students. The process involves the learner in a number of stages. Firstly, engagement with some form of organised mathematical activity. Secondly, objectification of some mathematical problem, i.e. the distancing of oneself from the problem so that it is clearly seen as the object of study. Thirdly, critical reflection upon the purpose and consequences of studying this problem in relation to wider values.
>
> (Abraham and Bibby, 1988, page 6)

However, conflict and controversy will result from

> *politicising education*. This situation is not all ideal. We do not like to think about little ones as people who can actually have different and even conflicting interests. And we like to practise cooperation and harmony, and not bring the nasties of the outside world into the classroom. But somehow we have to cope with conflicts when they are real. As educators most of us face situations which include dilemmas such as (competition versus

> cooperation) and (ideology x versus ideology y). We have to face rather than disregard such problems.
>
> (Mellin-Olsen, 1987, page 76)

The purpose of the public educator mathematics curriculum is to contribute to social change in the direction of greater social justice (Damerow *et al.*, 1986).

> Together, instructors and students can develop ways of overcoming the effects of racism, sexism and classism on instructional methodologies and student achievement in mathematics. These efforts, by necessity, will be directed towards structurally transforming the negative educational and societal conditions in which the dialogue exists.
>
> (Frankenstein and Powell, 1988, page 6)

Although the public educator group is the newest of the five ideological groups in mathematics education, it has growing support among mathematics educators. Beyond those quoted above, further contributors to the public educator perspective in mathematics education in Britain can be cited (for example: Burton, 1986; Ernest, 1986; Evans, 1988; Irvine, Miles, and Evans, 1979; Joseph, 1987; Lerman, 1988; Maxwell, 1984; Noss *et al.*, 1990).

Because of the novelty and controversiality of the public educator aims, few attempts at a public educator mathematics curriculum are published. Perhaps the two best known examples are the Mathematics in Society Project (Rogerson, undated, 1986) and the Radical Maths course for adult returners to education (Frankenstein, 1989). Further examples are discussed in Abraham and Bibby (1988) and Mellin-Olsen (1987).

D. The Public Educator Ideology of Mathematics Education

Aims of mathematics education

The aims of the public educator perspective are the development of democratic citizenship through critical thinking in mathematics. This involves empowering individuals to be confident solvers and posers of mathematical problems embedded in social contexts, and thus the understanding of the social institution of mathematics. At a deeper level, it involves assisting learners to become engaged in mathematical activity, which is embedded in the learner's social and political context (Mellin-Olsen, 1987). These aims stem from a desire to see mathematics education contribute to the furtherance of social justice for all in society.

Theory of school mathematical knowledge

School mathematical knowledge must reflect the nature of mathematics as a social construction: tentative, growing by means of human creation and decision-making, and connected with other realms of knowledge, culture and social life. School mathematics must not be seen as externally imposed knowledge from which students feel alienated. Instead it is to be embedded in student culture and the reality of their situation, engaging them and enabling them to appropriate it for themselves. In this way, knowledge of mathematics is to provide a way of seeing as well as thinking-tools (Mellin-Olsen, 1987). It provides an understanding of and power over both the abstract structures of knowledge and culture, and the mathematized institutions of social and political reality.

Theory of learning mathematics

The theory of mathematics learning of this perspective is that of the social construction of meaning, stemming from the theory of the social origins of thought of Vygotsky (1962) and the activity theory of Leont'ev (1978) and others. According to this theory, the child's knowledge and meaning are internalized 'social constructions' resulting from social interactions, the negotiation of meaning and engagement in 'activity'. This view has been explicitly proposed by supporters of the public educator position such as Bishop (1985), Cobb (1986) and Mellin-Olsen (1987), and is subsumed under social constructionism.

This theory sees children as needing to actively engage with mathematics, posing as well as solving problems, discussing the mathematics embedded in their own lives and environments (ethnomathematics) as well as broader social contexts. Learner (and teacher) conceptions and assumptions need to be articulated, confronted with other perspectives, and challenged, to allow the development of critical thinking. This leads to conflict, which is necessary for the accommodation and growth of new conceptions.

Theory of mathematical ability

Mathematical ability is viewed largely as a social construction, with the impact of the social context having a major role in individuals' development, and in particular on the manifestation of 'ability'. Individuals are understood, according to this perspective, to be far more comparable (equal) in characteristics and abilities at birth than after years of socialization in varying environments. Thus 'abilities' are conferred on students by their experiences and by the way that they are perceived and 'labelled' by others (Krutetskii, 1976; Meighan, 1986; Ruthven, 1987).

Theory of teaching mathematics

The theory of teaching includes a number of components:

1 genuine discussion, both student-student and student-teacher, since learning is the social construction of meaning;
2 cooperative groupwork, project-work and problem solving, for confidence, engagement and mastery;

3 autonomous projects, exploration, problem posing and investigative work, for creativity, student self-direction and engagement through personal relevance;
4 learner questioning of course contents, pedagogy and the modes of assessment used, for critical thinking; and
5 socially relevant materials, projects and topics, including race, gender and mathematics, for social engagement and empowerment.

In addition the teaching needs to be as democratic and open as the power asymmetries of the classroom allow, but with explicit recognition of this asymmetry. The teacher needs to play the role of neutral chairman or devil's advocate in discussions, but should also honestly reveal his or her views on controversial issues. The teacher also has a major responsibility to prepare students well for external assessments, as part of the real social context that surrounds the school situation. Above all, it is recognized that conflict has an essessential part to play, and cannot be smoothed away.

Theory of resources in mathematics education

This is based on the view that learning should be active, varied, socially engaged and self-regulating. Consequently the theory of resources has three main components:

1 the provision of a wide variety of practical resources to facilitate the varied and active teaching approaches;
2 the provision of authentic materials, such as newspapers, official statistics, and so on for socially relevant and socially engaged study and investigation;
3 the facilitation of student self-regulated control and access to learning resources.

Theory of assessment of mathematics learning

The theory of assessment is concerned to find measures of competence and positive achievement in mathematics without stereotyping students by ability, or presupposing an hierarchical model of mathematics. Above all, it values *fair* assessment of competence, regardless of gender, race, class or other social variables, as well as the reduction of competition. Thus a variety of forms of assessment can be used, including profiles or records of achievement, extended projects and examinations. Assessment tasks and outcomes should be open to pupil discussion, scrutiny and negotiation where appropriate (as in records of achievement), and student choice of topic for investigation and project-work. The content of assessment tasks, such as projects and examination questions, will include socially embedded mathematical issues, requiring critical thinking about the social role of mathematics.

This perspective is aware of the social importance of certification in mathematics, so it must provide a thorough preparation for examinations and external assessment. This is an essential part of the teacher's responsibility towards students, although it should be attained as a by-product of the 'social rationale' for mathematics (Mellin-Olsen, 1987). The possibility of a conflict of values here needs to be acknowledged.[1]

Theory of social diversity in mathematics education

The theory of social diversity reflects the underlying values and epistemology. Thus the mathematics curriculum should reflect its diverse historical, cultural and geographical locations and sources; its role in non-academic contexts (ethnomathematics) and its embeddedness in all aspects of the social and political organization of modern life (the social institution of mathematics). The mathematics curriculum must be 'friendly' to females, ethnic minorities, and other social groups, and positive action including anti-sexism and anti-racism are needed to enhance the mathematical education and social outlook of all, not merely to counter the problems of disadvantaged groups. The curriculum must be screened to remove obstacles to the success of all, such as language, stereotyping or narrow pedagogy which limit the full engagement, participation or development of all social segments. An overt discussion of the role of mathematics in the reproduction in social disadvantage is appropriate. Overall, social diversity is recognized, accommodated and celebrated as central to the nature of mathematics.

E. A Critical Evaluation of the Public Educator Perspective

Strengths

First of all, the public educator perspective, of those considered, most embodies social justice and the furtherance of democracy in its aims for mathematics education. It is the only ideology with the explicit aim of enhancing the self-realization of the learner both as an autonomous human being and as a member of society. It is also the only ideology fully committed to social justice, with regard to the social and political implications of providing 'mathematics for all', or better 'mathematics by all', especially for underprivileged social groups (Volmink, 1990). It promotes a vision of 'socialist mathematics education' teaching mathematics to all, for 'citizenship in a technological society' (Swetz, 1978, page 3). Thus, the public educator aims concern mathematics education based on democratic socialist principles and values.

This should be distinguished from mathematics education in a 'socialist society', described by Swetz, because of the contrast 'theoretical ideology *v.* practical realisation' (Howson, 1980). For apart from the fact that no socialist country has ever fully espoused the public educator aims, there is also the risk that in such countries the interests of the individual may be lost or submerged in an education driven by some notion of the 'collective good'.

Secondly, this is the only perspective to accommodate a fallibilist or social constructivist philosophy of mathematics, representing the leading edge of contemporary thought. Consequently the public educator mathematics curriculum reflects the nature of mathematics as a social institution, with all the powerful educational implications of this perspective. The roles of different races, countries and of women in creating mathematics are recognized, leading to the rejection of the myth of white European male ownership of mathematics. Also, history and the human context of mathematics become of central importance, leading to a less alienating and mystifying image of mathematics, and giving rise to one that is more humanistic and welcoming. The recognition of the fallibility of mathematics denies the centrality of the concept of student correctness or error in the mathematics, which is a powerful contributor to negative attitudes and mathephobia.

Overall, this perspective has strengths in terms of both its ethical and epistemological bases, and the translation of these into educational aims.

Weaknesses

The public educator perspective and aims suffer from a number of weaknesses, largely to do with the problems of implementation, but also due to a number of contradictions within the ideology.

Making mathematics education controversial

First of all, there is the problem of the controversiality of the public educator perspective and its implications for education. This has been recognized by the ICMI seminar on school mathematics in the 1990s in a discussion of the social role of mathematics which distinguishes two options, and possible negative outcomes.

> *Alternative 1* Mathematics is neutral, and is best taught in isolation from contentious social issues.
>
> . . .
>
> *Alternative 2* Since mathematics underpins both technology in all its manifest forms, and the policies that determine how it is used, its teaching should deliberately be related to these issues.
> Consequences: 1 It is very difficult to do. Indeed many — if not most — mathematics teachers will not see it as part of their job to touch on social and contentious issues.
>
> 2 Governments are likely to respond adversely. This has already happened in some countries which have attempted to include a 'social responsibility' component in the teaching of physics.
>
> (Howson and Wilson, 1986, page 5)

The public educator aims, and their underpinning values, represent a 'politicization' of mathematics education (Noddings, 1987). To spectators from the other four ideological groups, such aims are inconsistent with their own absolutist philosophies of mathematics, which view the discipline as unproblematic, neutral and value-free. These perspectives, therefore, deny that social and political values can enter into mathematics teaching purely for epistemological and educational reasons. Instead, the public educator approach, by deliberately treating contentious social and political issues, is at risk of being seen as an attempt to subvert mathematics education into a propagandist activity.[2] In the extreme case, it may be interpreted as an attempt by Marxists to subordinate mathematics education to a political ideology, with purely political (as opposed to educational) ends in mind. Such perceptions by parents, educational administrators or politicians, if sufficiently widespread, are likely to lead to intervention in the mathematics curriculum.

These considerations cannot be ignored by proponents of a public educator mathematics curriculum. The fact that implementation is likely to be met with controversy and opposition must be anticipated. Two strategies which have been proposed to reduce the risk of this are to offer such a course purely as an option (Abraham and Bibby, 1988), and to offer it outside the mainstream schooling to adult re-entrants to education (Frankenstein, 1989). These responses avoid confrontation by marginalizing the public educator approach.

This first set of problems raises the question: does any politicial system really want a public educator curriculum to educate its citizens to critically question its published statistics, and the mathematical assumptions and models underpinning its political decision making?

Introducing conflict into the classroom

The second problem area also concerns controversy and conflict, but within the classroom. The introduction of contentious social and political issues, and the encouragement of student questioning of subject matter, pedagogy and assessment will by design lead to conflict and controversy in the classroom. Beyond the problems described above, this may differ radically from the mode of teaching learners have experienced previously or elsewhere and may be unsettling and disturbing to learners. Controversy, conflict and rational argument are not only missing from much of educational practice, but also are alien to many of the cultural backgrounds of learners. Thus this aspect of the public educator approach could cause dissonance and crises for learners. This raises the question: how ethically justified is this conflictual approach, given the consequences for learners that can be anticipated?

The teachers has a professional responsibility in Britain to act *in loco parentis*, to provide personal, social and moral education as a tutor and counsellor, and by personal example. Thus, for example, a significant number of children from emotionally unstable backgrounds might derive a sense of security from the stability of their relationships with teachers. The question raised is will a public educator mathematics curriculum threaten the security of some children and hence lead to destructive as opposed to constructive outcomes? Naturally any answer must depend on the specifics of context and implementation. However, it may be that a public educator mathematics curriculum should not be fully implemented until the later years of schooling, and that for the younger learner controversy and conflict should be avoided.

On the other hand, conflict between young learners over answers to mathematical problems is a very effective learning strategy (Cobb, 1987, Yackel, 1987).[3]

Classroom propaganda

Thirdly, the admission of social, cultural and political issues into the mathematics curriculum opens the door to influences on or the overt manipulation of the mathematics curriculum by commercial and political groups. From the perspective of its own ideology the public educator aims for mathematics education are seen to be democratic, empowering and impartial. However this evaluation is unlikely to be shared by other ideological positions, who may feel that other sets of values are those that need to be promoted. Given the opportunity afforded by the incorporation of social and political issues into the mathematics curriculum, such groups, who are likely to be more powerful than the public educator grouping, may subvert the aims to their own ends.

An extreme case is provided in science, where a serious attempt has been made by Christian fundamentalist groups in the United States to replace or balance the teaching of the theory of evolution in schools with that of 'creation science', although it has no academic standing. In Britain, classroom materials produced for schools by some commercial groups, for example building societies, have been criticized for promoting the interests of their sector at the expense of offering learners an overall balanced view of the commercial factors involved (National Consumer Council, 1986). Thus direct political and commercial pressures on the school curriculum already exist. Such interest groups are likely to outweigh the public educator grouping in terms of both political power and financial resources. Thus there is the risk that a public educator mathematics curriculum will be subverted and exploited by political and commercial interest, to their own ends.[4]

Contradictions

Fourthly, there are a number of contradictions within, or likely to arise from a public educator mathematics curriculum. Some of these have already been discussed, such as *conflict versus stability*, *confidence versus threat*, and *critical thinking versus indoctrination*. But some powerful contradictions remain.

Personal empowerment versus examination success: Although referred to above, it must be recognized that this is a source of serious conflict. It parallels the distinction between the S-rationale and the I-rationale in education (Mellin-Olsen, 1987), and that between Relational and Instrumental understanding (Mellin-Olsen, 1981; Skemp, 1976). The use of examples, procedures, and strategies that are not directly applicable in the context of externally imposed assessment is likely to be attacked by some students, parents, and others within and without the education system. Furthermore, the public educator has the duty to address both these conflicting goals.

Ethnomathematics vs abstract mathematics: There is a conflict between the location of mathematics in the world of the student's experience, and the need to teach theoretical mathematics to provide the powerful thinking tools of abstract mathematics. This parallels the conflict between socially embedded and relevant applications of mathematics and academic mathematical structure and theory. A number of authors have pointed to the dangers of a restricted 'ghetto' curriculum (Dewey, 1966; Layton, 1973; Abraham and Bibby, 1988; Jones, 1989). Indeed Gramsci (1971) argues that the narrow cultural experience of working class children is an obstacle to the development of abstract and critical thought. The problem is to move from socially or concretely embedded mathematical situations to their theoretical content, without the loss of meaning and the switch into a new, disconnected realm of discourse. However, observations in Johnson (1989) suggest that these negative outcomes are common, even in the planned move from practical to formal mathematics within a single classroom.

There is no way to avoid these conflicts. They have to be recognized and addressed in any public educator curriculum.

2. A Critical Review of the Model of Ideologies

A tentative overall model of the ideologies underlying the mathematics curriculum in Britain has been presented in both philosophical and socially located terms. It is critically reviewed in this section, beginning with a consideration of its weaknesses.

A. Criticism of the Model

Arbitrarily joined philosophies, values and groups

The first criticism of the model is that of arbitrariness in the selection of the primary component types, their links, and their identification in each of the five ideologies. This has some foundation. The model is speculative and inter-disciplinary, drawing together elements of philosophy, psychology, sociology and history, both in and out of mathematics education. Whilst the constituent parts are well grounded in different theoretical disciplines, the overall synthesis is admittedly conjectural. Consequently no finality is claimed for the list of components in the model, which are joined together by associations of plausibility rather than logic.

Simplifying assumptions of the model

The model, of necessity, depends on many simplifying assumptions. It is assumed that a single ideology and interest group maintains its identity over the course of time treated, despite the large-scale changes in knowledge, society and education. Within each group different segments may form, band together in alliances, dissolve, or break away giving an overall picture of flux and change. It assumes that the five groups are represented both within and outside of mathematics education, and does not distinguish between divergent segments within each group. It also assumes that the five positions are discrete, and that individuals or sectors can be uniquely assigned to one of them. On the contrary, it may be that the aims of two or more of the positions are adopted by individuals or groups. Each of these represents a simplifying assumption. On the other hand, without such assumptions, no global model is possible.

The planned versus taught mathematics curriculum

The model concerns the ideologies of mathematics education, and does not consider the differences between the planned, implemented and learned mathematics curriculum. Recent research, both theoretical and empirical, has emphasized the gaps that exist between these three curriculum levels. The present model treats only the top level, the aims and ideology underpinning the planned curriculum in mathematics. Thus the question remains: which forces or factors intervene between these aims, intentions and ideologies and their implementation in the mathematics curriculum?

B. Strengths of the Model

Theoretically grounded model

Mathematics education has been criticized by a number of authors for being an atheoretical endeavour (see, for example, Bauersfeld, 1979). The present model combines a number of theoretical foundations. These include the philosophy of mathematics, theories of intellectual and ethical development, and the sociological-historical theory. By combining sets of ideas from these sources the model has the virtue of being theoretically well grounded.

The model accommodates complexity

By distinguishing five ideologies and interest groups, the model is able to accommodate some of the complexity of the history of the mathematics curriculum. It represents an advance on previous models, and due to its more refined characterization, is better able to accommodate the complexity of the ideologies and interests underpinning different sets of aims for the mathematics curriculum. It acknowledges that conflicts of aims and interests may lie behind different educational developments, and thus represents an improvement on accounts that assume consensus.

The applicability of the model

The model provides a critical tool for identifying the different aims and ideologies implicit in mathematics curriculum projects, reports and reforms. It is used this way, below. It should also be applicable outside mathematics to other areas of the school curriculum, since it is based on five interest groups whch transcend mathematics education.

Notes

1 A strategy for reconciling these conflicting aspects of the theory of assessment is, via alliances with others, to change the modes and content of external examinations in mathematics. This strategy was adopted by the ATM in developing an all coursework GCSE mathematics examination with the Southern Examining Group, although the aims reflected in early proposals for this examination were progressive, not public educator (Association of Teachers of Mathematics, 1986).

2 Indeed it is an attempt to subvert the implicit social aims of the other ideological groups, for they use education as a means to maintain the existing social hierarchy and to further their own interests, however hotly they would dispute this.

3 Opposition can be expected from progressive educators on this point, for they are concerned to shelter and protect children, and will ideologically oppose their exposure to conflict.

4 On the other hand, the increasing incidence of commercial and military biased materials or propaganda in schools means that a public educator approach is even more important.

10

Critical Review of Cockcroft and the National Curriculum

1. Introduction

A. Curriculum Theories

The model developed in this book represents one theoretical approach to the mathematics curriculum and the identification of its aims. It is multidisciplinary, resting on philosophy, sociology and history. In the literature, three types of approach can be distinguished, depending on which of these disciplines is the foundation.

First of all, there is the philosophical approach to the mathematics curriculum, employed by Confrey (1981), Lerman (1986) and Nickson (1981). This uses the philosophy of mathematics, and in particular, contrasting positions such as absolutism and fallibilism, as a basis for identifying the philosophy underlying the mathematics curriculum. Like the present approach, these authors recognize the significance of different philosophies of mathematics for its aims and pedagogy. However, the consideration of philosophical perspectives without locating them socially means that the interests served by aims are not identified.

Secondly, there is the sociological approach, used by Moon (1986) and especially, Cooper (1985). The underlying sociological model is that of competing social groups, with differing missions and interests, who form temporary alliances, irrespective of ideological differences, to achieve shared goals. This approach is strong in describing the social determinants of change, and the aims of competing groups.

Another sociological approach is that of the neo-marxists, who base their theories of education on the complex of relationships between culture, class and capital, derived from the work of Marx and others, such as Gramsci (1971) and Althusser (1971). Williams (1961) belongs to this group, as do other theorists including Apple (1979, 1982), Bowles and Gintis (1976), Gintis and Bowles (1980), and Giroux (1983). Their theories are beginning to be applied to the mathematics curriculum, in Mellin-Olsen (1987), Cooper (1989) and Noss (1989, 1989a). Such accounts offer powerful models of the relationship between schooling, society and power, beyond rhetoric and surface explanations. However, a common weakness is the lack of discussion of the nature of mathematical knowledge, necessary for a full account of the curriculum and its aims. A strand of thought which may compensate for this deficiency is that of Critical Theory (Marcuse, 1964; Carr and Kemmis, 1986), which is being applied to the mathematics curriculum (Skovsmose, 1985).

Thirdly, there is the historical approach to the mathematics curriculum, employed by Howson (1982, 1983) and Howson *et al.* (1981). This traces the history of innovations through key persons (Howson, 1982) or curriculum projects (Howson *et al.*, 1981; Howson, 1983). The former approach is individualistic, and risks losing sight of group ideology and philosophy, and the role of aims in serving the group interests. The latter approach is more relevant here, as it offers a model for classifying mathematics curriculum projects into five types (due to Keitel, 1975):

1. *New Math*, concerned largely with the introduction of modern mathematical content into the curriculum, pure or applied.
2. *Behaviourist*, based on behaviourist psychology, the analysis of content into behavioural objectives, and in some cases, the use of programmed instruction.
3. *Structuralist*, based on the psychological acquisition of the structures and processes of mathematics, typified by the approaches of Bruner and Dienes.
4. *Formative*, based on the psychological structures of personal development (e.g. Piaget's theory).
5. *Integrated-environmentalist*, an approach using a multi-disciplinary context, and using the environment both as a resource and motivating factor.

The scheme is descriptive, and is not based on a single, overall epistemological theory. Instead it draws upon the most distinctive feature or underlying theory of each project. Thus the project types are characterized by the theory of school mathematical knowledge (1); the theory of learning (3, 4); both (2); or a mixture of epistemology, and theories of learning and resources (5). This lack of systematic basis is a weakness, in that projects are not accounted for comparably, and there is no underlying theoretical foundation, beyond history. (It could be seen as a strength, keeping close to historical phenomena, and aiding the identification of the influences of one project on another.) A further weakness is that the model is 'internalist', not taking into account the aims and interests of the social groups associated with projects, nor their location in the power structures of society.

The current model can be applied to the scheme, at least concerning the aims of projects. Thus the New Math developments represent the old humanist or technological pragmatist aims, according to the balance between pure and applied mathematics in the curriculum. The behaviourist style of curriculum combines industrial trainer, technological pragmatist and possibly old humanist aims, because of its structured mathematical content coupled with a training delivery-format. The structuralist, formative and integrated-environmentalist type all embody variants of the progressive educator aims because of their child-centredness and emphasis of the processes of learning and discovery, the child's development, or the child's experience of the environment, respectively. However, structural type curricula also include old humanist aims, because of the emphasis on mathematical structure.

None of the project types reflect the public educator aims, presumably because such a curriculum has not been implemented. Bishop (1988) proposes the notion of a 'cultural approach' as a sixth type of curriculum project, which comes closer to the public educator view, but does not fully embrace its social change aims.

This analysis, superficial as it is, suggests that most patterns of mathematics curriculum development identified by Howson, Keitel and Kilpatrick (1981) originate with progressive educators, who are a powerful group within the education community, although not in society in the large.

B. Methodological Considerations

The methodology employed below consists of the analysis of the stated aims of a curriculum document, coupled with a reconstruction of the aims implicit in the text. This resembles the activity of the literary critic or sociologist searching for (or rather constructing) the deep structures of meaning in a document. The main conceptual tool used is the model of educational ideologies. Viewed through this lens, the pedagogical assumptions and models in the documents provide the main evidence of their underpinning ideology.

A focus on aims means that attention is restricted to the planned, as opposed to the taught and learned mathematics curriculum. Consequently, the scope is narrower than in empirical studies, such as Robitaille and Garden (1989), which explore all three of these dimensions, because of their difference in practice.

2. The Aims of Official Reports on Mathematics Education

The focus of this section is the Cockcroft Report (1982), but to give an indication of its impact on the intellectual climate we also consider two reports which pre- and post-date it.

A. 'Mathematics 5–11' (Her Majesty's Inspectorate, 1979)

This report is the publication of Her Majesty's Inspectorate, who represent a central mechanism for externally imposed evaluation ('quality control') in education. Lawton (1984) describes their ideology as consisting of professionalism, with values concerning 'quality' in education, and a 'taste' for impressionistic evaluation.

The report discusses the purpose, aims and objectives of teaching mathematics to primary school children (5 pages) and the content and sequence of the mathematics curriculum (58 pages). The purpose of school mathematics is said to concern its broad utility, cultural aspects and the training of the mind. However, both cultural purposes and the training of the mind are more or less identified with broad utility, so the purpose is stated largely in technological pragmatist terms.

The stated aims of teaching mathematics emphasize affective aims and attitudes, creativity and appreciation, as well as the mastery of mathematical knowledge, skills and applications. Thus the over-riding emphasis is progressive educator. However there are some comments that are redolent of the industrial trainer view.

> [C]hildren . . . have to learn to be neat and tidy, for muddled working may produce a wrong answer. They need to be careful and they ought to learn to be discriminating.
>
> (Her Majesty's Inspectorate, 1979, page 4)

This contradicts the aim of fostering creativity, which often results in untidy representations of children's methods and divergent thought. An emphasis on care, neatness and tidiness is part of the social training aim of the industrial trainer.

There are twenty objectives which are mathematics-centred, and the bulk of the report concerns the content and sequence of the primary school mathematics curriculum. The emphasis is on the structure of the topics, purportedly showing the stages of children's development, but in fact representing increasing mathematical complexity and sophistication. There is also some treatment of pedagogical considerations, such as the use of apparatus. The aims implicit in this section are therefore those of the old humanist and the utilitarian ideologies, being centred on mathematical content in a hierarchical form. However the presentation of pedagogical concerns indicates a concession to the progressive educator aims.

In summary, what is striking about this document is the inconsistency between the overt aims (progressive educator) and rationale (utilitarian), and the implicit aims of this document (utilitarian and old humanist alliance, with touches of industrial trainer). The progressive educator aims appear to be little more than window dressing, for the emphasis is on a content-centred, hierarchically structured mathematics curriculum.

B. *'Mathematics Counts' (Cockcroft, 1982)*

In 1978 the Cockcroft Committee was set up as a government response to a perceived, widespread concern with the levels of numeracy of school leavers, reflected in its terms of reference:

> To consider the teaching of mathematics in primary and secondary schools in England and Wales, with particular regard to the mathematics required in further and higher education, employment and adult life generally, and to make recommendations.
>
> (Cockcroft, 1982, page ix)

The committee was mostly made up of education professionals, particularly mathematics educationists and teachers. The committee took its brief very seriously, and had numerous meetings, attended to a great deal of opinion and evidence, and commissioned research on the mathematical needs of employment and adult life, and a review of research on the mathematics curriculum, its teaching and learning and its social context (Howson, 1983; Bell *et al.*, 1983; Bishop and Nickson, 1983).

The main body of the report is divided into three parts, the first devoted to the aims of mathematics education and the mathematical needs of adult life, employment and further and higher education (pages 1–55). The emphasis is clearly utilitarian, representing the technological pragmatist aims. In the second section, one whole chapter is devoted to discussing and endorsing the calculator and the computer. This is consistent with the aims of technological pragmatism. However, the endorsement is not unqualified. The emphasis is on these resources as aids to enhance mathematics teaching and learning, and the study of computing for its own sake within mathematics, is condemned.

An important feature of the report is the specification of a broad range of learning outcomes, based on the review by Bell *et al.* (1983), distinguishing facts and skills, conceptual structures, and general strategies and appreciation. This provided the basis for the most often quoted recommendation of the report, namely that the teaching and learning of mathematics at all levels should include problem solving, discussion, investigational and practical work, as well as exposition and practice (paragraph 243). This provides both an endorsement and a rationale for the progressive educators' aims and practices. A further indication of the learner-centredness of the report is the attention devoted to children's attitudes to mathematics, to the improvement of the mathematics curriculum for low attaining pupils, and to the relatively lower attainment of girls. These concerns indicate the strong influence of the progressive educator aims. (There is no indication that equal opportunity issues arise from a public educator perspective.)

Two critical aspects of the report are (1) a critique of 'top down' mathematics curriculum developments, and (2) a critique of mathematics examinations for 16 year olds, also condemned for being 'top down', although not in these words. These critiques amount to an overt attack on the old humanist domination of mathematics curricula and examinations at secondary schools. More implicitly, they amount to an attack on utilitarian ideologies, which use schooling and examinations to order children hierarchically, in preparation for work. In addition, the explicit rejection of rote learning and authoritarian teaching represent a rejection of the industrial trainer aims by the Cockcroft Committee. An anomaly, therefore, is the specification of a 'Foundation List' of basic mathematical skills, serving industrial trainer aims. This is probably best interpreted in terms of the model of mathematical ability assumed by the report. A basic skills curriculum thus serves the mathematically 'least able'.

Elements relating to public educator aims occur. Multicultural issues are treated briefly, and gender issues are treated more substantially. However the gender issues are relegated to an appendix, which marginalizes them (Berrill, 1982). The report is also criticized for making implicit assumptions about the nature of 'mathematical ability': that it is a fixed or abiding characteristic of children, and that it is unitary, susceptible to measurement along a single dimension (Goldstein and Wolf, 1983; Ruthven, 1986).

These assumptions amount to a rejection of the public educator model of childhood. Other key issues such as the social or fallible nature of mathematics, and the aim of teaching for critical awareness are neglected.

Overall, the Cockcroft Report (1982) can be seen as embodying the progressive educator aims together with the technological pragmatist aims. The aims of the other perspectives are rejected, except where they overlap with these two 'progressive' ideologies.

The report was influential, and its wide acceptance is partly due to the moderate and well founded basis of its progressive educator aims (research based). However, more significant is the pairing of technological pragmatist and progressive educator aims. This is consistent with the progressive 'new vocationalism' becoming widespread in education, especially post–16, typified by the work of the Further Education Unit (Hodkinson, 1989).

The report influenced educational thought and policy (Ernest, 1989). For example, the School Examinations Council (chaired by Cockcroft), was instituted. This replaced traditional assessments at 16+ by the GCSE Examination, addressing many of the criticisms of the Cockcroft Report. It shifted assessment from norm- to criterion-referencing, and incorporated novel oral and coursework modes. The impact of Cockcroft (1982) is shown by the inclusion of mathematical projects and investigations for assessment (School Examinations Council, 1985), reflecting the progressive educator aims.

C. *'Mathematics from 5 to 16' (Her Majesty's Inspectorate, 1985)*

Half of this document concerns the aims and objectives of mathematics teaching, with the remainder devoted to criteria for the choice of content and principles underpinning pedagogy and assessment. The stated aims concern the uses of mathematics (as a language and a tool), appreciation of mathematical relationships, in depth study, creativity in mathematics, and most of all, personal qualities (working systematically, independently, cooperatively and developing confidence). The objectives expand the analysis of learning outcomes given in Cockcroft (1982). Thus the progressive educator aims dominate this part of the document.

The criteria for the selection of content give weight to the technological pragmatist, and to a lesser extent, the old humanist aims, in addition to those of the progressive educator. The remainder of the document is broadly in keeping with the progressive educator aims, both in terms of teaching and assessment strategies.

Overall, the document is representative of the progressive educator aims, and the technological pragmatist aims to a lesser extent. In particular, computers and calculators are strongly endorsed. One overall indicator is the treatment of the school mathematics curriculum over the whole period of compulsory schooling (ages 5 to 16 years) as a unity. The continuity of learner development is given more weight than the organizational division of schooling, signifying a learner-centred, progressive view of the curriculum.

A summary of the responses of the education community to the document was published (Her Majesty's Inspectorate, 1987). A repeated criticism was the failure to treat the nature of mathematics.

> We concede that more could have been said explicitly on this matter... although the question 'What is mathematics?' is addressed implicitly throughout the document. It is also worth noting that the relevant section in *Mathematics Counts* has gained wide acceptance.
>
> (Her Majesty's Inspectorate, 1987, page 2)

In fact neither document (Cockcroft, 1982; Her Majesty's Inspectorate, 1985) answers the quoted question. The utility and personal uses of mathematics are discussed in Cockcroft (1982), but no discussion distinguishes or asserts either absolutist or fallibilist views of the nature of mathematics. My conclusion is that an absolutist view of mathematics is assumed, and that any philosophical debate is regarded as irrelevant to school mathematics. This is consistent with the progressive educator philosophy of mathematics (it is also consistent with all but the public educator ideologies).

A further criticism reported was the lack of attention to cultural diversity and equal opportunities.

D. Trends in the Official Publications, 1979–1987

The documents give an indication of the changing aims and perspectives on mathematics education of one sector of the establishment, most notably Her Majesty's Inspectorate. However, Lawton distinguishes three groups within the central authority in education, each with different beliefs, values and tastes:

> 1 The politicos (ministers, political advisers, etc);
> 2 bureaucrats (DES officials);
> 3 the professionals (HMI).
>
> (Lawton, 1984, page 16)

Consequently, it should not be assumed that there is a single unified establishment view. Two shifts in Her Majesty's Inspectorate thinking can be discerned over the period. The first is a shift from the old humanist/technological pragmatist perspectives and aims to those of the progressive educator. This is quite marked, with the emphasis shifting from the structure and content of the mathematics curriculum (Her Majesty's Inspectorate, 1979), to the development and realization of the child's individual potential through mathematical activity (Her Majesty's Inspectorate, 1985). One feature remains constant: the social utility of mathematics education is stressed throughout the period. The second shift concerns information technology. In 1979 there is no mention of the import of electronic calculators and computers, whilst by 1985 this receives major separate emphasis. Thus there is the addition of the pro-technology aspect of the technological pragmatist aims. This need not be seen as a revision of aims, but as a reflection of social and educational change.

One feature thrown up by the comparison is the lack of depth in Her Majesty's Inspectorate (1979, 1985, 1987) compared with Cockcroft (1982). The former documents consist of unjustified directives based on unanalyzed assumptions, with little or no theoretical basis. For all its flaws, Cockcroft (1982) offers an analysis of a number of previously unquestioned assumptions, includes novel facts, and is founded on a substantial body of theory and research.

3. The National Curriculum in Mathematics

The National Curriculum is part of the most far-reaching change in education in England and Wales for the past two decades, affecting all state schooling for 5 to 16 year olds. The government has taken direct control of education and is prescribing both the content and assessment of the school curriculum. At the same time, schooling at the regional, local and institutional level is being fundamentally reorganized. These are radical changes, and only in retrospect will their full significance become clear. Only then will we be able to evaluate the implemented changes in schooling and their outcomes. Meanwhile, we can critically examine the planned developments. The goal of this section is to provide a critique of the aims, ideologies and conflicts underlying the mathematics component of the National Curriculum.

A. The General Context

The Industrial Trainer Ideology and Interests

The National Curriculum needs to be seen first of all, within its national political context. The industrial trainers, represented by the Prime Minister Mrs. Thatcher, have been in power in Britain since 1979, in alliance with others of the political right. The ideology of this group includes a strictly hierarchical view of society, a moral vision demanding strict authoritarian regulation of the individual, coupled with a social philosophy based on the metaphor of the market place and 'consumer choice'. A key element underpinning this ideology is self-interest. The group represents upwardly mobile elements of society, the *petit bourgeoisie*, small and self-made business people, and ultimately, the *nouveaux riches*. This is expressed in policies which redistribute wealth upwards, rewarding the upwardly mobile in society.

During their period of office, the Thatcher government has implemented a range of policies concerning industry, commerce and social services based on the market place metaphor. The market place is discussed utilizing a rhetoric of 'freedom' and 'consumer choice'. This metaphor focuses on the disposal of income or wealth. However its deep structure is to invalidate discussion of the generation and distribution of wealth, and to mask a deep shift in values

The market place metaphor presupposes that all individuals have the income or wealth appropriate to their social station, and denies the validity of discussing or questioning this distribution. For the central concerns are seller competition and buyer income-disposal. It also represents a shift in values, from those of the welfare state, in which all individuals have an equal right to have their needs met, with the underlying values of equality, cooperation and caring. In the market place egalitarianism and sentiment are repudiated, and financial considerations alone rule. The values are those of competition and individual self-interest, with the better-off having no responsibility to care for others. An inevitable outcome is the erosion of equality and the hierarchical ordering of individuals into a 'pecking order' in terms of opportunity and wealth, and serves to reproduce the social distribution of wealth. Overall, the market place metaphor masks the naked self-interest of the groups with power and wealth in their favour (Bash and Coulby, 1989).

The Market Place and Social Policy[1]

In terms of social policy, the market place metaphor leads to the commodification of services. All goods, utilities or services are commodities 'manufactured' by 'workers', to be bought and sold in the market place. The commodities are subject only to minimal quality regulation, and consumer choice is the final arbiter of value. Market forces and competition ensure that only the fittest survive.

This metaphor leads to the rolling back of state and local government involvement, to allow the market to operate unchecked. Extreme free marketeers such as Hayek, Friedman and Libertarians support unchecked market forces (Lawton, 1988). However, the moral dogma and self-interest of the industrial trainer group cannot allow this. For certain social groups, such as teachers, educationists and other professionals (Aleksander, 1988), oppose or threaten the values of the market and the upward redistribution of wealth which it conceals. Thus the imposition of centralized control is also necessary, to coerce individuals and institutions impeding the interests of the industrial trainers to participate and conform to the values of the market. This regulation ensures that the services or 'commodities' offered to the market by professional groups meet minimum standards and are properly priced. Overall, two sets of contradictory forces are at work (Bash and Coulby, 1989).

B. The National Curriculum

This is the political context of the Educational Reform Act of 1988, the legal instrument for the reorganization of education, including the imposition of a National Curriculum. The two contradictory forces are also at work in educational policies (Maw, 1988). Both serve to shatter any semblance of equality between schools and equality of opportunities for students. Consequently, both aid the formation of a hierarchy and 'pecking order' of schools, and hence among students.

Rolling back state involvement Consumer choice and power is promoted by encouraging diversity (opting out, City Technology Colleges, discretionary charging and private schools), consumer power (increased parent and local business community representation on governing bodies), and consumer information (publication of test and exam results). Deregulation is at work in schools' new freedom to opt out of state control, in the Local Management of Schools, and in the independence of polytechnics. Much of this rolling back of state involvement is in fact the abolition of regional accountability and control by local education authorities. Overall, this encourages differences between schools.

Imposition of central control Centralized control is imposed differentially, with private education trusted to be self-regulating, subject only to the market. State education is subjected to strict central regulation, to permit market forces to work. This is seen in the imposition of tight conditions of service on teachers, and the imposition of the National Curriculum. The teachers are subject to strictly regulated conditions of service, as befits workers producing goods for the market place, to ensure productivity and the delivery of products of at least minimum standards. This is the 'proletarianisation of teachers' (Brown, 1988; Scott-Hodgetts, 1988). The National Curriculum imposes quality control and consumer labelling of educational products, like manufactured foodstuffs. Traditional school subjects correspond to recognized ingredients. Assessment results in the differential labelling of student and school achievement, allowing customers (parents) to choose schools according to their market value and their own means. This is the commodification of education (Chitty, 1987). The purpose is to emphasize and amplify differences between schools. All the new policies erode whatever equality of provision exists, and aid the formation of a hierarchy of schools, according to market forces. This hierarchy mirrors the stratification of society, which it serves to reproduce.

Constraints on Mathematics in the National Curriculum

Within this context, the National Curriculum for mathematics is limited, by the imposition of severe constraints (Department of Education and Science, 1987, 1988a):

1. Traditional subject boundaries, contrary to modern curriculum thinking and primary school practice (for example, in Her Majesty's Inspectorate, 1977).
2. A single fixed assessment model presupposing a unique hierarchical structure for subjects. (This carries with it assumptions about the fixity of social stratification, individual ability, as well as a disregard for cultural differences and needs.)
3. An assessment-driven curriculum, requiring the greatest degree of definition for core subjects (mathematics, English and science) in terms of a hierarchy of objectives specified as discrete items of knowledge and skill.
4. A very short timescale for development and implementation.
5. Severely limited terms of reference for the National Curriculum Working Groups restricting them to formulating clearly specified objectives and programmes of study.

Prior to the commencement of the design of the National Curriculum in mathematics, the industrial trainers have determined the form that the management and organization of schooling shall take, and have placed severe constraints on the nature of the school curriculum by making sure it is assessment driven. The only concession is that the curriculum model admits more sophisticated content than the industrial trainer minimum basic skills view.[2]

C. The National Curriculum in Mathematics

In Summer 1987 a Mathematics Working Group for the National Curriculum was constituted. It was made up of nine mathematics educationists, three head teachers, four educational administrators, two academics, one industrialist and one member of the New Right, although not all served for the full term. On 21 August 1987 the Secretary of State for Education, K. Baker informed the chair of the group of their report dates (30 November 1987 and 30 June 1988) and task: to design an assessment-driven mathematics curriculum for the age range 5–16 years, specified in terms of discrete items of knowledge and skill. Severe limits were imposed on what the group could discuss, permitting an initial consideration of 'objectives and the contribution of mathematics to the overall school curriculum', before focusing on 'clearly specified objectives' and a 'programme of study' (Department of Education and Science, 1988, pages 93–94).

On 7 September 1987 one of the mathematics advisers in the group circulated a key document including the following statement to the group.

> *Global statement*
> The mathematics curriculum is concerned with:
> (a) tactics (facts, skills, concepts)
> (b) strategies (experimenting, testing, proving , generalising, . . .)
> (c) pupil morale (pupil work habits, pupil attitudes)
>
> The treatment within the NMC [National Mathematics Curriculum] of what is contained in this statement is, to me, the most important issue facing the working group. There are many possible scenarios, but I will confine myself to two:
>
> **Scenario A**. The NMC deals relatively thoroughly with mathematical facts, skills and concepts (what I am calling the tactics of mathematics). But then it makes only superficial references to strategies and pupil morale, perhaps devoting, say, 5% of the statement of these aspects.
>
> **Scenario B**. The NMC starts off with a clear statement on pupil morale. This is followed by a detailed statement on general strategies which are the essence of mathematical thinking. Finally, it deals with mathematical tactics. Within this scenario it is strongly emphasised that pupil morale is paramount, followed by mathematical strategies and then mathematical tactics (concepts, skills and facts) — strictly in that decreasing

> order of importance. In simplistic terms this is based on the obvious principle: forgetting a fact (such as 7x8 = 56) can be remedied in a few seconds, but bad work habits and poor attitudes are extremely difficult to correct and may, indeed, be irreversible.
>
> (Mayhew, 1987, page 7)

This is a clear statement of both the mathematics-centred view of the old humanists (and technological pragmatists) (A), and the child-centred progressive view (B), but strongly endorsing the latter. This statement clearly sets out the ideological boundaries of the struggle to come. It excludes two views, those of the industrial trainers and the public educators.

The statement begins by assuming the definition of the outcomes of mathematics learning identified by Bell *et al.* (1983) and endorsed by Cockcroft (1982), as reformulated by Her Majesty's Inspectorate (1985). This replaces the notion of 'the appreciation of mathematics', which leaves open the possibility of an awareness of the social role and institution of mathematics, and hence public educator aims, by 'pupil morale' with its child-centred progressive connotations.

The internal struggle between the old humanists/technological pragmatists and the progressives was apparently won by the latter. The Interim Report (Department of Education and Science, 1987) was a clear statement of the progressive view of mathematics, following scenario B above. Simultaneously the sole member of the New Right on the working group (S. Prais) sent a note of dissent to K. Baker, complaining at length that the group were largely 'sold on progressive child-centred maths' instead of concentrating on essential basic skills (Prais, 1987, 1987a; Gow, 1988). He resigned shortly afterwards.

Baker dissociated himself from the report by not having his critical letter of acceptance published with it (unlike all the other Interim Reports), and the chair of the working group was replaced. His letter expressed his 'disappointment' and redirected them to 'deliver age-related targets' with 'urgency' and to make 'faster progress' and required the new chair to report progress at the end of February 1988 (Department of Education and Science, 1988, pages 99–100).

He also attacked the progressive philosophy of the report and instructed the group to give the greatest priority and emphasis to attainment targets in number. He pointed out 'the risks . . . which calculators in the classroom offer' and stressed the importance of pupil proficiency in computation and the 'more traditional paper and pencil practice of important skills and techniques'. This attack embodies the back-to-basics view of the industrial trainers.

On 30 June 1988 the final report of the Mathematics Working Group (Department of Education and Science, 1988) was sent to Baker. Its proposals represent a compromise between the old humanist, technological pragmatist and the progressive educator views of mathematics, as is reflected in the brief discussion of the nature of mathematics (pages 3–4). The old humanist and technological pragmatist impact on the proposals is evident from the range, depth and spread of mathematics content in the proposals. Mathematical content makes up two of the three profile components and is given a 60 per cent weighting. The impact of the progressives is shown in the third component comprisng mathematical processes and personal qualities, given a 40 per cent weighting. The technological pragmatist influence is shown in the name given to this components: 'Practical applications of mathematics', and in the attention given to technology. The public educator views are nowhere to be seen. Although lip-service is paid to social issues concerning equal opportunities, the need for multi-cultural mathematics is repudiated. Likewise the back-to-basics view of the industrial trainers is repudiated.

Baker accepted the report, because it contained the objectives for assessing mathematics that he needed. That was the specification of mathematics content in terms of twelve broad attainment targets each defined at ten age-related levels (and programmes of study based on this). This represented the old humanist/technological pragmatist part of the proposals. He rejected the third profile component, particularly personal qualities, which represented the heart of the progressive part of the proposals. He allowed for a token representation of the processes of mathematics, if they could be incorporated with the content targets under a technological pragmatist banner (Department of Education and Science, 1988, pages ii–iii).

Baker instructed the National Curriculum Council to prepare draft orders on the basis of these recommendations, and the industrial trainer view that 'pencil and paper methods for long division and long multiplication' needed to be included (National Curriculum Council, 1988, page 92). The Council carried out its instructions and published its report in December 1988 (National Curriculum Council, 1988). The part of the proposal reflecting the progressives view was marginal, comprising two out of fourteen attainment targets applying mathematical processes to the content areas defined by the two profile components. Even this concession is phrased in technological pragmatist rather than progressive educator language. The old humanists, however, managed to have an impact on the proposals, through the addition of extra mathematics content at the higher levels of the attainment targets.[3]

Scenario A (above) had been enacted, representing the triumph of the old humanist and technological pragmatist alliance, with marginal influences of the progressive educators, but within a framework dominated by the industrial trainers. This is reflected in the final form of the National Curriculum in mathematics (Department of Education and Science, 1989). This can be said to embody the aims of three groups. The curriculum represents a course of study of increasing abstraction and complexity, providing a route for future mathematicians, and meeting the old humanist aims. It is a technologically orientated but assessment driven curriculum, meeting the technological pragmatist aims. It is one component of a market-place approach to schooling, and an assessment-driven hierarchical curriculum with traces of the progressive educators expunged, meeting some of the aims of the industrial trainers. The overall range of content exceeds the basic skills deemed necessary by the industrial trainers. However, the underlying assessment framework ensures that below average attaining students will study little more than the basics, in keeping with industrial trainer aims. Overall, the outcome is largely one of victory for the industrial trainer aims and interests, together with their allies, despite the progressive climate of professional opinion since Cockcroft (1982).

4. Conclusion

Over the period of time considered, official publications on the mathematics curriculum shifted from the hierarchical content-driven view of the old humanist and utilitarian ideologies towards the progressive educator emphasis on the nature of the learner's mathematical experience. The National Curriculum in mathematics has reversed this trend, and negated the gains since Cockcroft (1982) (from a progressive educator perspective). Beyond this, the imposition of national testing at ages 7, 11 and 14 years and an assessment driven curriculum amounts to a reversal of the egalitarian 'comprehensivization' policy of the 1960s and 1970s. This has long been the goal of the industrial trainers and some old humanists (Cox and Dyson, 1969, 1969a, 1970; Cox and Boyson, 1975). An expected outcome of this shift is the hierarchical ordering of schools and pupils into a market-driven 'pecking order', eroding what equality of education provision exists.

The National Curriculum in mathematics provides an instructive case study of the powerful impact of social and political interests in the development of a curriculum. The professionals in the Mathematics Working Group tried to remain true to their largely progressive educator aims, even within the constraints first imposed on them. However, strong external pressures forced them to concede their position, and compromise with utilitarian and old humanist aims. There is nothing shameful in seeking compromise in the face of power. However, this meant that the reactionary central authority was able to unravel their compromise, extract the purely utilitarian and old humanist components, and make them serve industrial trainer purposes. Thus the professionals of the Mathematics Working Group were manipulated and exploited. Through such manipulations, the industrial trainers have wholly succeeded in imposing a market forces model on education, including its central feature: the imposition of an assessment driven curriculum, with traditional subject areas such as mathematics represented purely as an hierarchy of assessment objectives.

Not only have education professionals been used as unwitting agents in these anti-educational developments, but few have publicly dissented. In mathematics education a few critical voices are beginning to be heard, such as Scott-Hodgetts (1988), Ernest (1989e) Noss (1989, 1989a) and others in Noss *et al*, (1990).

Notes

1 The market-forces view of educational and social policy is widespread, for example in R. Pring (personal communication), J. Straw (quoted in Gow and Travis, 1988), and Chitty (1987).
2 Even this is under question, for Margaret Thatcher has stated that a basic core of Mathematics,

Science and English is enough for a national curriculum, reasserting the industrial trainer aims *Sunday Telegraph*, 15 April 1990; *Times Educational Supplement*, 20 April 1990. This is reinforced by strong indications that only these core subjects will be tested at age 7.

3 Additions of higher mathematical content were made to the upper parts of Attainment Targets, for example, to AT 7 at Level 7, AT 10 at Levels 9 and 10, AT 11 at Level 6, AT 12 at Levels 9 and 10, AT 13 at Level 10 and AT 14 at Level 10. In addition, the consultation process reported in National Curriculum Council (1988) attracted old humanist comments, for example the Mathematical Association's (1988a) response to the Final Report stated 'We wish . . . proof . . . [to] be included . . . Is there a case for the inclusion of matrices . . . ?' (page 5), and implicitly attacked the progressive philosophy 'AT 15 is of a different nature . . . [personal qualities] . . . are not specific to mathematics' (page 5).

11

Hierarchy in Mathematics, Learning, Ability and Society

1. Hierarchy in Mathematics

A theme of previous chapters is the assumption that mathematics has a unique fixed hierarchical structure. Analogues of this thesis include the assumptions that mathematics learning is best organized this way, that mathematical ability is structured this way, and that society has a more or less fixed hierarchical structure, which education needs to reflect. These are assumptions of profound social and educational significance, warranting a chapter to themselves.

A. Does Mathematics have a Unique Hierarchical Structure?

This question can be analyzed in two parts, concerning the existence and the uniqueness of a hierarchical structure for mathematics. Thus we have two subsidiary questions: does an overall hierarchical structure of mathematical knowledge exist? And if so, is this a unique and fixed hierarchical structure?

A hierarchy can be defined for any body of mathematical knowledge with an overall structure. Whether it is an axiomatic structure, based on axioms and rules of inference, or a definitional structure, based on primitive terms and further defined terms, then a hierarchy is definable, as follows. The primitive expressions of the hierarchy (axioms or primitive terms) comprise its lowest level (0). Any other expression E in the structure can be reached in some minimum number n of rule applications (rules of inference or definition) from an expression of level 0. This number n defines the level of expression E in the hierarchy.[1] So every expression is assigned to a unique level in the hierarchy. Thus any body of mathematical knowledge can be given a canonical hierarchical form provided that it constitutes a single mathematical system or structure, linked by inferential or definitional relationships.[2] Of these, inferential relationships are the most appropriate to consider, for they reflect the justificatory links between mathematical propositions and formulas, providing the structure of deductive axiomatic theories.

Using the distinction between the levels of formal, informal and social discourse of mathematics, we see that for an appropriate formal mathematical theory, a hierarchy can be defined. For a realm of informal mathematical inquiry, this may not be possible. For an axiomatic basis may not be fully specified, and the logical relationships between informal mathematical propositions may not be conclusively established. Thus in the following we will focus only on formal mathematical theories, or those informal mathematical theories which are ready for formalization. For otherwise the conditions for establishing a hierarchy may not be met.

We are now ready to consider the two questions. First of all: does an overall hierarchical structure of mathematical knowledge exist? We have seen that for formal mathematical theories, with a fixed set of axioms, there is a hierarchical structure. The choice of the axiom set, together with the specification of the rules of inference and the background formal language, determines a hierarchical mathematical theory. However, mathematics is made up of many different theories, many of which have many different axiomatic formulations. Axiomatic set theory, for example, has a number of quite distinct axiomatizations such as Zermelo-Fraenkel Set Theory and Godel-Bernays-von Neumann Set Theory (Kneebone, 1963). Beyond this, many mathematicians further vary the axiomatic set theory they are studying by appending further axioms (Jech, 1971; Maddy, 1984).

Consequently, there is no overall structure to formal mathematics, since it is made up of a myriad of different theories and theory formulations, each with its own structure and hierarchy. Furthermore, virtually every one of these axiomatic theories is incomplete, according to Godel's (1931) theorem. Thus there are truths of the theory which do not have a place in the deductive hierarchy. As we saw in earlier chapters, the attempts by some of the great mathematicians of this century to establish mathematical knowledge in a single foundational system whether logicist, formalist or intuitionist, all failed. Thus the results of meta-mathematics force us to recognize that mathematics is made up of a multiplicity of distinct theories, that these cannot be reduced to a single system, and that no one of these is adequate to capture all the truths even in its limited domain of application.

It follows that the question as to the existence of an overall mathematical hierarchy must be answered in the negative. This is irrevocable. However, in fairness, we should also consider a weaker question. Does a large and comprehensive informal structure of mathematics exist, even if it fails to satisfy the stringent criteria required to give an unambiguous structure to mathematics? Such a structure can be found in the *Elements* of Bourbaki (Kneebone, 1963). Bourbaki provides a systematic account of mathematics, beginning with set theory, and developing one after the other the major theories of pure, structural mathematics. Although the Bourbaki structure cannot be said to be complete (in the informal sense), for it leaves out the computational and recursive aspects of mathematics, it represents an informal codification of a substantial portion of mathematics. Does this provide an affirmative answer to the much weakened question? If we concede that it does, then the following caveats must be borne in mind:

1 a significant portion of mathematical knowledge is omitted;
2 the system is not formally well enough defined to allow a fixed hierarchy of mathematical knowledge to result;
3 the whole system depends on the assumption of classical set theory as the foundation of mathematics;
4 the whole system is culture-bound, reflecting mid-twentieth century structuralism.

Thus only in a very weak form can we assert that there is an overall structure to a significant part of mathematics.

The second question is as follows. Given the assumption that there is an overall structure to mathematical knowledge is it a unique and fixed structure on which a hierarchy can be based? This question again has two parts. The first is concerned with the uniqueness of the structure of mathematics. The second concerns the definability of a precise hierarchy in terms of this structure. We have already seen that this second part is untenable. Even if the structure provided by Bourbaki is conceded to be unique, it is informal and therefore does not suffice for the precise definition of a hierarchy. Thus in any strict sense, we can already assert that there is no unique hierarchy to mathematics.

But let us turn to the uniqueness of the structure of mathematics. This uniqueness would seem to depend on agreement as to the foundations of mathematics. Bourbaki assumes set theoretical foundations. Ignoring the differences between different set theories, can set theory be said to provide a unique, universally agreed foundation to mathematics? This question must be answered in the negative. We have already seen that the foundationist claims that mathematics rests on a unique foundation fail. At least two alternatives to the set theoretic foundations of mathematics exist. First of all, it has been claimed that Category Theory can provide an alternative foundation of mathematics, in place of set theory (Lawvere, 1966). This claim has not been fully justified, but it nevertheless constitutes a challenge to the uniqueness of set theoretic foundations. Indeed, there is a branch of category theory (Topos theory) to which both intuitionistic and classical logic can be reduced (Bell, 1981). Since axiomatic set theory is expressible in first order classical logic, it can be reduced to category theory.

Secondly, intuitionist logic provides foundations for mathematics. Although not all classical mathematics is expressible in terms of this basis, much of the intuitionist programme has been realized for analysis, by Bishop (1967) and others. Furthermore, intuitionistic logic accommodates combinatorial mathematics, unlike the set-theoretic foundations of classical mathematics. Thus on the basis of these two arguments, the claim that there is a unique structure to mathematics is refuted.

In fact, the history of mathematics teaches us the opposite lesson. Throughout its development mathematics changes through the fundamental restructuring of mathematical concepts, theories and knowledge (Lakatos, 1976). Thus although structures play a central role in organizing mathematical knowledge, they are multiple structures which form, dissolve and reform over the passage of time. There are no grounds for assuming that this process will ever cease, or for assuming that alternative theories and reformulations will ever be exhausted. Such a view is central to social constructivism, and to other philosophies of mathematics which acknowledge its historical basis. Thus not only is it untrue that at any one time mathematics can be described by a single unique hierarchical structure, but also over time whatever structures are present change and develop.

In refuting the claim that mathematics has a unique hierarchical structure, attention has been restricted to the logical, that is deductive structure of mathematical theories. As we have seen hierarchies can be defined in other ways, most notably, as hierarchies of terms and definitions. Whilst these are not nearly as significant in mathematics as deductive structures, the same arguments can be transposed to this realm. For the deductive structure of any theory carries with it a hierarchy of definitions, and almost as many definitional structures as deductive ones exist. Thus there is no unique hierarchy of definitions either. No further, global hierarchies are in use in mathematics. Within individual theories or domains some hierarchies certainly do exist, such as the Turing degrees (of unsolvability) in recursion theory (Bell and Machover, 1977). But these in no way structure even a significant fraction of mathematical knowledge. Thus it can be asserted unequivocally that mathematics does not have an overall hierarchical structure, and certainly not a unique one, even when the claim is interpreted generously and loosely.

Is mathematics a set of discrete knowledge components?

There is a further assumption concerning the nature and structure of mathematical knowledge which deserves examination because of its educational import. This is the assumption that mathematics can be analyzed into discrete knowledge components, the unstructured sum (or rather set) of which faithfully represents the discipline. This assumption requires that mathematical propositions are independent bearers of meaning and significance.

Distinguishing between the formal, informal and social discourses of mathematics, it is evident that this claim is best made for formal mathematics. For the other two domains presuppose meaning contexts, as will be argued below. Since structure is one of the characteristics of mathematical knowledge, this claim may also rest on the unwarranted assumption that there is a unique structure to mathematics. This may be needed so that when the discrete 'molecules' of knowledge are re-combined, a fixed and predetermined whole (the body of mathematical knowledge) results. We have disposed of this second assumption above. However, the pre-supposition that mathematical propositions are independent bearers of meaning and significance also fails. First of all, formal mathematical expressions derive their significance from the axiomatic theory or formal system in which they occur. Without this context they lose some of their significance, and the structure imposed by the theory collapses.

Secondly, the expressions of formal mathematics explicitly derive their semantic meanings from the interpretation or class of intended interpretations associated with the given formal theory and language. Such semantics has been a standard part of formal logic since Tarski (1936). This notion has been extended to the treatment of formal scientific theories by Sneed (1971), who adds the class of intended interpretations to the formal structure of the theory. Thus the separation of the expressions of mathematics into isolated and discrete parts denies them much of their significance and all of their semantic meaning. Such expressions consequently have little claim to be regarded as the 'molecular' components of mathematical knowledge.

Even more than the above, the expressions of informal mathematics discourse have implicit meanings associated with the overall background theory and context. For the rules and meanings which govern such expressions do not have precise formal stipulations, but depend even more on implicit rules of use (Wittgenstein, 1953). Models of the semantics of both formal and informal languages increasingly draw on the context of utterances (Barwise and Perry, 1982). Whether expressed in formal or informal language, the expressions of mathematics cannot be regarded as free-standing, independent bearers of meaning. Thus mathematics cannot be represented simply as a set of 'molecular' propositions, for these do not represent the structural relationships between propositions, as well as losing their context-dependent meanings.

B. Educational Implications

The fact that the discipline of mathematics does not have a unique hierarchical structure, and cannot be represented as a collection of 'molecular' propositions, has significant educational implications. However, first the relationship between the discipline of mathematics, and the content of the mathematics curriculum needs to be considered.

The relationship between mathematics and the curriculum

Two alternative relationships are possible. (1) The mathematics curriculum must be a representative selection from the discipline of mathematics, albeit chosen and formulated so that it is accessible to learners. (2) The mathematics curriculum is an independent entity, which need not represent the discipline of mathematics. Most curriculum theorists reject this second possibility, arguing the general case that the curriculum should reflect both the knowledge and processes of inquiry of the subject disciplines (Stenhouse, 1975; Schwab, 1975; Hirst and Peters, 1970). A form of Case 2 is devastatingly satirized by Benjamin (1971).

Studies of curriculum change have documented how developments in mathematics give rise via the pressures exerted by mathematicians to changes in the school mathematics curriculum reflecting these developments (Cooper, 1985; Howson *et al.*, 1981). More generally, in mathematics education it is accepted that the content of the curriculum should reflect the nature of the discipline of mathematics. Such acceptance is either implicit, or explicit, as in Thwaites (1979), Confrey (1981) and Robitaille and Dirks:

> the construction of a mathematics curriculum . . . [results from] a number of factors which operate on the body of mathematics to select and re-structure the content deemed to be most appropriate for the school curriculum.
>
> (Robitaille and Dirks, 1982, page 3)

An international seminar on the future of mathematics education explicitly considered the possibility that 'real mathematics' would not form the basis of the mathematics curriculum for everyone (the majority would study only 'useful mathematics'). However, this was contradicted by the three other options considered, including the most widely accepted view that a differentiated but representative curriculum is needed (Howson and Wilson, 1986).

Of the five ideologies distinguished in this book, all but the industrial trainers strongly endorse case 1. As a consequence of this brief survey, it can be said that the principle that the mathematics curriculum should be a representative selection from the discipline of mathematics represents the consensus of experts.

If the mathematics curriculum is therefore to reflect the discipline of mathematics faithfully, it must not represent mathematics as having a unique, fixed hierarchical structure. There are multiple structures within any one theory, and no one structure or hierarchy can ever said to be ultimate. Thus the mathematics curriculum should allow for different ways of structuring mathematical knowledge. In addition, the mathematics curriculum should not offer a collection of separate propositions as constitutive of mathematics. For the components of mathematics are variously structured and inter-related, and this should be reflected in the mathematics curriculum.

These educational implications allow us to criticize the National Curriculum in mathematics on epistemological grounds. For the mathematics curriculum is represented as a unique hierarchy of fourteen 'topics' (attainment targets) at ten levels (Department of Education and Science, 1989). Furthermore, at each level, the topic is represented by a number of propositions or processes, and mastery of the discipline of mathematics is understood to result from the mastery of these disparate components. Thus the National Curriculum misrepresents mathematics, contrary to the accepted curriculum principle. It embodies a hierarchy that is unjustified in terms of the nature of mathematics, as well as portraying mathematical knowledge as a set of discrete facts and skills.

A possible defence is that the mathematics curriculum might fail to represent the discipline of mathematics in order to fulfil a psychological purpose, such as to represent a psychological hierarchy of mathematics.

2. Hierarchy in Learning Mathematics

A. The View that Mathematics Learning is Hierarchical

It is often claimed that the learning of mathematics is hierarchical, meaning that there are items of knowledge and skill which are *necessary* prerequisites to the learning of subsequent items of mathematical knowledge. Such views are embodied in Piaget's theory of intellectual development. Piaget postulates a sequence of four stages (sensori-motor, pre-operational, concrete operational, formal operational) which form a hierarchy of development. The learner must master operations at one stage before she is ready to think and operate at the next level. However the rigid hierarchical aspect of Piaget's theory has been criticized (Brown and Desforges, 1979). Indeed Piaget coined the term 'decalage' to describe hierarchy-transgressing competencies.

Another psychologist who proposes that learning is hierarchical is Gagne. He argues that a topic can only be learned when its hierarchy of prerequisites have been learned.

> [A] topic (i.e., an item of knowledge) at a particular level in the hierarchy may be supported by one or more topics at the next lower level... Any individual will not be able to learn a particular topic if he has failed to achieve *any* of the subordinate topics that support it.
>
> (Gagne, 1977, pages 166–7)

Gagne claims that in empirical testing, in none of his topic hierarchies has there ever been more than 3 per cent of contrary instances.

Thus two influential psychologists representative of the developmental and neo-behaviourist traditions assert that learning is hierarchical. Furthermore, both these psychologists have made special studies of mathematics. Within mathematics education, there has been empirical research purporting to uncover learning hierarchies in mathematics. An influential British project, Concepts in Secondary Mathematics and Science, proposed a number of 'hierarchies of understanding' in some of the major areas of school mathematics (Hart, 1981). This study offers up to eight hierarchical levels in each of the topics studied.

The cited theories and empirical work are a small selection of the research concerned with identifying hierarchies in the learning of mathematics. Such research, possibly coupled with absolutist-foundationist views of the nature of mathematics, have led to the widely held belief that the learning of mathematics follows a hierarchical sequence. For example, this view is articulated in the Cockcroft Report.

> *Mathematics is a difficult subject both to teach and to learn.* One of the reasons why this is so is that mathematics is a hierarchical subject... ability to proceed to new work is very often dependent on a sufficient understanding of one or more pieces of work, which have gone before.
>
> (Cockcroft, 1982, page 67, original emphasis)

The hierarchical view of learning mathematics has its ultimate expression in the

National Curriculum in mathematics, as we have seen (Department of Education and Science, 1989). This is a fixed hierarchical specification of the mathematics curriculum at ten levels, constituting the legally required basis for the mathematics studies of all children (in English and Welsh state schools) from the ages of 5 to 16 years.

B. Criticism of the Hierarchical View of Mathematics Learning

The hierarchical view of mathematics learning rests on two assumptions. First of all, that during learning concepts and skills are 'acquired'. Thus prior to some particular learning experience a learner will lack a given concept or skill, and after an appropriate and successful learning experience the learner will possess, or have acquired, the concept or skill. Secondly, that the acquisition of mathematical concepts or skills necessarily depends on the possession of previously learned concepts and skills. This relation of dependence among concepts and skills provides the structure to the learning hierarchy. Thus to learn a concept of level $n+1$, the learner must already have acquired the appropriate subset of concepts of level n (but not necessarily all of that level). Consequently, according to this view, mathematical knowledge is organized uniquely into a number of discrete levels. Each of these two assumptions are problematic, and open to criticism.

Hierarchical dependency relationships between concepts

One assumption is that there are fixed hierarchical relations of dependence among concepts and skills, resulting in a unique hierarchy of concepts and skills. Two main criticisms can be advanced against this assumption. First, it presupposes that a concept or a skill is an entity which is possessed or not possessed by a learner; this is the second assumption, criticized below. But without this assumption it cannot be claimed that a concept of level $n+1$, depends on the possession of concepts of level n. For to make this claim it must be possible to claim that a learner determinately has, or has not, concepts or level n or $n+1$.

A more substantive criticism is that the uniqueness of learning hierarchies is not confirmed theoretically or empirically. Resnick and Ford (1984) conclude their review of research on learning hierarchies with the warning that they must be used with caution, and cite Gagne's comments of 1968 as remaining valid: 'A learning hierarchy . . . cannot represent a unique or most efficient route for any given learner.' (page 57).

A number of studies comparing the effects of instruction following different sequences of concepts from a proposed hierarchy (Phillips and Kane, 1973) or matching individual learners' knowledge to a learning hierarchy in a fine grained way (Denvir and Brown, 1986) confirm that no one hierarchy best describes the sequence or structure of every learners' knowledge acquisition. Although many authors report the effectiveness of learning hierarchies for sequencing instruction (Bell *et al.*, 1983; Horon and Lynn, 1980), the fact is that equally effective alternative strategies such as 'advance organizers', 'adjunct questions' and the 'deep end principle' deliberately thwart their hierarchical ordering assumptions (Begle, 1979; Bell *et al.*, 1983; Dessart, 1981). Thus such teaching studies do not tell us how learners' knowledge is structured.

The widely held view of cognitive scientists and psychologists is that the organization (and nature) of learners' knowledge is idiosyncratic, and that it cannot be subsumed to a single fixed hierarchy. Hence learners' concepts or conceptual structures have been termed 'alternative conceptions' or 'alternative frameworks' (Easley, 1984; Gilbert and Watts, 1983; Pfundt and Duit, 1988). Whilst such differences are on the micro-scale, the notion that learners' understanding across different mathematical topics can be equated in an overall mathematical hierarchy is also rejected (Ruthven, 1986, 1987; Noss *et al.*, 1989).

Concepts as entities that are acquired

The remaining assumption concerns the nature of mathematical concepts and skills, but a treatment of concepts alone suffices to establish the argument. The term 'concept' has two psychological meanings. Its narrow sense is that of an attribute or set of objects. It can be defined intensively, by means of a defining property, or extensively, in terms of the membership of the set. A concept in this sense allows discrimination between those objects that fall under it, and those that do not. Concepts in this sense are simple, unitary mental objects. The broader sense of 'concept' is that of conceptual structure, comprising a number of concepts (in the narrow sense) together with the relationships between them (Bell *et al.*, 1983). Conceptual structures are also called schemas, or 'concepts with interiority' (Skemp, 1979). Almost all that is referred to as a concept in the psychology of mathematics, such as the concept of place value, or even the concept of ten, has this broader sense of conceptual structure, since subsidiary components can be distinguished within each of the concepts.

Given these distinctions, three major objections can be raised against the assumption that concepts are acquired all at once, or are either 'possessed' or 'lacking' by a learner. First of all, given that most concepts are in fact composite conceptual structures, it is evident that their construction must be an extended process of growth, rather than an all or nothing state of affairs. In view of the complex interconnections between concepts, the acquisition of a concept can be an almost lifelong affair.

Second, a learner's possession of a concept can only be manifested indirectly, through its use, since mental structures are theoretical entities which cannot be directly observed. But a learner's use of a concept must necessarily be within some context, so the concept is linked to its contexts of use. To abstract the 'essence' of a concept from its contexts of use, and claim that this 'essence' represents the concept is presumptive. Current thinking in psychology points to the contextually situated nature of cognition (Brown *et al.*, 1989; Lave, 1988; Solomon, 1989; Walkerdine, 1988). Indeed, there is a substantial body of research which shows that a learner's use of a mathematical concept or skill in different contexts varies greatly (Carraher, 1988; Evans, 1988a). Thus the learner's grasp of a concept grows according to the range of contexts of use that are mastered, again undermining the notion that its acquisition is an all or nothing process.

Third, the notion that a concept is a uniquely specifiable objectively existing entity, is open to both philosophical and psychological criticism, as Chapters 4 and 5 have shown. It is widely accepted that individuals construct unique personal meanings (Novak, 1987). To claim that different individuals both possess the same concept, is not to say that some identical objective entity, albeit abstract, is 'owned' by both of them. This would be to reify a purely hypothetical theoretical entity. Such a claim is merely a *facon de parler*, meaning that two individuals' performances are comparable. Since acquiring a concept is the process of effecting an idiosyncratic personal construction, it is no longer valid to claim that a learner determinately possesses or does not possess a particular concept.

Overall, we see that the claim that the learning of mathematics follows a unique learning hierarchy cannot be sustained. The individual construction of concepts and their relationships is personal and idiosyncratic, even if the outcome may be shared competencies. A Vergnaud puts it:

> [T]he hierarchy of mathematical competences does not follow a total order organization, as the theory of stages unfortunately suggests, but rather a partial order one: situations and problems that students master progressively, procedures and symbolic representations they use, from the age of 2 or 3 up to adulthood and professional training, are better described by a partial-order scheme in which one finds competences that do not rely upon each other, although they may all require a set of more primitive competences and [may] all be required for a set of more complex ones.
>
> Vergnaud (1983, page 4)

Consequences for the National Curriculum in Mathematics

This discussion has consequences for hierarchical curriculum frameworks, and hence for the National Curriculum in mathematics (Department of Education and Science, 1989). Most importantly, there is no psychological justification for imposing a unique, fixed hierarchical structure on the mathematics curriculum for all children from ages 5 to 16. The empirical results reported above mostly concern small portions of the mathematics curriculum and restricted age and attainment ranges. Even under these favourable restrictions, the conjecture that a single hierarchy accurately represents mathematics psychologically must be rejected. Beyond this, we have seen that there are powerful theoretical reasons why a fixed hierarchy cannot describe student learning. Coupled with the earlier epistemological rejection, the outcome is a powerful condemnation of the framework in principle, without a detailed scrutiny of its contents.

It is also worth noting that virtually all of the arguments used in this criticism can be transposed to other areas of the curriculum, since detailed reference to the content of the National Curriculum has not been made.

When the detailed contents of the National Curriculum in mathematics are brought into the discussion, a possible justification can be anticipated. Namely, that although the curriculum does not have any epistemological or psychological necessity, nevertheless it might reflect the best available knowledge about children's overall achievements in mathematics.

There is a substantial amount of such knowledge available from large scale achievement testing in Britain and other countries, such as in Assessment of Performance Unit (1985), Hart (1981), Keys and Foxman (1989), Carpenter *et al.* (1981), Lindquist (1989) and Lapointe *et al.* (1989), Robitaille and Garden (1989), and Travers and Westbury (1989). Such information is inevitably a cultural product, reflecting the outcomes of mathematics curricula mediated by the institutional structure of schools and the system of assessment. Nevertheless, it provides a baseline, albeit pragmatic, against which a proposed hierarchical mathematics curriculum can be validated. The information need not wholly constrain a new curriculum, for there might be a clear rationale for altering aspects of past practice. However, given this caveat, any serious, large-scale curriculum development must carry out a minimal check of areas of intended agreement and disagreement with empirical research, and justify and anticipate any major deviations.

The National Curriculum in mathematics has ignored such issues, and does not reflect the current state of knowledge. Keohane and Hart (1989) and Hart (1989) show that a single level of the planned curriculum includes contents on which there has been greatly varying facilities. Level four is included in the programmes of study for children of ages 8–16. In a study of a large sample of 11 year olds (Hart, 1981), there were facility rates spread from 2 per cent to 95 per cent on items corresponding to level four statements of attainment.

Not only does the National Curriculum in mathematics lack any parity with, or reference to, the results of empirical research. The Mathematics Working Group was instructed by its chair, D. Graham, not be concerned with such matters.

> [T]he group is not expected to come up with water-tight research-based recommendations; it is expected to reflect good practice in a pragmatic way.
>
> (Nash, 1988, page 1)

This illustrates the fact that no attempt was made to develop the National Curriculum on the basis of research, let alone to test it empirically. Instead, it was put together by a committee, working as three sub-committees, in a matter of a few weeks. Overall, it has been shown to lack any epistemological or psychological validity, in its hierarchical assumptions. Given its status, and the resources available, this is highly remiss of its originators (the government).

3. The Hierarchy of Mathematical Ability

A. The Hierarchical View of Mathematical Ability

General intelligence has been regarded by psychologists as a fixed, inborn mental power, as the following quotation from Schonell shows.

> General intelligence may be defined as an inborn, all-round mental power which is but slightly altered in degree by environment although its realization and direction are determined by experience.
>
> (Tansley and Gulliford, 1960, page 24)

Although widespread, this view is not shared by all modern psychologists (Pigeon, 1977). Nevertheless, since 'mathematical ability' has been identified as a major factor of general intelligence (Wrigley, 1958), it may well have contributed to the widespread perception that the mathematical ability of an individual is fixed and enduring. In a penetrating analysis Ruthven (1987) shows that this perception is widespread, and is commonly seen by teachers and others as the main cause of different levels of attainment in mathematics. He uses the term 'ability stereotyping' for the tendency of teachers to entertain stable perceptions of pupil ability together with expectations of their achievement, even in the face of contrary evidence.

> In effect, individual pupils appear to be subject to a form of stereotyping in which teachers characterise them in terms of a summary, global judgement of cognitive capability and entertain correspondingly overgeneralized expectations of them.
>
> (Ruthven, 1987, page 252)

One consequence of ability sterotyping is that, in the extreme case, observed differences in performance on particular tasks are taken as indicative of the 'mathematical ability' of individual learners. A well known example is the 'seven year difference' of Cockcroft (1982). This is discussed after a characterization of the numerical attainment of 'average', 'much below average' and (implicitly) 'much above average' children

> **[T]here is a 'seven year difference' in achieving an understanding of place value which is sufficient to write down the number which is 1 more than 6399.** By this we mean that whereas an 'average' child can perform this task at age 11 but not at age 10, there are some 14 year olds who cannot do it and some 7 year olds who can.
>
> (Cockcroft, 1982, page 100)

This quotation suggests that individual children's performances on a particular item on a particular occasion are linked with, and even taken as an indicators of an overall construct of 'mathematical ability'. The presupposition of an underlying and persistent global construct of individual 'mathematical ability', giving rise to enduring levels of attainment, is confirmed by the following quotations.

> Even if the average level of attainment can be raised, the range of attainment is likely to remain as great as it is at present, or perhaps become still greater, because any measures which enable all pupils to learn mathematics more successfully will benefit high attainers as much as, and perhaps more than, those whose attainment is lower.
>
> (Cockcroft, 1982, page 101).

> In the case of children whose low attainment in mathematics is associated with low general ability, the mathematics course needs to be specifically designed to build up a network of simple related ideas and their applications
>
> (Cockcroft, 1982, page 98)

Overall, there is a widespread assumption, clearly evident in Cockcroft (1982), that there is a fixed linear hierarchy of mathematical ability from the least able to the most able (or mathematically gifted); every child can be assigned a position in this hierarchy, and few shift their position during the years of schooling.

One important outcome of these stereotyped perceptions and expectations of pupils is the adoption of limited goals for the mathematical education of lower attaining pupils. Ruthven provides evidence of this, and concludes that

> the emphasis on repetitive activities, on instrumental learning, and on computation — reflects stereotyped perceptions of the cognitive capabilities of less successful pupils and the curriculum goals appropriate to them, and stereotyped expectations of their futures, both as learners and as members of society.
>
> (Ruthven, 1987, page 250)

B. Criticism of the Hierarchical View of Mathematical Ability

Ruthven (1987) provides a powerful critique of ability stereotyping, arguing on the one hand, that the consistency of student mathematical attainment is less than is supposed, varying across both topics and time. On the other hand, teacher expectations and stereotyping become self-fulfilling, and curriculum differentiation in mathematics which makes high and low cognitive demands of high and low attaining students, respectively, exacerbates any existing differences. This criticism can be supported by means of two theoretical perspectives: sociological and psychological.

The sociological argument for rejecting a fixed hierarchical view of ability in mathematics comes from labelling theory. A strong association between social background and educational performance of almost all types is one of the longest established and best supported findings in social and educational research (Department of Education and Science, 1988b). Specifically, there is extensive evidence in Britain of the correlation of educational life chances and social class (Meighan, 1986). Perhaps the best supported theoretical explanation of these effects is based on labelling theory, due to Becker (1963) and others. The key feature of labelling individuals as 'mathematical low attainers', for example, is that it is often self-fulfilling. Thus streaming by ability, which is widespread in the teaching of mathematics, although only loosely related to measured attainment, has the effect of labelling by ability, and thus affects achievement in mathematics, becoming self-fulfilling (Meighan, 1986; Ruthven, 1987).

The second theoretical basis for rejecting a fixed hierarchical view of ability is psychological. There is a tradition in Soviet psychology which rejects the notion of fixed abilities, and links psychological development with socially mediated experience. This development was expedited politically by the 1936 Soviet ban on the use of mental tests, which halted research on individual differences in ability (Kilpatrick and Wirszup, 1976). A seminal contributor to this tradition is Vygotsky (1962), who proposed that language and thought develop together, and that a learner's capabilities can be extended, through social interaction, across a 'zone of proximal development'. The interaction of personal development and social context and goals through 'activity' has been the basis of the Activity Theory of Leont'ev (1978) and others. Within this overall tradition, the psychologist Krutetskii (1976) has developed a concept of mathematical ability which is more fluid and less hierarchical than that discussed above. He first offers a critique of the relatively fixed views of mathematical ability stemming from the psychometric tradition in psychology. He then offers his own theory of mathematical ability based on the mental processes developed by individuals used in attacking mathematical problems. He acknowledges individual differences in mathematical attainment, but gives great weight to the developmental and formative experiences of the learner in realizing his or her mathematical potential.

> Of course, the 'potentialities' are not constant or unalterable. The teacher should not content himself with the notion that the children's varied performances — in mathematics say — are the reflection of their ability levels. Abilities are not something foreordained once and for all: they are formed and developed through instruction, practice and mastery of an activity. Therefore we speak of the necessity of forming, developing, cultivating and improving children's abilities, and we cannot predict exactly how far this development may go.
>
> (Krutetskii, 1976, page 4)

This Soviet psychological tradition is having an increasing impact on mathematics education (Christiansen and Walther, 1986; Crawford, 1989; Mellin-Olsen, 1987). It is increasingly acknowledged that the cognitive level of student response in mathematics is determined not by the 'ability' of the student, but the skill with which the teacher is able to engage the student in mathematical 'activity'. This involves the development of a pedagogical approach in mathematics which is sensitive to and relates to student goals and culture. Students labelled as 'mathematically less able' can dramatically raise their levels of performance when they become engaged in socially and culturally related activities in mathematics (Mellin-Olsen, 1987).

Another empirical confirmation of the fluidity of ability can be found in the *idiot savant* phenomenon. Here, persons labelled as 'ineducable' can perform at an astonishingly high level in domains in which they have become engaged (Howe, 1989).

Overall, there is a strong theoretical (and empirical) basis for rejecting the fixed hierarchical view of mathematical ability, and linking it much more to social development, stemming from the Soviet tradition. Coupled with the sociological arguments, these comprise a strong case against the hierarchical view of ability in mathematics.

C. The Hierarchical View of Ability in the National Curriculum

A hierarchical view of mathematical ability is evident in publications concerning the National Curriculum. The Task Group on Assessment and Testing was established to develop testing 'for all ages and abilities' (Department of Education and Science, 1987a, page 26) and its terms of reference included the giving of advice on assessment to 'promote learning across a range of abilities' (Department of Education and Science, 1988b). The Secretary of State for Education (K. Baker) wrote to the chair (P. Black) on ability and differentiation, as follows.

> I am asking the subject working groups [including mathematics] to recommend attainment targets for the knowledge, skill and understanding which pupils of a range of different abilities should normally be expected to achieve at the four age points; but so far as possible to avoid setting qualitatively different targets — in terms of areas of knowledge, skill or understanding — for children of different ability.
>
> (Department of Education and Science, 1988b, appendix B)

The final report of the mathematics working group (Department of Education and Science, 1988) also uses the language of ability stereotyping. The accompanying letter to the Secretaries of State states that the enclosed proposals are: 'appropriate for children of all ages and abilities, including children with special educational needs.' (page vi). Further examples picked at random from the report include: 'Teachers of infant children . . . will need to . . . refer to programme B in order to extend the work of their most able pupils.' (page 63) 'A time comes when even the brightest child needs to make an effort' (page 68) 'some of the least able 10 per cent of pupils may have difficulty with, for example, Level 1 at age 7 and Level 2 at age 11.' (page 83)

These quotations strongly suggest that the both official (governmental) view and that represented in the publications of the Mathematics Working Group are of a hierarchy of mathematical ability, in which individuals can generally be assigned a fixed and relatively stable position.

In addition, the National Curriculum in mathematics results in the restriction of the curriculum experience for low attaining pupils in mathematics. As the curriculum and assessment framework for the National Curriculum shows (Figure 11.1), the net result is a single curriculum in mathematics for all pupils, with those of 'low ability' restricted to the lower, simpler levels.

The outcome of these assumptions in the National Curriculum in mathematics is likely to be the exacerbation and exaggeration of individual differences in performance. As we have seen, this is almost certain to be self-fulfilling, denying success in mathematics to a very large number of school children.

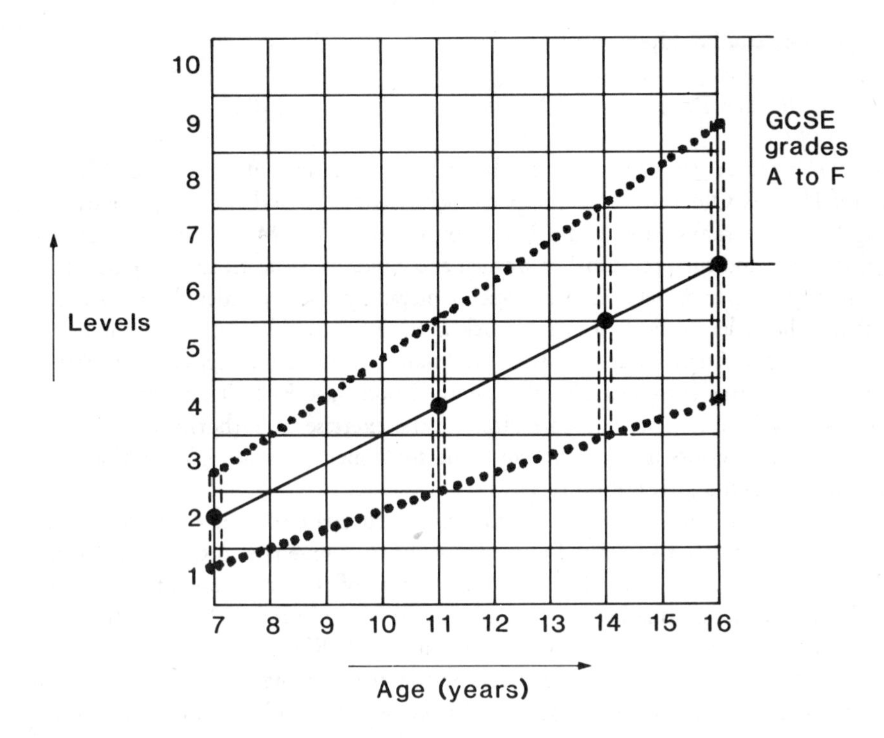

Figure 11.1: The Curriculum and Assessment Framework of the National Curriculum (Adapted from Department of Education and Science, 1988b)

The lines show the progress of students by ability: 'high ability' (90th percentile) — top dotted line; 'average ability' (50th percentile) — middle line; 'low ability' (10th percentile) — lower dotted line.

Of course ability stereotyping in mathematics is not only due to observed differences in attainment. There is incontrovertible evidence that class (as well as ethnic and gender) factors play a major part in such labelling (Meighan, 1986). The ability stereotyping built into the National Curriculum in mathematics assumes that every child can be assigned a position in the 'hierarchy of mathematical ability', and that few shift their position during the years of schooling. Consequently, working class, black and female children are likely to be placed lower, rather than higher, in this hierarchy, in conformity with stereotyped expectations. This is another anti-egalitarian feature of the National Curriculum, which will impose a fixed and hierarchical 'pecking order' by attainment on students.

4. Social Hierarchy

B. The Roots of Social Hierarchy

Social hierarchies have a long history, dating back to the ancient Hebrews and Greeks.[3] In the Hebrew Old Testment an implicit hierarchy places God at the top, followed by angels in their ranks, then come earthly prophets such as Moses, followed by tribal chiefs, men, and then presumably women and children. Beneath them are devils, and ultimately Lucifer or Satan himself. Such a hierarchy linearly orders humankind, but extends the ordering both above and below to limits or 'ideal points', analogous to projective geometry. Values are strongly linked with the hierarchy, the higher the better, with the extremes identified with Good and Evil. These values have a justificatory function, serving to legitimate the exercise of authority and power by superiors on inferiors in the hierarchy. The subsequent 'divine right of kings' is an example of such a justification of power.

In Chapter 7 this view was traced from an alternate source, Aristotle's view of nature, with which it was fused in medieval times to give rise to the Great Chain of Being (Lovejoy, 1936). Another important source of this tradition is the division of people into three stratified types, termed gold, silver and bronze (Plato, 1941). This is significant because of its link with education, in which different curricula were deemed appropriate for the three types, dictated by the needs of the different stations in life they would assume. This is the source of a theme that will be seen running through this section. We have also seen that the Greeks distinguished between the work of the hand and the work of the brain, giving rise to the association between pure knowledge and the more powerful classes (and its obverse).

The combined modern outcome of these traditions is a widely accepted pyramidal hierarchical model of society, with power concentrated at the top, legitimated and reinforced, if not reproduced, by the associated cultures and values. This model of society is seen by many to be the 'natural' state of affairs, as exemplified by humankind and groups of animals in the wild. Such biological roots are firmly rejected by the feminist re-analysis of history and anthropology, which sees pyramidal hierarchy as associated with male dominance in society, and rejects the claim that it is universal (Fisher, 1979). Indeed, such an unquestioned hierarchical view of society may be seen as part of the culture which sustains the existing structure of society, and hence male and upper/middle class dominance. The identification of pyramidal hierarchy as the 'natural' structure of society is an instance of the 'naturalistic fallacy', the false assumption that what *is* the case, is what *has to be*; contingency is mistaken for necessity.

When the power structure of society is physically threatened, force is likely to be brought into play to maintain it. However, what is more interesting, is the impact of perceived threat on the associated culture and values. According to Douglas (1966), social groups have 'group' boundaries, distinguishing members from outsiders, and 'grid' boundaries, distinguishing the different sectors or strata within the group.[4] Under threat, according to Douglas, the group becomes concerned with purity in its culture, and with strict group and grid boundaries. In this view, the purity associated with dominant class culture, becomes intensified, as does the strictness of the boundary definitions, including internal gradations in a hierarchy.

B. Education and the Reproduction of Social Hierarchy

Perhaps the most influential modern theorist on the structure of society is Karl Marx (1967). He argues that material conditions and the relations of production have a central determinative power over the structure and inter-relationships in society. In particular, society has an infrastructure, or economic base, which in the 'last instance' determines the two levels of its superstructure, the law and the state, and the associated ideologies. The state, through the 'repressive state apparatus' (police, prison, army, etc.) sustains and reproduces industrial production in favour of capital and the dominant class.

However this thesis can be interpreted in two ways concerning the constraining forces on the masses and society in the large. There is the 'hard' view that social conditions are highly determinative, and that humankind is imprisoned without the key of Marxist theory with which to penetrate false consciousness and oppression. There is also the 'softer' determinist position, that humanity is capable of reacting, and is everywhere able to 'create' social change (Simon, 1976). A comparable distinction is drawn by Giroux (1983) between 'structuralist' and 'culturalist' traditions in neo-marxist theory, which emphasize the importance of social and economic structure, or the culture and its relationship with human agents, respectively.

Hard determinism

An influential modern theorist in this tradition is Althusser (1971). He argues that in addition to the 'repressive state apparatus' social reproduction depends upon an 'ideological state apparatus', which includes education, religion, respect for the law, politics, and culture; and that no class can maintain power without extending hegemony or cultural domination over such areas. Education is the most powerful 'ideological state apparatus' in reproducing productive relations, that is instilling the acceptance of labour and the conditions of life of the masses.

Bourdieu and Passeron (1977) propose a theory of schooling and the reproduction of society which fits in this category. In this account linguistic culture (more generally 'cultural capital') is especially important in determining the social outcomes of education, in terms of class membership. They term 'symbolic violence' the cultural domination of the working classes which masks social reproduction.

An influential development of the hard deterministic thesis, which downplays the role of ideology is that of Bowles and Gintis.

> The current relationship between education and the economy is ensured not through the *content* of education but through its *form*: the social relations of

> the educational encounter. Education prepares students to be workers through a *correspondence* between the social relations of production and the social relations of education. Like the division of labor in the capitalist enterprise, the educational system is a finely graded hierarchy of authority and control in which competitition rather than cooperation governs the relations between participants... The hierarchical order of the school system admirably geared towards preparing students for their future positions in the hierarchy of production, limits the development of those capacities involving the exercise of reciprocal and mutual democratic participation and reinforces social inequality by legitimating the assignment of students to inherently unequal 'slots' in the social hierarchy.
>
> (Gintis and Bowles, 1980, pages 52–53)

Powerful as these arguments are, they suffer from two major flaws. First of all, they are too deterministic in shackling education to the conditions of production. In this, they do not allow for the exploitation of contradictory forces at work in the system, nor for human agency or resistance from within (Giroux, 1983). Secondly, especially in the case of Bowles and Gintis (1976), they ignore the nature of knowledge, which as we have seen previously, relates to ideology and class, and cannot be ignored.

Soft determinism

Many of the insights considered above remain valid for the softer and less deterministic view of reproduction considered here. However, beyond this structural determinism Gramsci (1971) argues that the domination of society by one class requires cultural hegemony. This is the domination of culture by one class legitimating, mystifying and reinforcing its power and prestige. Such hegemony saturates the 'common-sense' of the masses, and hence secures their unwitting consent and collusion.

Williams (1976) builds on the concept of hegemony, but argues that there are alternative and oppositonal forms of social life and culture in addition to the dominant class culture. These can coexist with the dominant culture which may incorporate alternative or even contradictory forms. This illustrates an important and more general point made by Williams, concerning the multiplicity of ideologies and cultures. It is all too easy to fall into the trap of moving from hegemony to a simplified and static view of culture. Williams stresses its complexity and dynamics.

Giroux (1983) acknowledges the complex nature of culture. He proposes that within school cultures there is resistance which is more than mere response to an authoritarian curriculum, and which reflects instead implicit alternative agendas. He argues alongside Freire and other public educators that through critical education, students can be emancipated from the reproductive forces at work in schooling.

Overall, according to this second grouping, the forces tending to reproduce the hierarchical structure of society are acknowledged, as is the importance of culture, ideology and knowledge. But these are seen to have a dual role, both as the essential means of domination, and also as the possible means for emancipation.

C. Reproducing the Social Hierarchy through Mathematics

A number of authors have applied one or other form of the above ideas to mathematics education, such as Cooper (1989), Mellin-Olsen (1987), Noss (1989, 1989a) and others in Noss *et al.* (1990). Both Noss and Cooper conclude that it is the form rather than the content of mathematics education (i.e., the 'hidden curriculum') which conveys its social aims.

Cooper argues that the hegemony of schooling exercises a negative power over primary school teachers binding them to traditional authoritarian and routinized approaches to mathematics, and to differentiated curricula which serve to recreate the social hierarchy. Elements of this account ring true, and it provides a valuable insight as to how cultural pressures follow the chain of command in the hierarchy of schooling. Nevertheless, the account is oversimple in failing to acknowledge the range of beliefs and ideologies of teachers and social pressure groups.

Noss presents a powerful case for the weak deterministic thesis in mathematics education, and identifies the National Curriculum in mathematics as serving a socially reproductive function which is 'profoundly and intentionally anti-educational' (Noss, 1989, page 1). He argues that there are contradictions in the system which allow it to be subverted to serve genuinely educational purposes. In particular, the low priority accorded to mathematical content means, in his view, that it can be exploited to serve empowering, democratic education. However he does not acknowledge that the hierarchical structure of the content of the curriculum serves to recreate an hierarchically stratified society, as will be argued below. (Although in Noss, 1989a, it is suggested that a basic-skills curriculum in mathematics serves to deskill labour for financial exploitation.)

Mellin-Olsen (1987) acknowledges the existence of reproductive forces in mathematics education and society, and builds them into a theoretical account drawing also on social psychology, anthropology and psychology. He stresses, following Giddens (1979), that individuals create ideologies as well as living in them. In particular, he identifies resistance to hegemony with the production of alternative ideologies by means of activity.[5] He argues that an empowering mathematics education must grasp such opportunities: critical mathematics education should provide learners with 'thinking tools' to engage in activity which challenges the implicit ideologies of schooling.

This brief account cannot do justice to Mellin-Olsen's theory, supporting arguments and links with practice. However, it can be said that it shares two areas of weakness identified earlier in accounts of social reproduction. First, it does not adequately distinguish the different ideologies and social interest groups at work in the mathematics curriculum. This may not seem necessary for the general argument put forward by Mellin-Olsen, but it is necessary before the implicit ideologies of schooling can be challenged. Second, it does not explore the elements of ideology adequately, and above all, does not consider views of the nature of mathematics, which are so central to mathematics education, according to the thesis of this book.

Overall, there is widespread support for the thesis that education helps to reproduce the hierarchical structure of society, serving the interests of the wealthy and privileged. However, this thesis needs to be understood in a way that recognizes the complexity of the relationships within society, and which modifies the deterministic character of its original formulation. This modified reproduction thesis depends a great deal on ideology, and so it is appropriate to explore its relationship with the model of educational ideologies of this book.

The industrial trainers

In terms of the social station of the masses, the industrial trainer aim is directly reproductive. Thus the social training of the masses through mathematics is part of a preparation for a life of obedient labour. Drill, rote, practice, the dualistic demarcation between correct and incorrect, and the strict hierarchical authority of the teacher help to inculcate the appropriate expectations and values to discipline future workers for an unquestioning role in society, whilst the future higher strata of society are not so regulated. Low level training also ensures that the masses become cheap labour (Noss, 1989a). The central goal of this group is to contain and define the (lowly) place of the masses. The petit-bourgeois ideology derives from many of the group having 'bettered themselves', and this ideology involves keeping their social origin group 'in their place'.

The old humanists

The old humanist aim focuses on development of the mathematically able and gifted and the inculcation of pure mathematical values. This serves the nurture and reproduction of the body of mathematicians, who represent a portion of the professional, middle class élite, with a middle class purist culture. This can be traced back to the divisions between the work of the hand and the brain, and the concomittent cultural and class distinctions (Restivo, 1985). This group have traditionally exercised much power over the content of the mathematics curriculum, making it 'top down' (serving the interests of the group) rather than 'bottom up' (serving first the interests of all). By focusing on the needs of an élite, and its survival, this ideology seeks to reproduce the class structure of society.

These two groups both focus on the preservation of a group and its boundary. The old humanists are part of a middle class professional group with economic and political power, and with a culture whose purity serves to define and defend the group boundary. Douglas (1966) has argued in general that purity serves to defend group boundaries in this way, on the basis of extensive anthropological work. The purist aims and ideology of this group fit with this pattern. The industrial trainer aims for mathematics education are not purist, and yet also serve to preserve the group boundary around the masses, and hence their own group boundary. This would seem to be inconsistent with the findings of Douglas. However their dualistic moral values centre on moral purity in the Judeo-Christian tradition ('cleanliness is next to

Godliness', 'original sin'), as opposed to the epistemological purity of the old humanists. Thus Douglas' conception of cultural (and bodily) purity as a response to group boundary threats applies here, too.

The technological pragmatists

The technological pragmatist aims are not so strictly concerned to preserve class boundaries, and hence not so strictly reproductive. Society is seen as based on wealth and progress, following technological innovation and progress. Mathematics education is part of the overall training of the population to serve the needs of employment, and the overt social aims are meritocratic. Social mobility on the basis of merit or technological attainment is part of this perspective, since industry and other sectors are continually expanding and needing technologically trained personnel. However, the existing social stratification by class is not questioned, and consequently various factors and expectations largely serve to reproduce social divisions and stratification.

The progressive educators

The progressive educator aims for mathematics concern the realization and fulfilment of human beings through mathematics as a means of self expression and personal development. The emphasis of this perspective is very individualistic. Whilst it is directed at the advancement of individuals in a number of ways, it does not locate them in a social matrix, nor does it recognize the conflicting forces at work in society which undermine the efficacy of progressive education. Thus the perspective although socially progressive, does not seriously undermine the reproductive forces at work in society and school. Factors such as the unequal resourcing of schools and teacher stereotyping of students are not challenged. Socially, the progressive educator aims concern the amelioration of the conditions of individuals, not social change to provide emancipatory conditions.

Of these two ideologies, that of the progressive educator is the most concerned to develop and empower individuals, and hence to facilitate meritocratic social progress. That is, it is the more progressive of the two ideologies. Despite this, both perspectives are blind to the social context and its impact on social advancement. Both attribute this to the endeavours and achievements of individuals, against the backdrop of an assumed and unquestioned social hierarchy. Neither perspective questions the fact that different sectors are socialized to have different educational expectations, and receive differing forms of education in line with their class origins. Nor do they acknowledge that consequently the hidden curriculum tends to reproduce stratification by employment and wealth. As Mellin-Olsen (1981) argues, working class and middle class students expect and are conditioned to be taught mathematics instrumentally or relationally (respectively).

Only one of these two meritocratic perspectives has a purist ideology. This is the progressive educator view, which emphasizes pure child-centredness and creativity, in opposition to utility. The romanticism and focus on the pure interests of the child, provides a group defining ideology, protecting the privileged middle class position of professional educators. It also serves to elevate the progressive educator into the privileged parental role in their relations with children, and analogously in society, as a middle class professional. Thus the purity of this ideology can be seen, *pace* Douglas, to safeguard the group boundary and interests.

The public educators

The public educator aims concern the empowerment of learners, through mathematics, to be autonomous, critical citizens in a democratic society. The public educator mathematics curriculum is intended to be emancipatory through teachers' integration and public discussion of mathematics in its social and political contexts, through students' freedom to question and challenge assumptions about mathematics, society, and their place in it, and their empowerment through mathematics to understand and better control their own life situations. The perspective fully acknowledges the impact of the social context on education and indeed sees education as a means of achieving social justice. Thus there is a concern with the unequal allocation of resources and life-chances in education, and a concern to combat racism, sexism and other obstacles to equal opportunities. Of all five ideologies, this alone is a social change perspective, acknowledging the injustice of our stratified and hierarchical society, and seeking to break the cycle which reproduces or recreates it through education.

5. Inter-relating Mathematical, Ability and Social Hierarchies

This chapter has explored different contexts in which hierarchies appear in mathematics education, arguing in each case that fixed and inflexible views of hierarchy have negative outcomes for education. Beyond this, I wish to argue that the components of each group ideology work together to serve the social aim of the group, and that hierarchies plays a central part in this process. In particular, there is a loose correspondence between the structures of mathematics, ability and society in each of the ideologies. In terms of the five ideologies, three loose groupings may be distinguished.

Rigidly hierarchical ideologies

Two of the ideologies, the industrial trainers and old humanists, are strictly reproductive in intent. It serves their interests to reproduce the hierarchical structure of society. Basic skills training is intended to produce quiescent workers, whilst higher mathematical study is reserved largely for the future middle classes. In each ideology, a rigid hierarchical theory of mathematics is translated into a rigid hierarchical mathematics curriculum; it is associated with a fixed hierarchical view of ability in mathematics; also with a strictly hierarchical view of society, class and employment. In both views, there is a correspondence between the levels of the hierarchies: low level practical mathematical knowledge is deemed to be the appropriate curriculum for students labelled as having lower mathematical ability and intelligence, who are prepared for the lower level occupations and strata in the social hierarchy. Higher level theoretical mathematics is considered the appropriate curriculum for students of 'higher mathematical ability' who are expected to obtain higher level occupations and social positions, possibly in the professions. This is illustrated in Figure 11.2, which also shows the fairly rigid class barriers which separate the two main tracks, allowing social mobility only in exceptional cases.

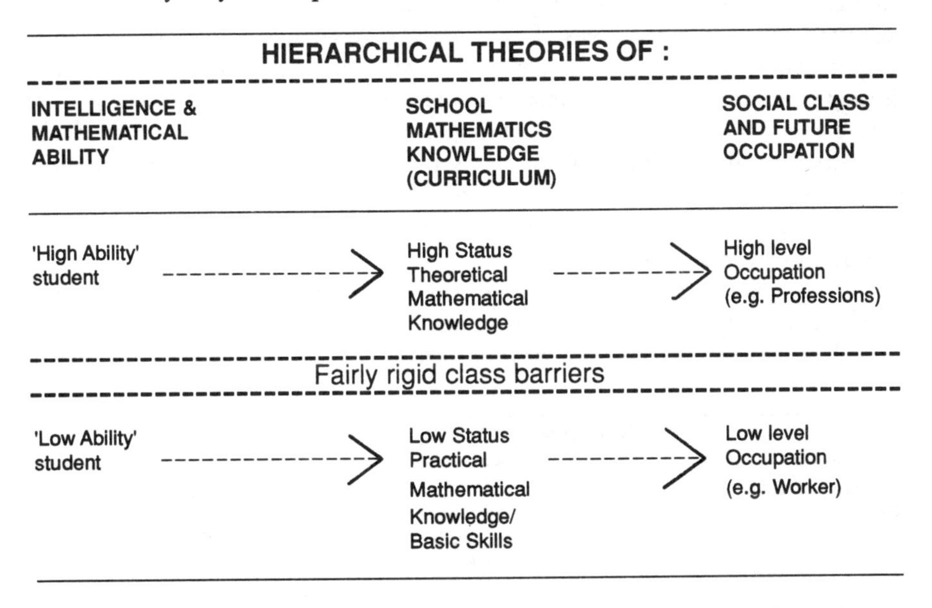

Figure 11.2: Correspondence between Rigid Hierarchical Theories of Mathematics Curriculum, Ability and Social Class/Occupation

Naturally, this model is greatly simplified. It shows only two discrete levels of hierarchy. In addition, pedagogy is not shown. This comprises most of the secondary elements of the ideologies, and plays a crucial role in the reproduction of social hierarchy. This is accommodated by identifying the industrial trainer ideology, including its pedagogical elements, more with the lower track. This brings in the social training and instrumental aspects of the pedagogy (Mellin-Olsen, 1981). The old humanist ideology is identified more with the upper track, with its focus on pure theoretical mathematics, and its relational emphasis in pedagogy. Such specialization has some substance, for the industrial trainers are happy to leave élite education in the hands of the élite, and the old humanists are most concerned with this strand, anyway. Thus the two ideologies complement each other with their emphases on different strata of the hierarchies. The net result is that they work to be more reproductive than the other ideologies.

Progressive hierarchical ideologies

Two of the perspectives, the technological pragmatists and the progressive educators are each, for different reasons, reproductive of the social hierarchy, but less strictly and rigidly so than those above. They both view themselves as meritocratic by permitting or encouraging social mobility within the assumed pyramidal hierarchy of society. The technological pragmatists aim to serve the needs of industry, employment and society, through the upward social mobility of the technologically skilled. The progressive educators aim to serve the needs of individuals, by encouraging them to develop as persons. The outcome of this is potential upward social mobility, for those who develop self-confidence, and marketable knowledge and skills. The two ideologies are thus to a varying degree progressive, seeking improvement to society and individuals. However behind these overt intentions is the unquestioning acceptance of social stratification. Consequently the two positions tend to loosely recreate the hierarchical structure of society, with slight progressive modifications, namely the upward movement of a small sector of society.

Once again, in their underpinning ideologies, hierarchical theories of mathematics and the mathematics curriculum are associated with hierarchical views of ability in mathematics and a hierarchical view of society, class and employment. As above,

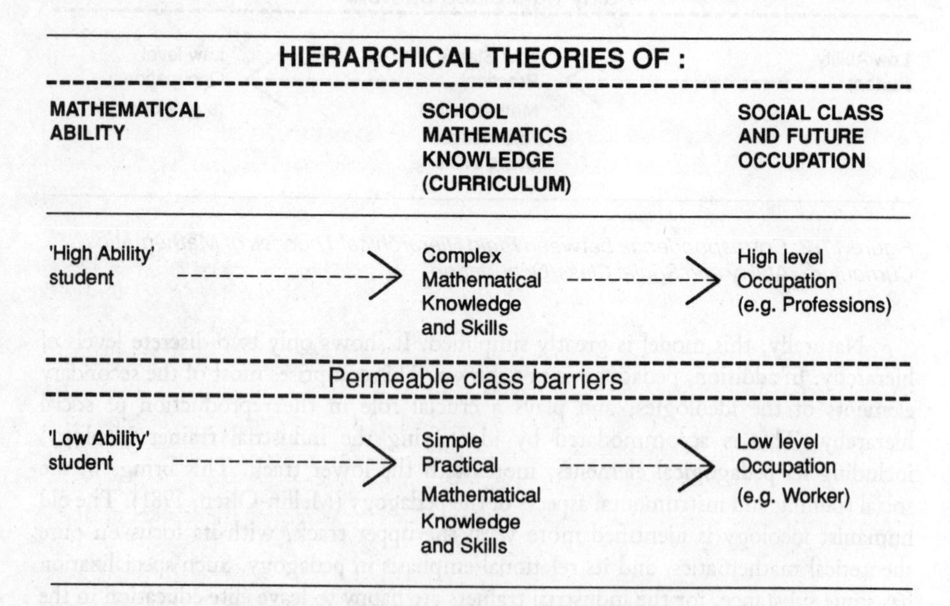

Figure 11.3: Correspondence between Progressive Hierarchical Theories of the Mathematics Curriculum, Ability and Social Class/Occupation

there is a correspondence between levels: simple, practical mathematical knowledge and skills are considered to make up the appropriate curriculum for 'low ability' students, who are presumed to be largely destined for lower level occupations and social strata. More complex mathematical knowledge and skills make up the curriculum for 'high ability' students, presumed destined for higher level occupations and social positions. This is shown in figure 11.3, which also indicates the permeable division of hierarchies into strata corresponding to class divisions. The permeability of this barrier is due to both ideologies viewing themselves as progressive and meritocratic, although any progress due to merit presupposes the hierarchical framework.

Once again, the model is greatly simplified, showing only two levels of hierarchy and not indicating pedagogy, which plays a central role in reproduction. Unlike the previous case, there is not so sharp a distinction between the pedagogy (and the secondary ideological elements) at the two different levels of the hierarchy. Neither the technological pragmatists nor the progressive educators overtly distinguish between the pedagogical approach considered appropriate for different levels of the hierarchies. However, in practice different pedagogical approaches are imposed on students at the different levels, as a consequence of the underlying hierarchical theories. Lower level students tend to be trained instrumentally, and upper level students tend to be taught relationally, and encouraged to become independent learners. Whilst this is an oversimplified account of a more complex phenomenon which varies according to ideology, it shows that the hierarchical theories of the two ideologies serve to recreate the hierarchical structure of society, if not so rigidly as the previously considered two.

The social change ideology

Lastly, there is the social change ideology of the public educators. This perspective acknowledges the existence and inequality of the pyramidal social class hierarchy, but seeks to change it to achieve social justice. It seeks to break the reproductive cycle in education, whether rigid or progressive, by openly acknowledging its existence and promoting emancipatory education. Once again, mathematics, ability and society are linked in this ideology, but by their fluidity, malleability and rejection of fixed hierarchical structures. The theory of mathematics is a conceptual change theory, social constructivism. This is translated into a flexible theory of school mathematical knowledge adaptable to serve learners and their social context; it is associated with a fluid theory of mathematical ability, concerned with the the zone of proximal development instead of stereotyped 'ability' levels; also with a social change theory of society, class and employment. Thus there is a correspondence between all of these theories, but it is one which acknowledges change and rejects fixed hierarchiacal structures, be they rigid or progressive.

Overall, it is suggested that in each of the five perspectives many of the ideological elements work harmoniously together to achieve their social aims, overt or convert. The theories of mathematics, curriculum, pedagogy, ability and society are all similar, be they hierarchical or change orientated. The form of mathematics education plays a central part in the reproduction (or challenging) of the social hierarchy, but it is only one of several such elements, which included the philosophies of mathematics and the theories of school mathematical knowledge. Thus epistemology and the content of education plays a crucial role in recreating or changing the social hierarchy. Beyond this, it has been shown that a single model of social reproduction will not serve. Different ideological groups strive to reproduce social relations differently. In practice the outcome is a complex amalgam of these forces, according to their strengths, and according to other forces arising from the education system or society. A novelty in this account is the linking of hierarchical conceptions in apparently disparate realms, and showing that they work together to serve social aims and interests.

Finally, turning to the National Curriculum in mathematics we can identify within it hierarchical conceptions of school mathematics and ability, and infer such views of mathematics and society, the ingredients of a reproductive position. Given that the ideologies underpinning the development are a mixture of rigid and progressive hierarchical positions, it is evident that the purpose of the curriculum, whether implicit or explicit, is social reproduction. Since this involves denying opportunity and the realization of human potential, it is, to paraphrase Noss, profoundly anti-educational in intent.

Notes

1 If in a formal language we allow infinitary rules of inference or definition, it is a simple matter to extend the hierarchy to include levels of transfinite ordinality.

2 Any other similar relation on the set S of formulas or terms of mathematics will also impose an hierarchical structure on mathematics. All that is needed to specify it is a triple <S, B, R> , where R is a well-founded partial ordering of S, and B is the subset of S which is minimal with respect to R.

3 They were undoubtedly found in other ancient societies, including Egypt, Sumer/Babylon and China.

4 I also draw on the account in Bloor (1983).

5 This refers to the activity theory of Leont'ev (1978).

12

Mathematics, Values and Equal Opportunities

1. Mathematics and Values

A. The View of Mathematics as Neutral and Value-Free

Absolutist philosophies are committed to a belief in the absolute objectivity and neutrality of mathematics, as are a range of personal philosophies of mathematics. However, despite this belief, the view of mathematics they promote is itself value-laden. For, as we have seen, within mathematics there are implicit values. Abstract is valued over concrete, formal over informal, objective over subjective, justification over discovery, rationality over intuition, reason over emotion, general over particular, theory over practice, the work of the brain over the work of the hand, and so on. These constitute many of the overt values of mathematicians, as well as being shared by much of British and Western scientific culture.

Having identified these values, the question is, how can so overtly a value-laden view of mathematics claim that it is neutral and value-free? The answer of the absolutists is that these values concern mathematicians and their culture, and not the objective realm of mathematics itself. It is claimed that the content and methods of mathematics, by their very nature, make it abstract, general, formal, objective, rational, theoretical and concerned with justification. That is the nature of theoretical scientific knowledge, including mathematics. There is nothing wrong with the concrete, informal, subjective, particular, or the context of discovery, according to this view. It is just that it is not science, and certainly not mathematics (Popper, 1979).

What I wish to claim is that the values of the absolutists are smuggled into mathematics, either consciously or unconsciously, through the definition of the field. In other words, all that the absolutist perspectives will admit as *bona fide* mathematical knowledge must satisfy these values, and that anything that does not is rejected as inadmissible. Mathematical propositions and their proofs, the products of formal mathematical discourse, are admitted as legitimate mathematics. Mathematical invention, the practices of mathematicians and other products and processes of informal and professional mathematical discourse are not.

Once the rules of demarcation of the discipline are established in this way, then it can legitimately be claimed that mathematics is neutral and value free. For in place of values there are rules which determine what is admissible. Preferences, choices, social implications and all other expressions of values are all eliminated by explicit and objective rules. In fact, the values lie behind the choice of the rules, making them virtually unchallengeable. For by legitimating only the formal level of discourse as mathematics, it relegates the issue of values to a realm which is definitionally outside of mathematics.

If this criticism is accepted, at the heart of the absolutist neutral view of mathematics is a set of values and a cultural perspective, as well as an ideology which renders them invisible.

Having identified above which values and whose culture, there is a further question to ask. Whose interests do they serve? Britain and the West are largely ruled by white males from or in the upper strata of society. Most sectors of employment and power have a pyramidal hierarchical structure, with the upper strata dominated by this group. Thus for example, among university mathematicians, the group who serve to define the subject, it is white males of the middle and upper classes who overwhelmingly predominate.

The values of mathematicians have developed as part of a discipline with its own powerful inner logic and aesthetics. So it would be absurd to claim that these values do nothing but explicitly serving the social interests of the group. Nevertheless, whether accidentally or not, the fact is that these values do serve the interests of a privileged group. They advantage males over females, whites over blacks, and middle classes over lower classes, in terms of academic success and achievement in school mathematics. This promotes the interests of the more privileged in society, because of the special social function of mathematics as a 'critical filter' in terms of access to most well paid professions (Sells, 1973, 1976). Thus the covert values of mathematics and school mathematics serve the cultural domination of society by one sector.

The absolutist response to this charge is that mathematics is objective and neutral and value-free. Any so called 'values' implicit in mathematics do not represent choices or preferences but are essential to the nature of the enterprise. Mathematics is the science of the abstract, the formal and the objective, it is primarily concerned with generality, with theory and with justification. Of itself, mathematics has no social preferences. It just so happens that certain sectors of the population, namely whites, males and members of the middle-classes are intrinsically better equipped for the demands of mathematical study. Their cognitive styles embody the properties described as mathematical values. Furthermore, according to this perspective, this is supported by historical evidence, since virtually all great mathematicians have belonged to this group.

This argument can be criticized at several points. First of all, there is the premise that mathematics is neutral. Secondly, even if this premise were to be granted, there is the hidden assumption that mathematics teaching is also neutral, and cannot compensate for the nature of mathematics. In contrast, I have argued that all teaching is intrinsically value-laden and can be made to serve egalitarian (or other) principles.

Thirdly, there is the assumption that the underparticipation of various social groups in mathematics is a consequence of their intrinsic character. This is shown below to be the unwarranted assertion of certain ideological perspectives. Lastly, there is the historical argument. This can be refuted on the grounds that under-representation in the history of mathematics by groups who have been denied access to it, is to be expected.

B. The View that Mathematics is Value-Laden and Culture-Bound

Social constructivism views mathematics as the product of organized human activity, over the course of time. All the different fields of knowledge are the creations of human beings, interconnected by their shared origins and history. Consequently, mathematics like the rest of knowledge is culture-bound, and imbued with the values of its makers and their cultural contexts.

The history of mathematics records its creation

The history of mathematics is the record of its making, not merely the track left by mathematics as it approaches the truth ever more closely. It records the problems posed, and the concepts, propositions, proofs and theories created, negotiated and reformulated by individuals and groups to serve their purposes and interests.

A consequence of this view, since absolutist philosophies have dominated the field, is that the history of mathematics needs to be rewritten in a non-teleological, non-eurocentric way. Absolutist views of mathematics as necessary truth implicitly assume that its discovery is virtually preordained and that modern mathematics is the inevitable outcome. These need correction, for modern mathematics is no more the inevitable outcome of history than modern humankind is the inevitable outcome of evolution.

Many histories of mathematics, such as Eves (1953), promote a simplified Eurocentric view of its development. Scholars such as Joseph (1987) have criticized these histories, and indicated how much more widespread and numerous the traditions and foci of mathematical research and development are, in the centres of culture and civilization throughout world history.

A social constructivist history of mathematics needs to show what mathematical, philosophical, social and political forces drive particular creations, or block them. For example, Henry (1971) argues that the creation of the calculus was within Descartes' grasp, but that he avoided the issues because to approach the infinite would be blasphemous. Less speculatively, an increasing number of studies, such as Restivo (1985), MacKenzie (1981) and Richards (1980, 1989) show the interplay of forces at work in the social history of mathematics, which depend on the social positions and interests of the participants, rather than on purely objective and rational criteria.

All fields of human knowledge are interconnected

Social constructivism begins from the premise that all knowledge is generated by human intellectual activity, providing an underlying genetic unity to all fields of human knowledge. Social constructivism also rests the justification of all knowledge on a shared foundation, namely human agreement. Thus both in terms of its origins and its justificatory foundations, human knowledge has a fundamental unity, and all fields of human knowledge are thereby interconnected. Consequently, according to social constructivism, mathematical knowledge is indissolubly linked with other fields of knowledge, and through its shared roots, it is also value-laden, as other fields of knowledge are acknowledged to be, since it is connected with them.

This is in direct opposition to the Anglo-American tradition in epistemology, according to which the justificatory bases of the different branches of knowledge are wholly distinct. For example, Hirst and Peters (1970) and Hirst (1974) argue that knowledge is divided into distinct autonomous 'forms', each with their own characteristic and unique concepts, relationships, truth tests and verification criteria, and methodologies and procedures. Thus, according to this view, there are quite distinct methods of justification, applied in the different fields of knowledge. However, even this view acknowledges that there is a shared substratum to the genesis of knowledge in different fields, for, according to Hirst:

> The various forms of knowledge can be seen in low level developments within the common area of our knowledge of the everyday world. From this there branch out the developed forms which, *taking certain elements in our common knowledge as a basis*, have grown in distinctive ways.
>
> (Brown *et al*, 1981, page 230, emphasis added.)

Thus even traditional epistemological approaches acknowledge the shared origins of all human knowledge in our common culture, even if the means of justification vary in the different branches of knowledge.

Less conservative parallels of the social constructivist views of knowledge are found in other areas of inquiry, including branches of philosophy, sociology and psychology, as we saw in Chapter 5. One such parallel is to be found in the modern continental 'post-structuralist' or 'post-modernist' philosophers, such as Foucault (1972) and Lyotard (1984). These authors take the existence of human culture as a starting point. Foucault argues that the divisions of knowledge accepted today are modern constructs, defined from certain social discourses. Lyotard (1984) considers all of human knowledge to consist of narratives, each with their own legitimation criteria. What both these thinkers exemplify is a new intellectual tradition which affirms that all human knowledge is interconnected through a shared cultural substratum, as social constructivism asserts.

Mathematics is culture-bound and value-laden

Since mathematics is linked with all human knowledge, it is culture-bound and imbued with the values of its makers and their cultural contexts. Consequently it pervades social and cultural life (Davis and Hersh, 1988). This means that a basis for the cultural location of mathematics is needed.

Shirley (1986) proposes that mathematics can be divided into formal and informal, applied and pure mathematics. Combining these distinctions leads him to four categories of mathematical activity, each including a number of different practices. These are:

1 formal-pure mathematics, including university research mathematics, and much of the mathematics taught in schools.
2 formal-applied mathematics, carried out both in educational establishments, and beyond, such as with statisticians working in industry.
3 informal-pure mathematics engaged in outside the social institutions of mathematics, which might be termed 'cultural' pure mathematics.
4 informal-applied mathematics, comprising the vast range of mathematics embedded in everyday life, crafts, customs or work.

Dowling (1988) offers a richer model of the contexts of mathematical activity than this, building on the work of Foucault and Bernstein. He distinguishes four fields as one dimension of his model. These are the fields of Production (creation), Recontextualization (teacher rhetoric and pedagogic representation), Reproduction (classroom practice) and Operationalization (application and implementation of mathematical knowledge). The second dimension consists of four 'careers' or social locations of mathematical practices. These are Academic (higher education), School, Work and Popular (consumer or domestic). The result is a detailed model of the different social spaces, practices and discourses of mathematics (sixteen in all), which acknowledges the legitimacy of many aspects of non-academic mathematics.

By including this variety of contexts of mathematics we are admitting what D'Ambrosio (1985, 1985a) terms 'ethnomathematics'. According to the thesis of Bishop (1988, 1988a), such culturally embedded mathematics, in particular the activities arising from counting, locating, measuring, designing, playing and explaining, are the cultural roots of all mathematics. Dowling (1988) claims that the identification of such cultural invariants is illusory. Nevertheless, there is agreement that more than the traditional academic mathematics is legitimate.

An outcome of this view of mathematics is a challenge to the cultural domination of abstract, white, male mathematics. For if ethnomathematics is admitted as genuine mathematics, then mathematics is no longer the province of a privileged élite. On the contrary, mathematics is a universal human characteristic, which like language, is the cultural birthright of all peoples.

As a part of a society's culture, mathematics contributes to its overall goals. That is to help people make sense of life and the world, and to provide tools for dealing with the full range of human experiences. As a part of culture, mathematics serves these overall goals. But in each different culture mathematics may be assigned different roles and parts to play, as a contribution to these goals. Thus the purposes of mathematics in a culture may be religious, artistic, practical, technological, study for its own sake, and so on. Whichever it is, the mathematics of each culture presumably serves its own purposes efficiently and well, since it has evolved to meet certain needs and survived. Consequently, the mathematics of each culture is equally valuable, since all cultures are equally valid.

An objection to this argument can be anticipated. To claim that all cultural forms of mathematics are equally valuable, is to deny the power of what might be termed Western academic mathematics. This is the discipline which lies at the heart of modern science and technology, and industry and production. Great progress has been achieved in these fields, and Western mathematics has made a critical contribution. In harnessing the forces of nature and enhancing industrial production the mathematical component of Western culture is immensely efficient, and has no rival.

But it is a fallacy to argue from this that consequently Western academic mathematics is more valuable or efficient than the mathematics of any other culture. For claims of the value or efficiency of mathematics assume a system of values. Each culture has values that are a part of its view of the world, its overall goals, and the purposes it gives to its members. Each culture, like each individual, has the right to integrity. Thus, the system of values of each culture are *ab initio*, equally valid.[1] In absolute terms, there is no basis for asserting that the system of values of one culture or society is superior to all others. It cannot be asserted, therefore, that Western mathematics is superior to any other forms because of its greater power over nature. This would be to commit the fallacy of assuming that the values of Western culture and mathematics are universal.

The acknowledgment of the culture-bound nature of mathematics leads inevitably to the acknowledgment of its value-laden nature. There is a growing literature which recognizes the values implicit in mathematics, and the need for their critical examination. Bell *et al.* (1973) raised the issues of military involvement in mathematics, and the moral issues raised. More recently authors such as Maxwell (1984), Restivo (1985), Ernest (1986), Bishop (1988) and Evans (1988) have raised the question of the values implicit in mathematics, mostly from an educational standpoint.

C. The Educational Implications of Social Constructivism

The social constructivist view of mathematics alone does not entail a particular approach to education; only when combined with a set of values and principles. Therefore we need to add a set of values and principles concerning education.

A set of values

The set of values adopted here are those of the public educator ideology. These are also largely the values of Western liberalism. These comprise the following:

> liberty, equality and fraternity as the basic values; respect for the integrity and individuality of all persons;
>
> the extension of these values to all groups of people and cultures, according them equal value and status;
>
> the right of all individuals to equal opportunities regardless of sex, race, creed, social class, sexual orientation, age, disability, or other discriminatory characteristics;
>
> the acceptance of democracy, the right of persons to collectively determine their life circumstances, as a means of enacting these values politically.

There is no necessity behind this choice of values, for no values are strictly necessary. However, they are more or less the values endorsed by the charters, bills and universal declarations of human rights, as well as in the laws of Western liberal democracies.

The educational implications of social constructivism

Educational principles are also required. We adopt two: (1) schooling and the curriculum should embody and respect the above values as much as is possible, and (2) the mathematics curriculum should be a representative selection, reflecting the nature of the discipline itself.

In accordance with these principles, school mathematics should reflect the following features of mathematics.

1. Mathematics consists primarily of human mathematical problem posing and solving, an activity which is accessible to all. Consequently, school mathematics for all should be centrally concerned with human mathematical problem posing and solving, and should reflect its fallibility.
2. Mathematics is a part of human culture, and the mathematics of each culture serves its own unique purposes, and is equally valuable. Consequently, school mathematics should acknowledge the diverse cultural and historical origins and purposes of mathematics, and the real contributions of all, including women and non-European countries.
3. Mathematics is not neutral but laden with the values of its makers and their cultural contexts, and users and creators of mathematics have a responsibility to consider its effects on the social and natural worlds. Consequently school mathematics should explicitly acknowledge the values associated with mathematics, and its social uses. Learners should be aware of implicit social messages in the mathematics curriculum and should have the confidence, knowledge and skills to be able to understand the social uses of mathematics.

2. Anti-racist and Multicultural Mathematics Education

The value-laden nature of mathematics has consequences for anti-racist and multicultural education.

A. What is the Problem?

The first issue to be addressed is the nature of the problem concerning the underparticipation of racial and ethnic groups in mathematics. Briefly, three aspects of the problem can be identified from the literature.

The educational underachievement of racial minority students

The most immediate aspect of the problem is the educational underachievement of students from certain ethnic minority groups. The report of the committee of inquiry into the education of children from ethnic minority groups concluded that the achievement of pupils from certain ethnic groups were significantly below average levels in Britain at 16+ (Swann, 1985).

> For example, in the DES 1981/82 Survey of five multi-racial LEAs, 6% of West Indians gained 5 or more higher grade passes at CSE and O Level exams compared to 17% of Asians and 19% of all others (the national average for England as a whole was 23%).
>
> (Commission for Racial Equality 1985, page 3)

More recent findings confirm that pupils from certain ethnic minority groups underachieve educationally. For example, a study of 3000 pupils in four urban areas found that children in some but not all black groups scored substantially below average in mathematics on entering secondary school (Smith and Tomlinson, 1989).

However such findings must not be generalized to entire ethnic groups. For the problems of underachievement vary greatly from school to school and region to region (Smith and Tomlinson, 1989), as well as with socio-economic status and other factors (Commission for Racial Equality, 1985). What the results do very strongly suggest is that many ethnic minority students are not offered equal opportunities to learn and achieve in British schools.

Institutional racism in education

A second dimension of the problem is that of institutional racism in education. According to the US Civil Rights Commission, racism is 'any attitude, action or institutional structure which subordinates a person or group because of their colour . . . Racism is not just a matter of attitudes; actions and institutional structures can also be a form of racism.' Institutional racism, according to most authorities, is largely to blame for the first aspect of the problem (Commission for Racial Equality, 1985; Meighan, 1986; Gill and Levidow, 1987; Shukla, 1989). In conjunction with unconscious attitudes and actions, from which it is inseparable, institutional racism is manifested in a number of ways in education, including the following, culled from the literature:

1 *The cultural content of the curriculum* Mathematics is presented as absolutist and the product of white middle-class males, with decontextualized, abstract and formal mathematics most prized. This alienates learners, erects linguistic and cultural barriers, and devalues other cultural heritages.
2 *Assessments* The use of culture-biased tests, such as IQ tests, and modes of assessment, such as competitive tests, serves as a cultural filter to differentially exclude blacks, other ethnic minorities and women. They also result in ability labelling in schools, resulting in a disproportionate number of blacks labelled as 'remedial' and 'SEN'.
3 *Biased texts and worksheets* The use of biased texts, reflecting and presupposing the dominant cultural context, and ignoring or stereotyping representatives of minority groups, serves to reinforce and build prejudice and stereotyping, in black and white students alike.
4 *The modes of teaching* A widespread emphasis on written individualistic reproductive work, as opposed to oral, cooperative or creative mathematics, culturally discriminates and disadvantages a variety of social groups including females.
5 *School organization* Selection, tracking, setting or grouping all often disadvantage black students, since they are often based on culturally biased norms and assessments.
6 *Lack of positive black role models among school staff* This serves to reinforce the stereotypes and prejudices of all concerning the power and authority, even the superiority of whites, and the inferior social status of blacks.
7 *Unconscious racism among teachers* This results in unconscious discriminatory behaviour patterns towards black students, as well as stereotyped expectations, reducing their opportunities to learn.

Institutional racism in education results in directly reduced educational opportunities and consequent life-chances. It also has a negative impact on the beliefs, attitudes and self-concepts of students, both black and white. Both groups learn to see blacks as inferior in view of the dominance of white culture and the relatively low position of blacks throughout school and society.

Racism in society

The third and most sinister aspect of the problem is the widespread existence of racism in society, in both overt and institutional forms. As Lord Scarman said:

> Black Britons are suffering from a discrimination and racial disadvantage

> deriving partly from a deep unconscious prejudice in the rest of the community, which we've not yet exposed and killed. I think this is true of almost everybody.[2]

Racism in society is manifested in a number of important forms, including the following.

1. *Overt racist behaviour and beliefs* These are widespread in Britain today, are directed at black people throughout their lives, and are particularly damaging during the vulnerable pre-school years (Commission for Racial Equality, 1990).
2. *Cultural domination* This denies the equal validity of the cultures that co-exist in multicultural Britain, especially black cultures, and keeps in place and legitimates institutional racism in education.
3. *Structural institutional racism* This discriminates against blacks, denying them equal access to housing, education, employment, justice, political representation, and ultimately to power and wealth, thus reproducing social inequality.

Institutional racism in society plays an important part in the white middle-class male hegemony, something akin to the 'ideological state apparatus' of Althusser (1971), which keeps in place the distribution of wealth and power. The original black immigrant groups (and other such groups) in the main originally came to take up employment in the lowest social strata. This hegemony serves to confine the British descendants of immigrants, and those identified with them on account of race, to these lowest strata.

Attempts to question this cultural domination in society are met with swift reactionary responses, illustrated by Thatcher's (1987) attack: 'children who needed to count and multiply were learning anti-racist mathematics — whatever that might be', and media assault on local education authorities implementing anti-racist policies, such as ILEA and Brent (Jones, 1989).[3] In the USA, when the governing body of Stanford University voted to modify the 'Western Civilization' course to include multicultural elements, former Education Secretary W. Bennett violently attacked the move in public (Reed, 1988).

Epistemology is central to these problems, for it plays a major role in transmitting institutional racism through schooling. Philosophical presuppositions about the nature of mathematics underpin the content of the mathematics curriculum, and are presupposed in assessment and pedagogy. Such factors all contribute to the cultural domination of society through an epistemology, with an absolutist view of mathematics as its centrepiece.

It is evident that the problem of black and ethnic minority underparticipation in mathematics cannot simply be solved through the dissemination of information in schools, or even through public education. There are powerful forces at work keeping cultural domination and institutional racism in place, for it serves the interests of capital and the politically powerful.

B. Ideological Views of the Problem and its Solution

Each of the five ideologies of education has a different perception of the problem of mathematics education in a multicultural society, and its solution.

Industrial trainers

The industrial trainers deny the existence of the problem. On the contrary, they assert that anti-racism, anti-sexism and even multiculturalism are (1) illegitimate since mathematics is viewed as neutral, and (2) are themselves the problem because they represent political interventions aimed at undermining British culture and values (Cox *et al.*, 1987; Palmer, 1986). The solution to the problem as redefined by the industrial trainers is (1) to strenuously deny that there are problems in the education provided for ethnic minorities in Britain, (2) to actively oppose anti-racism and anti-sexism (Palmer, 1986; Brown, 1985; Thatcher, 1987), (3) to vociferously assert that mathematics and science are value-free and unrelated to social issues (Cox and Boyson, 1975; CRE, 1987; Cox *et al.*, 1987), and (4) to promote a monocultural view of Britain and its history (Brown, 1985).[4]

These responses stem from the ideology of the industrial trainers which asserts the neutrality of mathematics and is anti-egalitarian, monoculturalist and crypto-racist. Equality and hence the principle of equal opportunities are explicitly denied by this perspective (Cox and Dyson, 1970). Its hierarchical view of humankind assigns a lower place to blacks than whites, and uses statistics of the lower achievement of ethnic minorities to support this view of their inferiority (Flew, 1976). This racism is usually denied, so it is termed crypto-racism. However, it is becoming more overt, with a government decision permitting choice of schools on purely racial grounds, contrary to the Race Relations Act of 1976 (Hughill, 1990). The myth that Britain is a monoculture has been refuted elsewhere (Ernest, 1989).

The industrial trainer views of knowledge, culture and race are deeply repugnant, because of their denial of fundamental principles of natural justice, and their false conceptions. However, the question arises, what are the roots of these views? Part of their ideology, it has been argued, includes the need to defend the boundaries of their social group,[5] and to strictly demarcate the boundary of the working class. Douglas (1966) links perceived threat to the group boundary with a concern for purity, and sure enough we find such concerns in (1) keeping mathematics neutral and free from social issues ('even mathematics ... [suffers from the] gradual *pollution* of the whole curriculum', Cox *et al.*, 1987, page 3, emphasis added), and (2) keeping the culture and national identity pure, unbesmirched by 'alien' influences (Palmer, 1986; Brown, 1985). These impulses therefore can be attributed to deep unconscious feelings of insecurity and threat (making them all the more dangerous).

Old humanists

The old humanists do not acknowledge that there is any knowledge-related educational problem concerned with race or ethnic groups. They can be said to make a fetish of the purity of middle-class culture, and mathematics is viewed as pure and unrelated to social uses and issues (the above interpretation of purity applies here). This group is certain of the superiority of its own culture, and sees mathematics in other cultures as lacking the power, abstraction and purity of Western academic mathematics. Consequently issues of race, gender or social diversity are deemed irrelevant to mathematics, and any attempt to accommodate such social issues by making knowledge and learning more accessible stands to dilute standards and threaten the purity of the subject. In short, this view actively promotes the cultural domination of absolutism, which is the epistemological basis of institutional racism.

Technological pragmatists

This perspective perceives a problem in the educational underachievement of ethnic minority students, as they affect the utilitarian needs of employment and vocational training. Beyond this, the diversity of culture, race and gender in society are seen as immaterial. Thus although this perspective's perception of the problem is limited, there is a concern to improve education to bring ethnic and other minority students onto the labour market. Thus, multicultural initiatives and awareness are encouraged strictly with this end in mind, as is illustrated by Shukla (1989), sponsored and published by the Further Education Unit (see also Further Education Unit, 1984; 1985).

However, this perspective sees mathematics as neutral, unless it is applied to industry and technology. Thus multicultural mathematics is seen as an irrelevance, since applications do not concern cultures but occupational uses. Although purportedly meritocratic, technological pragmatism is essentially reproductive of the social hierarchy, and most ethnic minority students are expected to find occupations at the lower end of the job market. Consequently, if not overtly racist, this perspective contributes to institutional racism in education and society.

Progressive educators

The progressive educator perspective is person-centred, concerned to meet the needs of all learners, including those of ethnic minority students. This perspective perceives a problem in the underachievement and alienation of black students, including in mathematics. The solution from this perspective is multiple, and involves (1) reacting to individual instances of prejudice in the classroom, (2) working with ethnic minority students to enhance self-esteem and confidence, (3) adding a multicultural dimension to education, to accommodate the ethnic diversity of the school population, and in particular (4) adding multicultural elements to mathematics, such as introducing rangoli patterns and mentioning the Hindu-Arabic origins of our number system (Cockcroft, 1982). This is part of the general strategy to give a human face to mathematics, and bring it closer to the interests and experiences of all learners, to enable them to be creative and to develop positive attitudes towards it, although the viewpoint remains (progressive) absolutist with regard to mathematics. This perspective is also concerned to shield children from conflict, and is often over-protective concerning controversial issues such as racism and other socio-political concerns (Carrington and Troyna, 1988).

The progressive educators treat the problems of ethnic minorities as they perceive them, and consequently the solutions proposed are only partial, with a number of weaknesses, including the following. (1) The culture-bound nature of knowledge, mathematics in particular, is not acknowledged, and so the solution fails to address the cultural domination of the curriculum. (2) There is insufficient recognition of institutional racism in education and the overt and institutional racism in society, and so these root problems are not addressed. (3) The problems of overt and institutional racism are avoided in the classroom, with the aim of protecting the sensibilities of the learners. The outcome is a denial of these problems and their importance, despite their impact on children even before they begin schooling (Commission for Racial Equality, 1990). (4) Multicultural education is seen to be the solution to the problems of black children, and is not seen as a necessary response to the nature of knowledge and to the forms of racism that exist in society, and hence of importance for all learners, teachers and members of society. (5) Through this limited perception and response to the problems, this perspective is palliative, and tends to reproduce cultural domination and the structual inequalities in society. Despite these weaknesses, this is the most concerned and responsive of the perspectives considered so far.

Public educators

On the basis of its epistemology and values, the public educator ideology incorporates an awareness of the problem of black students in education as described at the beginning of this section. It acknowledges that overt and institutional racism are the causes of some black student' underparticipation and underachievement in mathematics, and their consequent alienation from school and society. In particular, it is recognized that absolutist views of mathematics provide the epistemological foundations of institutional racism.

This view is reflected in the public educator strategies for addressing the problem. These work at three levels, concerning first, explicit discussion and consciousness-raising, second, the selection of the content of the mathematical curriculum, and third, the modes of pedagogy and organization employed.

The first four strategies are directed at epistemological and political education in mathematics: to communicate a social constructivist view, and to reveal and question the cultural domination of our society by absolutism, as part of the 'ideological state apparatus' which sustains the structures and existing inequalities, including the inferior position of blacks.

(1.1) There is explicit discussion of mathematics as a social construction, and the fallibility of mathematical knowledge is communicated through its teaching, preventing student invalidation through failure to attain correctness (Abraham and Bibby, 1989).

(1.2) The multicultural origins of mathematics are discussed, and explored. The validity of the mathematics of all cultures is affirmed and exemplified (Zaslavsky, 1973; Wiltshire Education Authority, 1987; Mathematical Association, 1988).

(1.3) Racism at all levels and institutional racism, with its cultural domination and denial of opportunities, are discussed and condemned (Cotton, 1989).

(1.4) The social contexts of schooling are discussed, including the hierarchies of power relations in school and society, and the social problems facing all citizens, especially blacks, with the public educator aim of conscientization' through mathematics education (Frankenstein, 1983; Abraham and Bibby, 1988).

The next cluster of strategies focus on the content of teaching, aiming to make school mathematics meaningful to the learners' experiences and lives; de-mystifying social statistics; and empowering all to make sense of and take control of their lives.

(2.1) 'Installed mathematizations' in society are exemplified, analyzed and discussed, including social statistics concerning the life situations of ethnic and minority groups in Britain and globally (Abraham and Bibby, 1988; Frankenstein, 1989).

(2.2) Multicultural mathematics including ethnomathematics, games, number systems, and geometrical patterns is explored and related to its cultural origins and purposes (Gerdes, 1985; Bishop, 1988; Wiltshire Education Authority, 1987; Mathematical Association, 1988).

(2.3) Errors and the limits of confidence in mathematics are treated, such as in measurement, probability and statistics, to foster critical thinking about the uses of mathematics and confidence limits involved.

(2.4) Mathematics based controversy and value-rich issues, such as voting systems, the misleading use of statistics in advertisements and by government, historical controversies, are debated and argued (Dugas, 1989; Frankenstein, 1989; 1989a).

The goals addressed in these content issues depend essentially on the third cluster of strategies concerning pedagogical and school approaches. Their purpose is primarily to empower students to become autonomous problem posers and solvers and to develop a collateral ethos of cooperation, not competition; to link education with life in the community and oppose racism in both; and to contribute to students' development of awareness and engagement in all aspects of social life, without denying the problems and contradictions in society (Frankenstein and Powell, 1988).

(3.1) Varied learning approaches are used, including cooperative projects, problem solving and discussion, and the autonomous pursuit of mathematical investigations. This is applied to a thorough treatment of the mathematical content necessary for external assessments, for certification is a most important outcome of education for future life chances (Brown, 1984).

(3.2) The teacher adopts a classroom role that respects the responsibility and integrity of students whilst explicitly acknowledging the power asymmetries of schooling (Agassi, 1982).

(3.3) The teacher screens classroom materials for bias and stereotyping, for cultural barriers such as the use of language, and fosters an anti-racist classroom climate in which instances of bias, when found, are explicitly discussed (Inner London Education Authority, 1985).

(3.4) Outside the classroom the teacher participates in a whole school equal opportunities policy promoting anti-racist education, and also promotes links between home, school and the local community.

This ideological position is the most committed to the issues of equal opportunities and anti-racist mathematics teaching, and is the only perspective to perceive the problems of racial education fully, with their epistemological and socio-political dimensions. However, this approach can lead to problems and conflicts. These include, first of all, the conflict between instrumental goals (mastering the routine aspects of mathematics, preparing for external examinations) versus social goals (studying mathematics for its intrinsic power in providing thinking tools). This is not merely a conflict of classroom aims, but may also be a conflict between cultures of schooling (Mellin-Olsen, 1981; 1987; Crawford, 1989).

A second conflict is that between the aims and atmosphere of the anti-racist mathematics classroom and the rest of the school ethos, or with learners' previous school culture.

The third area of conflict is between the aims of anti-racist mathematics education and the aims of the other four ideologies. The most powerful groups in society wish for mathematics education to serve their aims and to largely reproduce the hierarchical structure of society, which is facilitated by docile blacks and workers who do not question or threaten the stability of society. Empowering and politicized anti-racist education has goals explicitly opposed to this, and is met with powerful and irrational opposition (Jones, 1989). When in 1986 a mathematics examination question presented candidates with real data on combined East-West military spending and asked what portion would meet basic world needs, it was headlined as 'sinister', and examiners were appointed to politically vet future questions.[6] The implementation of a public educator anti-racist mathematics curriculum will be opposed and may be compromised by the conflict.

The views of the progressive and public educators serve to distinguish the approaches of multicultural and anti-racist education, respectively. Multicultural education seeks to ameliorate the conditions of black learners in mathematics, and is concerned with individualistic concerns, such as prejudice, confidence and self esteem. Anti-racist education goes beyond this, focusing on racism, social structures and cultural domination, seeking to use mathematics to empower learners, to further social justice, and ultimately to change society (Gaine, 1987).

The public educators alone recognize the significance of epistemology, especially the philosophy of mathematics, in sustaining or challenging the hegemony of white, middle-class males. This includes the saturation of views of knowledge, curriculum, human nature and ability and society with hierarchical notions, and the associated valuing of their higher levels. Only the public educator perspective challenges these ideological assumptions, and recognizes that the problems concern schooling for all, not just for a minority.

3. Gender and Mathematics Education

A. What is the Problem?

A second equal opportunities issue is the disparity between men's and women's participation rates in mathematics. For two decades evidence has been accumulating that women come out of mathematics education disadvantaged relative to men (Fox *et al.*, 1977).[7] In Britain, Hilary Shuard documented this disparity fully at the beginning of the 1980s (Cockcroft, 1982). In purely descriptive terms, the problem has two components.

1. Underachievement of females in external examinations

There is overwhelming evidence that the proportions of females passing mathematics examinations at 16 and 18 years of age in Britain is less than that of males, and that the proportions of males at the higher grades outweighs that of females (Cockcroft, 1982; Burton, 1986; Open University, 1986).

2. Underparticipation of females in mathematics post 16

From this age onwards, at each decision point the proportion of females opting for mathematical study diminishes, relative to men.

Since mathematics is a 'gateway' to many areas of further study, and a 'critical filter' in employment, this is very important (Sells, 1973, 1976). It is a source of inequality, it closes many educational and career opportunities to women, and deprives society of the benefits of their talents.

However, the gender problem in mathematics goes much deeper than has been indicated. There are two further dimensions: institutional sexism in education, and sexism in society, which lie at the root of the problem (Cockcroft, 1982; Walden and Walkerdine, 1982; Whyld, 1983; Burton, 1986; Open University, 1986; Walkerdine, 1989; Walkerdine *et al.*, 1989). The problem almost exactly parallels that concerning ethnic minorities, and can be summarized similarly.

Institutional sexism in education

This is manifested in terms of:

- the cultural content of the curriculum (mathematics as a male domain);
- the forms of assessment used (competitive);
- gender-biased texts and worksheets (stereotyped);
- the modes of teaching employed (individualistic instead of oral and cooperative);
- the organization of schooling and selection;
- the insufficiency of positive female role models among mathematics teachers; and
- unconscious sexism among teachers.

Sexism in society

This is manifested in a number of powerful forms, including:

- overt sexist beliefs and behaviour;
- cultural domination (legitimating and reproducing gender-stereotyped roles and gender-biased areas of knowledge, including mathematics); and
- structural institutional sexism (which denies women equal opportunities, thus reproducing the gender inequalities in society).

The way that some of these factors are inter-related and contribute to the gender problem in mathematics can be shown as a reproductive cycle (Figure 12.1). This shows how girls' lack of equal opportunities in learning mathematics, from a variety of causes, leads to girls' negative views of their own mathematical ability, and reinforces their perception of mathematics as a male subject. A consequence is girls' lower examination attainments and participation in mathematics. Because of its 'critical filter' role in regulating access to higher level occupations, this leads to lower paid employment for women. The positioning of women disproportionately in the lower paid and lower status occupations reproduces gender inequality in society. This reinforces gender stereotyping, among both men and women. This in turn contributes an ideological component to institutional sexism in education, which reproduces the lack of equal opportunities for girls in mathematics, completing the cycle.

This cycle should not be taken as rigidly reproductive or understood too deterministically. It does illustrate how some aspects of the gender related problems in mathematics combine with other factors to reproduce social inequalities. It also suggests that any would-be solutions must be multiple, attacking each stage of transmission in the cycle, and that the problem is not merely educational, but also exists in the socio-political realm. For although three components concern mathematics education there are also three components that are essentially socio-political in nature, as the figure shows. The broader dimension means that it is a problem for all of society, not just for girls and women.

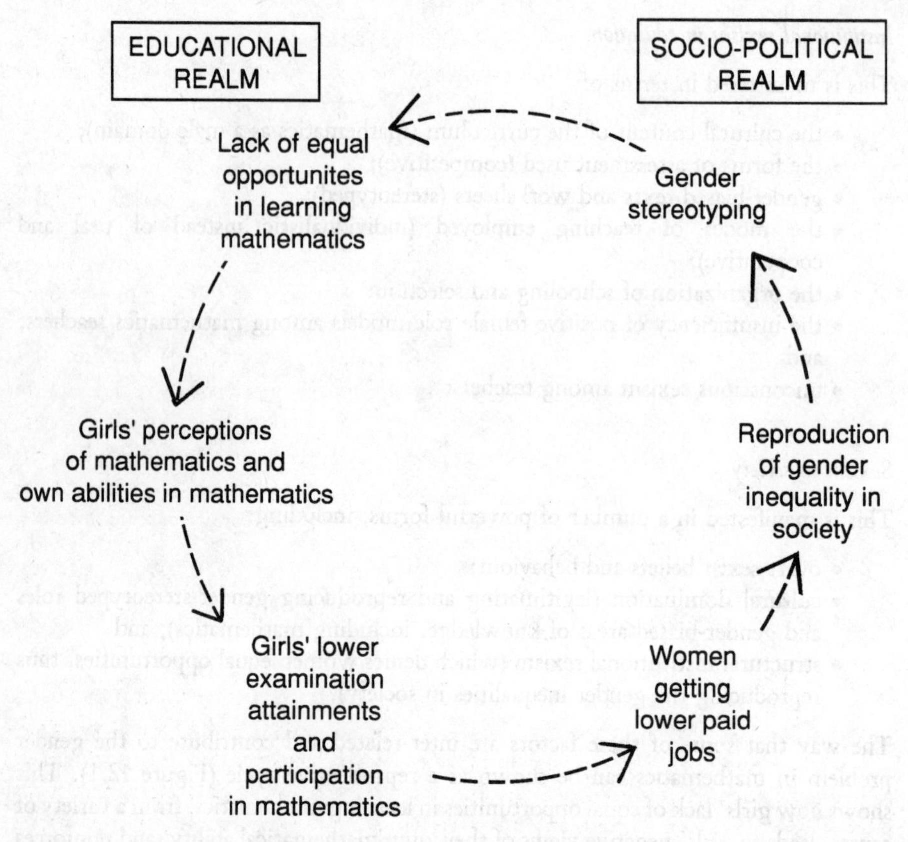

Figure 12.1: The Reproductive Cycle of Gender Inequality in Mathematics Education

B. Perceptions of the Problem and its Solution

Each of the five ideologies of mathematics education has a different perception of the problem of gender and mathematics, and its solution, paralleling their views on race.

The industrial trainers deny the existence of any problem, seeing the inequality of women as stemming from the intrinsically hierarchical nature of humankind ('Sex equality is an impossible dream', Campaign for Real Education, 1989, page 2). Mathematical ability is seen as fixed and inherited, and distributed in the same unequal fashion. The old humanists share this view, although they adopt a less reactionary stance than the industrial trainers, who actively oppose anti-sexist approaches to mathematics. Both of these ideologies help to sustain and recreate the gender inequalities in our hierarchically structured society.

The technological pragmatists see the problem in terms of obstacles to women joining the technological workforce, which they believe should be overcome through girl-friendly training. They do acknowledge that steps need to be taken to overcome gender bias in mathematical and technological education (see for example, Women's National Committee, 1985). However they do not perceive that mathematical knowledge itself may be gender-biased.

The progressive educators view the problem in terms of individual girls' underachievement and lack of confidence. According to this view, there are personal obstacles to girls achieving their potential, which may be exacerbated through insensitive or sexist teaching and materials. The progressive educator solution is to alleviate this problem by (1) making sure curriculum materials are not gender biased and providing good female role models in mathematics; and (2) helping girls to develop positive mathematical self-concepts and attitudes, through individual attention and the experience of success in mathematics.

This approach is individualistic, locating the problem in individuals, and seeking to ameliorate their conditions. It represents the deepest and most principled response considered so far. However, like the technological pragmatists, by not seeing that the problem has epistemological and socio-political roots, and failing to challenge the structural and institutional sexism in schools and society, it helps to reproduce existing gender inequalities.

The public educator view

The public educators perceive the problem of gender and mathematics in terms of its epistemological and socio-political basis, and even question the 'fact' of girls' underachievement in mathematics. This reconceptualization of the problem is supported by research, for prior to examinations at 16+ large scale achievement testing does not show the superior achievement of boys unequivocally. For example, the APU found little in terms of statistically significant differences favouring boys, at age 11.

> In only two subcategories have the differences been significant for each of the five surveys. These subcategories are Length, area, volume and capacity and Applications of number. Eleven year old girls have achieved higher mean scores than boys in every survey for the subcategory, Computation; whole numbers and decimals, two of the differences being significant.
>
> (Assessment of performance Unit 1985, page 698).

Thus in the largest scale British surveys the significant differences in attainment that favour boys are more or less balanced by those that favour girls. Furthermore, these results show a large amount of individual, institutional and regional variation, far outweighing the overall sex differences. The claim that boys outperform girls in mathematics during most of the years of schooling is not sustained by the published evidence. Indeed, in the '11+' examination used in the 1950s and 1960s for selection at age 11, girls consistently outperformed boys in mathematics (and language and verbal reasoning) so that differential pass marks were imposed to give boys equal pass rates.

At 16+ there are significant overall differences in examination performance in mathematics, with a higher proportion of boys passing and attaining higher grades (Cockcroft, 1982; Burton, 1986; Open University, 1986; HMI, 1989). However some of this difference appears to be due to the differentiated curriculum experiences of the genders. Sharma and Meighan (1980) compared the attainments of boy and girls in mathematics who also studied physics, technical drawing or neither. They found that collateral studies were far more significant statistical correlates of higher attainment in mathematics than gender. The highest, middle and lowest mean grades in mathematics were achieved by those also studying physics, technical drawing or neither, respectively, and in no case were the gender differences significant. However, the ratio of boys to girls taking 16 + examinations in physics and technical drawing in 1984 were 3:1 and 17:1 respectively (Open University, 1986), so girls have far less experience of these collateral studies. Although these results do not imply that simply studying physics and technical drawing will solve the problem, they do suggest that the inequalities are social artefacts.

After a sustained research programme the Girls and Mathematics Unit (1988) have concluded that underachievement is not the cause of women's underparticipation in mathematics, but that it is due in large part to institutional sexism mediated by teachers.

> [T]he failure of women to enter higher grade careers requiring mathematics can in no way be attributed to general poor performance . . . That girls as a sex attain well in school relative to boys is difficult to dispute, yet everywhere we are surrounded with gender divisions which represent girls as unreasoning, irrational and passive.
>
> We have concentrated on teachers' attributions of performance, but the teachers are not to blame in any simple sense. The discourses they use surround them, both in scientific ideas about children and in other cultural and social practices and institutions.
>
> (Girls and Mathematics Unit 1988, page 11)

Walkerdine argues that the powers of rationality and mathematical thinking are so bound up with the cultural definition of masculinity, and 'that the discursive production of femininity [is] antithetical to masculine rationality to such an extent that femininity is *equated* to poor performance, even when the girl or woman is performing well.' (Walkerdine, 1989, page 268)

Thus from the public educator perspective, the problem of women's underparticipation in mathematics is seen to be due to deeply entrenched cultural discourses which identify mathematics with masculinity and power, and the consequence of these definitions is to 'count girls out' of mathematics (Walkerdine *et al.*, 1989). Thus the problem is seen at base to be epistemological, and inseparably, socio-political in nature. For the cultural domination of rational and scientific knowledge by masculine values, serves to legitimate and sustain men's domination of the power, status and wealth, and hence political hierarchies in society.

The public educator solution is anti-sexist education, which sets out to (1) reveal and combat explicit and institutional sexism in teachers, texts, views of knowledge, and ultimately in the cultural definitions of gender; (2) to provide all with an empowering mathematics education. The aim of this is not merely to compensate girls for their disadvantage. The outcome must be a reconceptualization of the nature of knowledge, especially mathematics, as a social construction, and the restructuring of gender definitions and social divisions, in recognition of these insights.

The public educator ideology offers the broadest conceptualization of the problem of women's underparticipation in mathematics, which is a great strength. However its weaknesses are (1) that it is a controversial position likely to generate widespread opposition from positions of greater power (which it threatens); (2) by identifying the site of the problem as society-wide, it means that anything less than massive social change cannot count as full success. This is both unrealistic and dispiriting, and may draw attention away from what can be achieved in a more limited site, notably the classroom.

4. Conclusion

The five ideologies of education have very different perceptions of the problems of equal opportunities in mathematics, and their solutions. Only the public educator perspective accepts that the problems result from distortions of epistemology and relations in society, serving the interests of the dominant group. In mathematics, these distortions result in the myth of the neutrality and objectivity of mathematics, and the associated values. This myth is destructive in terms of human relations with mathematics, and results in the anxiety, fear, alienation and the 'irrational and impeditive dread of mathematics' (Lazarus, in Maxwell, 1989, page 221) felt by so many (Tobias, 1978; Buxton, 1981; Open University; 1986; Maxwell, 1989). Thus the problem of equal opportunities in mathematics is not merely that of lost opportunities for ethnic minority groups and women. The absolutist view of mathematics creates a problem for all.

Underpinning the neutral view of mathematics is the cultural perspective and values that dominate Western scientific culture. This is the culture of rationality, which values reason but denigrates feeling. It separates knower from known, and objectifies its perceptions, removing the knowing subject from the universe of discourse. It is a discourse of separation and power (Gilligan, 1982), which seeks to subjugate nature and demands certainty and security from the knowledge it legitimates (Walkerdine, 1989). It represents the aggressive masculine half of human nature, which has rejected the receptive and compassionate feminine half. Out of balance, it leads to assertions of power, ever more destructive armaments and conflicts, and the rape of the environment (Easlea, 1983).

The view of mathematics as male owned or as a joint social construction plays a central role in sustaining or challenging the male domination of Western culture. Success at dehumanized male mathematics may diminish our humanity, our ability to care, relate and feel. Sustaining the inferiority of ethnic minority groups and women through this view of mathematics does symbolic violence to all, and subtracts from our integrity as human beings.

Notes

1. This argument does not necessarily lead to moral relativism. To claim that the system of values of each culture are *ab initio* equally valid does not mean that we cannot adopt an overarching statement of universal human rights and values. It would still be relative to humankind or a significant portion of it, but would be pan-cultural.
2. Quotation in BBC1 TV programme *Anglo-Saxon Attitudes,* 5 September 1982; printed in *Times Educational Supplement*, 3 September 1982.
3. Documented instances of media and right-wing attacks on anti-racism include the Goldsmiths College Media Monitoring Unit refutation of the media fabrication that the rhyme 'Baa baa black sheep' was banned as racist in a London borough.
4. Thatcher intervened in the National Curriculum in History 'pressing for a more traditional approach to the curriculum', and 'insist[ing] on more British and less world history' (Judd, 1989).
5. This is illustrated by Thatcher's concern with whether individuals are 'one of us?' (Young, 1989).
6. Question from June 1986 CSE mathematics paper set by London Regional Examination Board. 'Sinister' sub-headline in *The Sun*; report of future vetting in *The Guardian*; both 14 June 1986.
7. The Fox paper cites 200+ publications, most treating gender and mathematics.

13

Investigation, Problem Solving and Pedagogy

1. Mathematics Results from Human Problem Posing and Solving

Social constructivism identifies mathematics as a social institution, resulting from human problem posing and solving. Mathematics is probably unique in the central place it gives to problems, which can remain unsolved but of great interest for thousands of years. But mathematical problems are more significant than long-lived challenges. Often the techniques devised to solve them represent major advances in mathematics. Thus problems also serve as growth points for mathematics.

A number of philosophers have identified problems and problem solving as lying at the heart of the scientific enterprise. Laudan (1977) explicitly proposes a Problem Solving Model of scientific progress. He argues that provided it occurs in contexts (or cultures) permitting critical discussion, problem solving is the essential characteristic of scientific rationality and methodology. In the philosophy of mathematics, Hallett (1979) proposes that problems should play a key role in the evaluation of mathematical theories. He adopts the 'Hilbert Criterion', that theories and research programmes in mathematics should be judged by the extent to which they aid the solution of problems. These approaches both acknowledge the importance of problems in scientific progress, but they both share a focus on the justification rather than the creation of theories. This is the 'context of justification', contrasted by Popper (1959) with the 'context of discovery', which he disparages.

Since the time of Euclid, or earlier, the emphasis in presentations of mathematics has been on deductive logic and its role in the justification of mathematical knowledge. This is one of the great achievements of mathematics. But the emphasis on theorems and proof, and in general on justification, has helped to buttress traditional absolutist views of mathematics. Recognition of the central place of problems and problem solving in mathematics reminds us of another tradition in the history of mathematics, one which emphasizes the context of discovery or creation.

From the time of the ancient Greeks, at least, it has been recognized that systematic approaches can facilitate invention in mathematics. Thus, for example, Pappus wrote a treatise which distinguished between the analytic and synthetic problem solving methods. The former involves separating the logical or semantic components of a premise or conclusion, whereas the latter involves bringing novel elements into play and attempting to combine them. This distinction has recurred throughout history, in recent times has been used by psychologists to distinguish different levels of cognitive processing (Bloom, 1956).

Since the Renaissance, a number of important methodologists of science have attempted to systematize creation in ways that are forerunners of mathematical heuristics. Bacon (1960) proposed a method of induction for arriving at hypotheses, which were then to be subjected to testing. In order to facilitate the genesis of inductive hypotheses, he proposes the construction of systematic tables of results or facts, organized to show similarities and differences. Such proposals, published in 1620, anticipate the heuristics of modern researchers on mathematical problem solving, such as Kantowski, who specified 'Heuristic processes related to planning . . . Searches for pattern . . . Sets up table or matrix' (Bell *et al.*, 1983, page 208).

In 1628 Descartes (1931) published a work embodying twenty-one 'Rules for the direction of the mind'. This proposes further heuristics, many explicitly directed at mathematical invention. These include the simplication of questions, the sequential enumeration of examples to facilitate inductive generalization, the use of diagrams to aid understanding, the symbolization of relationships, the representation of relationships by algebraic equations, and the simplication of equations. These heuristics anticipate many of the heuristics published 350 years later as aids to teaching problem solving, such as in Mason *et al.* (1982) and Burton (1984).

In the 1830s Whewell published 'On the philosophy of discovery', which gave an account of the nature of scientific discovery (Blake *et al.*, 1960). He proposed a model of discovery with three stages: (1) clarification, (2) colligation (induction), and (3) verification, each of which has a number of components and methods attached. Whewell was largely concerned with empirical science, although he believed, following Kant, that necessary truths occur in both mathematics and science. Nevertheless, there is a striking analogy between his model of discovery and that proposed by Polya (1945) for mathematics, a century later. If two of the stages of Polya's model are combined, the result is (1) understanding the problem, (2) devising a plan and carrying it out, and (3) looking back. There is now an exact parallel between the function of these stages and those in Whewell's model.

This, together with the previous examples, serves to show how much of the recent thought on mathematical invention and problem solving in psychology and education has been ancitipated in the history and philosophy of mathematics and science. Evidently the theory of mathematical invention has a history comparable to that of the theory of justification. However, it is not recognized in most histories of mathematics. On the contrary, this century, until Polya (1945), it would seem that writings on mathematical 'discovery' have tended to mystify the process. Thus, for example, Poincaré (1956) and Hadamard (1945) both emphasize the role of intuition and the unconscious in mathematical creation, implicitly suggesting that great mathematicians have a special mathematical faculty which allows them to mysteriously pierce the veil surrounding mathematical 'reality' and truth. This view of mathematical invention supports élitist, absolutist views of mathematics, by mystifying its human creation.

Such views are confirmed by the values attached to mathematics. Mathematical activity and discourse take place on the three levels of formal, informal and social discourse of mathematics. In western society, and in particular, in the culture of the professional mathematician, these are valued in descending order. The level of formal mathematical discourse is reserved for justificatory presentations of mathematics, which is accorded high value. Informal mathematical discourse takes place at a lower level, which is assigned a lower value. But mathematical activity and the creation of mathematics naturally takes place at the informal level, and this means that it has lower status (Hersh, 1988).

Such distinctions and valuations are social constructs, which can be scrutinized and questioned. In earlier chapters, an account of social constructivism is given which inter-relates the creation of subjective and objective knowledge in mathematics. This suggests that the contexts of 'discovery' (creation) and justification cannot be completely separated, for justifications, such as proofs, are as much the product of human creativity as concepts, conjectures and theories. Social constructivism identifies all learners of mathematics as creators of mathematics, but only those obtaining the critical assent of the mathematical community produce *bona fide* new mathematical knowledge, i.e. that which is legitimated (Dowling, 1988). The mathematical activity of all learners of mathematics, provided it is productive, involving problem posing and solving, is qualitatively no different from the activity of professional mathematicians. Non-productive mathematics does not offer the same parallel, because it is essentially reproductive as opposed to creative, comparable to 'frozen' mathematics (Gerdes, 1985).

2. Problems and Investigations in Education

Given that a major part of mathematics is human problem posing and solving, and that this is an activity which is accessible to all, then important consequences for education follow. These consequences, which also depend on the values and principles specified in the last chapter, include the following:

- School mathematics for all should be centrally concerned with human mathematical problem posing and solving.
- Inquiry and investigation should occupy a central place in the school mathematics curriculum.
- The fact that mathematics is a fallible and changing human construction should be explicitly admitted and embodied in the school mathematics curriculum.
- The pedagogy employed should be process and inquiry focused, or else the previous implications are contradicted.

One outcome of these principles is that mathematics *for* all becomes mathematics *by* all (Volmink, 1990).

A. Problems and Investigations: Some Distinctions

Problem solving and investigational work have been a widespread part of the rhetoric of British mathematics education since Cockcroft (1982). Worldwide, problem solving can be traced further back, at least to Brownell (1942) and Polya (1945), and probably earlier. By 1980, in a selective review of research in mathematical problem solving, Lester (1980) cited 106 research references, representing only a small proportion of what had been published by then. In British mathematics education, problem solving and investigations probably first appeared on the scene in the 1960s, in the Association of Teachers of Mathematics (1966) and the Association of Teachers in Colleges and Departments of Education (1967).

One of the difficulties in discussing problems and investigations is that these concepts are ill defined and understood differently by different authors. However, there is agreement that they both relate to mathematical inquiry. Thus, there are a number of preliminary distinctions which can be usefully applied to both of them. For it is possible to distinguish the object or focus of inquiry, the process of inquiry, and an inquiry based pedagogy.

1. The Object of Inquiry

The object or focus of inquiry is either the problem itself or the starting point of the investigation. One definition of a problem is 'a situation in which an individual or a group is called upon to perform a task for which there is no readily accessible algorithm which determines completely the method of solution . . . It should be added that this definition assumes a desire on the part of the individual or the group to perform the task.' (Lester, 1980, page 287). This definition indicates the non-routine nature of problems as tasks which require creativity for their completion. This must be relativized to the solver, for what is routine for one person may require a novel approach from another. It is also relative to a mathematics curriculum, which specifies a set of routines and algorithms. The definition also involves the *imposition* of a task on an individual or a group, and willingness or compliance in performing the task. The relationship between an individual (or group), the social context, their goals, and a task, is complex, and the subject of activity theory (Leont'ev, 1978; Cristiansen and Walther, 1986). In particular, the relationship between teacher and learner goals is complex, and it is not possible to simply transfer one to the other by command.

The concept of an investigation is problematic for two reasons. First of all, although 'investigation' is a noun, it describes a process of inquiry. Thus a dictionary definition of investigation is 'The action of investigating; search, inquiry; systematic examination; minute and careful research.' (Onions, 1944, page 1040). However, in mathematics education there has been a shift in meaning, or a fairly widespread adoption of a curtailed *facon de parler*, which identifies a mathematical investigation with the mathematical question or situation which serves as its starting point. This is a metonymic shift in meaning which replaces the whole activity by one of its components (Jakobsen, 1956). The shift is also teacher-centred, focusing on teacher control through 'setting an investigation' as a task, analogous to setting a problem, in contrast to a learner-centred view of investigation as a learner directed activity.

The second problem is that whilst investigations may begin with a mathematical situation or question, the focus of the activity shifts as new questions are posed, and new situations are generated and explored. Thus the object of the inquiry shifts and is redefined by the inquirer. This means that it is of limited value to identify an investigation with the original generating situation.

2. *The process of inquiry*

Contrasting with the object of inquiry is the process of inquiry itself, although these cannot be separated entirely, as we have seen in the case of investigations. If a problem is identified with a question, the process of mathematical problem solving is the activity of seeking a path to the answer. However this process cannot presuppose a unique answer, for a question may have multiple solutions, or none at all, and demonstrating this fact represents a higher order solution to the problem.

The formulation of the process of problem solving in terms of finding a path to a solution, utilizes a geographical metaphor of trail-blazing to a desired location. Polya elaborates this metaphor. 'To solve a problem is to find a way where no way is known off-hand, to find a way out of a difficulty, to find a way around an obstacle, to attain a desired end that is not immediately attainable, by appropriate means.' (Krulik and Reys, 1980, page 1). This metaphor has been represented spatially (Ernest, 1988a, Fig. 8). Since Nilsson (1971) it has provided a basis for some of the research on mathematical problem solving, which utilizes the notion of a 'solution space' or '*state-space* representation of a problem [which] is a diagrammatic illustration of the set of all states reachable from the initial state. A *state* is the set of all expressions that have been obtained from the initial statement of the problem up to a given moment.' (Lester, 1980, page 293). The strength of the metaphor is that stages in the process can be represented, and that alternative 'routes' are integral to the representation. However a weakness of the metaphor is the implicit mathematical realism. For the set of all moves toward a solution, including those as yet uncreated, and those that never will be created, are regarded as pre-existing, awaiting discovery. Thus the metaphor implies an absolutist, even platonist view of mathematical knowledge.

The geographical metaphor is also applied to the process of mathematical investigation. 'The emphasis is on exploring a piece of mathematics in all directions. The journey, not the destination, is the goal.' (Pirie, 1987, page 2). Here the emphasis is on the exploration of an unknown land, rather than a journey to a specific goal. Thus whilst the process of mathematical problem solving is described as convergent, mathematical investigations are divergent (HMI, 1985).

Bell *et al.* (1983) propose a model of the process of investigation, with four phases: problem formulating, problem solving, verifying, integration. 'Here the term "investigation" is used in an attempt to embrace the whole variety of means of acquiring knowledge.' (Bell *et al.*, 1983, page 207). They suggest that mathematical investigation is a special form, with its own characteristic components of abstracting, representing, modelling, generalizing, proving, and symbolizing. This approach has the virtue of specifying a number of mental processes involved in mathematical investigation (and problem solving). Whilst other authors, such as Polya (1945) include many of the components of the model as processes of problem solving, the central difference is the inclusion of problem formulation or problem posing, which precedes problem solving. However, whilst the proposed model has some empirical basis, there is little rationale or justification for the choice of the components or their relationships.

3. Inquiry based pedagogy

A third sense of problem solving and investigations is as pedagogical approaches to mathematics. Cockcroft (1982) endorsed these approaches under the heading of 'teaching styles', although the terminology employed does not make the distinction between modes of teaching and learning. One way of contrasting inquiry approaches is to distinguish the roles of the teacher and the learner, as in Table 13.1.[1]

Table 13.1 illustrates that the shift from guided discovery, via problem solving, to an investigatory approach involves more than mathematical processes. It also involves a shift in power with the teacher relinquishing control over the answers, over the methods applied by the learners, and over the choice of content of the lesson. The learners gain control over the solution methods they apply, and then finally over the content itself. The shift to a more inquiry-orientated approach involves increased learner autonomy and self-regulation, and if the classroom climate is to be consistent, a necessary accompaniment is increased learner self-regulation over classroom movement, interaction and access to resources.

Problem solving and mathematical investigation as teaching approaches require the consideration of the social context of the classroom, and its power relations. Problem solving allows the learner to apply her learning creatively, in a novel situation, but the teacher still maintains much of her control over the content and form

Table 13.1: A Comparison of Inquiry Methods for Teaching Mathematics

Method	Teacher's role	Student's role
Guided Discovery	Poses problem, or chooses situation with goal in mind. Guides student toward solution or goal.	Follows guidance.
Problem solving	Poses problem. Leaves method of solution open.	Finds own way to solve the problem.
Investigatory approach	Chooses starting situation (or approves student choice)	Defines own problems within situation. Attempts to solve in his/her own way.

of the instruction. If the investigational approach is applied so as to allow the learner to pose problems and questions for investigation relatively freely, it becomes empowering and emancipatory. However, the characteristics that have been specified are necessary but not sufficient for such an outcome. What is also needed is the communication of a progressive or fallibilist view of mathematics through classroom experiences. This de-emphasizes the uniqueness and correctness of answers and methods, and centres instead on humans as active makers of knowledge, and the tentative nature of their creations.

B. Different Perceptions of Problems and Investigations

One of the outcomes of the above distinctions is that different interpretations can be given to problems and investigations, and their role in the teaching of mathematics.

Rejection of problem solving and investigations

The strongest negative reaction to problems and investigations is their rejection as inappropriate to school mathematics. This is based on the perception that school mathematics is content orientated, and that its central function is to inculcate basic mathematical skills. In contrast problems and investigations are seen to be frivolous, a squandering of time that should be given over to 'hard work'.

This is the response from the industrial trainer group. In particular, investigational work is explicitly opposed (Froome, 1970; Lawlor, 1988). This group has a narrow view of mathematical content because of its dualistic epistemology. In addition, the industrial trainer theory of teaching is an authoritarian transmission model, and any move to increase learner autonomy is strongly opposed (Lawlor, 1988). Loss of power over learners and the encouragement of emancipatory educational strategies are anathema to this perspective.

The incorporation of problems and investigations as content

A second group of responses to problems and investigations is to treat them as additional content to be adjoined to a content driven mathematics curriculum. Thus they are perceived as objects of inquiry used to enrich teaching, and not in terms of the learners' processes or the pedagogical approach adopted for mathematics. In particular, investigations are not understood in terms of problem posing.

In their different ways, both the old humanist and technological pragmatist ideologies share this view. Both are based on absolutist philosophies of mathematics. Both largely see problems as ways of enriching the content of the mathematics curriculum, and more or less identify investigations with problems.

The old humanist perspective values problems as non-routine applications of knowledge, as an important means for the demonstration of learning, understanding and talent. However, this perspective is concerned to transmit the body of pure mathematical knowledge to learners, so investigations are not understood in terms of problem posing by learners.

The technological pragmatist perspective values and encourages applied problem solving, and mathematical modelling. Thus problem solving is understood in terms of practical ('real') problems, leading to tangible outcomes. Mathematical investigations are subsumed into this conception of problems, or understood as puzzles. Thus Burghes (1984), representing this perspective, categorizes investigations into (1) eureka investigations (which are puzzles), (2) escalator investigations ('process' or combinatorial problems), (3) decision problems and (4) real problems. This represents the identification of investigations with problems, for the problem posing dimension is ignored or denied. Overall, problems and investigations are identified with objects of inquiry, and treated as adjuncts to the content of the curriculum, except that the mathematical modelling is understood in terms of processes.

Problem solving and investigation as pedagogy

The third group of perspectives see problem solving and investigation as pedagogical approaches to the whole curriculum, and not just an addition. Such views arise from philosophies of mathematics which see it at least as a growing field of knowledge, if not as a social construction. These views are concerned with the role of human beings in the growth of knowledge, and hence reflect the processes of mathematical problem solving and investigation in the curriculum. The full incorporation of these processes into the curriculum, including problem posing, leads to a problem solving and investigational pedagogy.

The progressive educator perspective is concerned to facilitate individuals' creativity in mathematics, and problem solving and investigation are perceived to be central to this. Thus problem solving and investigation are understood both in terms of the learners' processes and the pedagogical approach adopted in the classroom. In support of this pedagogy learners are offered carefully structured environments and situations for mathematical exploration, encouraging them to formulate and pursue their own investigations. The role of the teacher is understood in ways that support this pedagogy, as manager of the learning environment and learning resources, and facilitator of learning. The range of subjects for investigation is likely to be restricted to pure mathematical situations, or thematic topics concerning 'safe' as opposed to political issues. In keeping with the overall ideology, the emphasis is on the individual student and their interests, and not the structural social context in which they live, study and will earn a living.

The public educator accepts much of the previous perspective's views of a problem solving and investigational pedagogy, but adds a socio-political dimension. Thus the pedagogy adopted by this approach will involve a number of features which facilitate investigational approaches, including cooperative groupwork and discussion, autonomy and student self-direction in problem posing and investigation. All this may be shared with the progressive educator perspective. However where the public educators go beyond this, is through the encouragement of critical thinking through learner questioning of course content, pedagogy and assessment; and the use of socially relevant problems and situations, projects and topics, for social engagement and empowerment of the learners. Thus problem solving and investigation will be partly based on authentic materials, such as newspapers, official statistics, and social problems. For the public educator, this pedagogy is a means to develop the skills of citizenship and social engagement amongst learners.[2]

C. The Relationship between Epistemology and Pedagogy

In recent years, a number of official and authoritative reports have been published recommending the incorporation of problem solving into the teaching of school mathematics. In Britain such reports have included Cockcroft (1982) and Her Majesty's Inspectorate (1985), and in the USA they have included National Council of Teachers of Mathematics (1980, 1989).

However, one obstacle to such curriculum reforms is the interpretation given to such recommendations. For the concepts of problem solving and investigation are assimilated into the perspective of the interpreter, and understood accordingly, as we have seen above. Teachers' very performance at problem solving, not to mention their teaching approaches, depend on their beliefs about mathematics (Schoenfeld, 1985). Empirical evidence suggests that teachers may interpret problems and investigations in a narrow way. Lerman (1989a), for example, describes how investigational work in school mathematics is subverted by the view that there is a unique correct outcome, betraying an underlying absolutist philosophy of mathematics.

A second obstacle is that of implementation. This involves the relationship between theories of teaching and learning, which embody the pedagogy of a particular perspective, and classroom practice. On the large scale, this is the difference between the planned and the taught curriculum. On the small scale, this is the difference between the teacher's espoused theories of teaching and learning, and the enacted versions of these theories. Several studies have revealed teachers who espoused a problem solving approach to mathematics teaching (typically, consonant with that of the progressive educators) but whose practices revolved around an expository, transmission model of teaching enriched by the addition of problems (Cooney, 1983; 1985; Thompson, 1984; Brown, 1986).

The main explanation for the disparity put forward in these studies is that the constraints and opportunities afforded by the social context of teaching cause teachers to shift their pedagogical intentions and practices away from their espoused theories (Ernest, 1989b, 1989c). The socialization effect of the context is sufficiently powerful that despite having differing beliefs about mathematics and its teaching, teachers in the same school are observed to adopt similar classroom practices (Lerman, 1986). Figure 13.1 provides a model of some of the relationships involved.

It shows how one primary component of the teacher's ideology, the personal

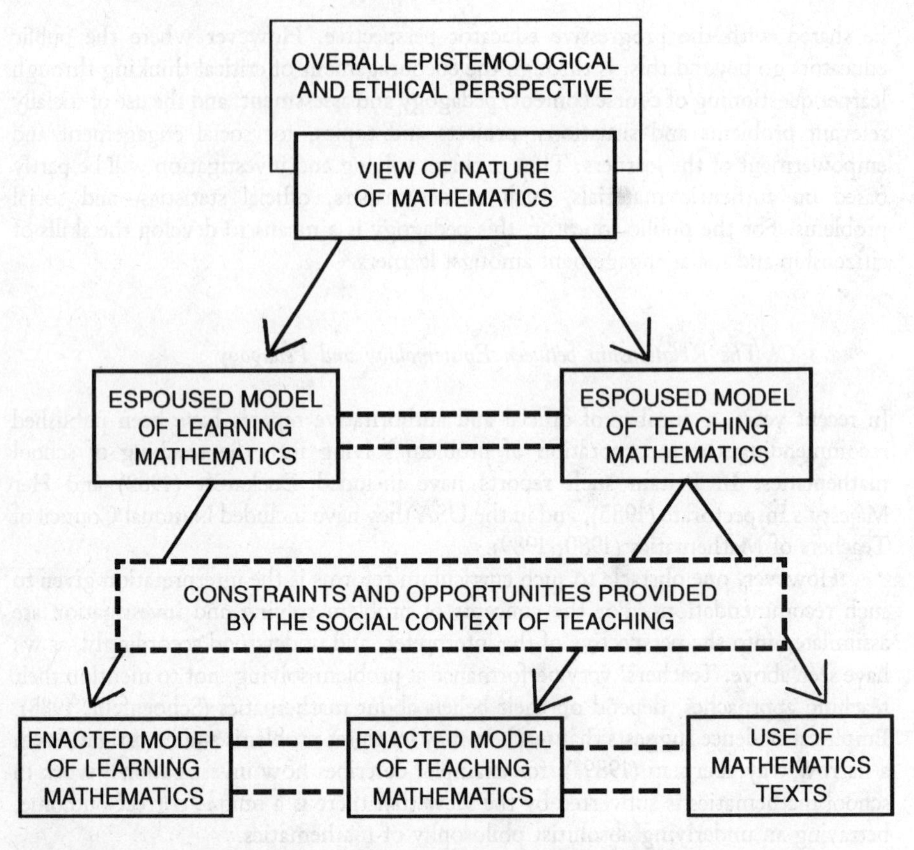

Figure 13.1: The relationship between espoused and enacted beliefs of the mathematics teacher

philosophy of mathematics, underpins two secondary components, the theories of teaching and learning mathematics. These in turn have an impact on practice as the enacted models of learning and teaching mathematics, including one use of resources which is singled out, namely the use of mathematics texts. This is important enough to be distinguished, for texts embody an epistemology, and the extent to which their presentation and sequencing of school mathematics is followed is a crucial determinant of the nature of the implemented curriculum (Cooney, 1988; Goffree, 1985). The downward arrows in the figure show the primary direction of influence. The content of the higher belief components is reflected in the nature of the lower components. Because the enacted models are inter-related, as are the espoused theories of teaching and learning, this is represented in the figure as horizontal links drawn between them.

However the impact of the espoused theories on practice is mediated by the constraints and opportunities provided by the social context of teaching (Clark and Peterson, 1986). The social context has a powerful influence, as a result of a number of factors including the expectations of others, such as students, their parents, fellow teachers and superiors. It also results from the institutionalized curriculum: the adopted text or curricular scheme, the system of assessment, and the overall national system of schooling. The social context leads the teacher to internalize a powerful set of constraints affecting the enactment of the models of teaching and learning mathematics.

Naturally the model illustrated in Figure 13.1 is greatly simplified, since the relationships are far more complex and far less mechanistic than they appear. Thus, for example, although the enacted beliefs are shown separate from the social context, they are embedded in it. Furthermore, all the beliefs and practices are part of an interactive system, and pressures at any point, such as in classroom practices, will feed-back and may influence all the other components.[3]

3. The Power of Problem Posing Pedagogy

A problem posing pedagogy, as in the public educator theory of teaching mathematics, and to a lesser extent, the progressive educator theory, represents a powerful emancipatory teaching approach, and when successfully implemented, empowers learners epistemologically. That is, it encourages active knowing and the creation of knowledge by learners, and it legitimates that knowledge as mathematics, at least in the school context.

It has been argued that it is the form and not the content of education that has the greatest impact, in general (Bowles and Gintis, 1976), and for mathematics (Noss, 1989). This view was challenged in Chapter 11, where it was argued that hierarchical views of knowledge as well as hierarchical forms of organization contribute to the re-creation, if not the reproduction, of social inequalities through education. The implications of this are that both the content and form of teaching matter, although it may be illusory to think that they can be separated. To reflect a social constructivist, or even a progressive absolutist view of mathematics, a problem posing pedagogy must include a treatment of content as well as a teaching approach

A. Countering Reproductive Forces in the Mathematics Curriculum

Four of the five ideologies of mathematics education presented above have aims that are socially reproductive, either in a hard or a soft sense. The hard sense corresponds to rigid class boundary definitions, whilst the soft sense corresponds to semi-permeable class boundary definitions, permitting limited social rise on merit, in a society seen as progressive and meritocratic. Of the five ideologies, only that of the progressive educators is a social change ideology. It seeks to empower learners to become aware and later to take control of their lives to challenge the reproductive forces at work in school and society. The central means of achieving this is through a problem posing pedagogy. This is reflected in the classroom empowerment of learners, at first epistemologically, and ultimately, socially and politically, through a critical awareness of the role of mathematics in society. This approach tries to minimize or make explicit the hidden power hierarchies exemplified in the classroom, which play an important part in reinforcing the tacit acceptance of fixed social hierarchies. The public educator perspective also challenges rigid hierarchies in the nature of mathematical knowledge, in the mathematics curriculum, and in the attribution of mathematical ability to learners. All of these hierarchies can serve to support and consolidate the reproduction of social hierarchy.

Furthering equal opportunities in mathematics

There are specific problems of social reproduction which concern the denial of equal opportunities in mathematics to ethnic minorities, notably blacks, and to women (Chapter 12). The proposed solution to these problems is the implementation of a problem posing pedagogy, based on the public educator ideology. To further the equal opportunities of blacks in mathematics, schooling and hence society requires anti-racist mathematics teaching. Likewise, to further the opportunities of women requires anti-sexist mathematics teaching. Both approaches rest on a problem posing pedagogy, which is proposed because it empowers *all* learners, not a deficient minority.

B. Subverting the Narrow Aims of the National Curriculum in Mathematics

In Chapter 11 the ideologies underpinning the development of the National Curriculum in mathematics were sketched. In simple terms, it was argued that the utilitarian ideologies are most central to this development, whilst the whole thrust of the educational reforms, including the reorganization of schooling and the curriculum and assessment frameworks, was driven by one of them, that of the industrial trainers.

In this section I wish to argue that these developments have in them ambiguities and contradictions which provide windows of opportunity through which the narrow aims of these ideologies can be countered or subverted. Furthermore, the main means of carrying this out is by means of the problem posing pedagogy of the public educators.

However, although this possibility exists, the obstacles to this solution should not be overlooked. One problem is that through a tight specification of the content and modes of assessment in mathematics, the National Curriculum is part of a move to 'deskill' and de-professionalize teachers (Giroux, 1983; Brown, 1988). The incorporation and institutionalization of problem solving and its assessment into the mathematics curriculum (Department of Education and Science, 1988; National Curriculum Council, 1989) serves to routinize strategic mathematical thinking, which robs it of its emancipatory power. Given that there are obstacles, we consider the opportunities for change.

Ambiguities

The concepts of problem solving and investigation have different meanings according to the perspective to which they are assimilated and interpreted, as we have seen. These meanings vary from reproductive applications in mathematics, to creative and epistemologically empowering approaches. Thus there is an important ambiguity underlying this range of meanings. A number of other terms used in educational discourse are ambiguous, and open to a comparable range of interpretations. Two will be considered here, the concepts of relevance and citizenship.

Relevance is a value laden term used to denote what is deemed apposite by a speaker. Thus utilitarian perspectives such as those of the industrial trainers and technological pragmatists see the 'relevance' of mathematics in terms of basic life skills (such as numeracy) or employment relatedness. Relevance, with the same meaning, is anathema to the purist ideology of the old humanists. The progressive educators, in contrast, understand relevance in terms of the perceived interests and needs of learners. To the public educators, relevance applies in mathematical contexts where learners are personally and socially engaged. This perspective draws upon activity theory to understand and inter-relate personal and social goals, which underpin this interpretation of the term (Mellin-Olsen, 1987). A 'relevant' curriculum, according to this perspective, is one which permits learners to engage mathematically with their social context, in an empowering way. It concerns a problem posing pedagogy, giving learners ownership of their learning.

Thus the concept of relevance is ambiguous, being relativistic in its meaning (Keitel, 1987). Evidently 'relevance' is a ternary relation which holds between a situation or object S, a person or group P, and a goal G when object S is deemed relevant by P to goal G.

Currently, there is an interest in 'citizenship in education', which extends across the political spectrum (Morrell, 1990). The National Curriculum Council has set up a Citizenship Task Group to report on how citizenship should be tackled within the curriculum. What is interesting, in the present context, is the ambiguity in the term citizenship. An interpretation favoured by the political right is that of the active citizen who participates in neighbourhood watch schemes, aids the policing of society, and carries out other activities perceived to be socially useful.[4] The centrist Commission on Citizenship understands the term more broadly, to include:

> The civil, political and social entitlements and corresponding responsibilities of the individual in the community or state ... the values associated with those rights and the institutions ... which give them effect. The entitlements ... are set out in the European Convention of Human Rights and ... other charters and conventions ... Citizenship also includes the furtherance of the public good, civic virtue and voluntary social obligation.
>
> (Morrell, 1990, page 38)

The more radical view of the left is that of the public educators who see critical citizenship as their aim (Giroux, 1983). This notion entails a critical view of social statistics and social structures, as well as positive action to redress inequalities and combat inequalities and combat racism and sexism (Frankenstein, 1990).

These are three different meanings for the same term. Like each of the above ambiguities this allows the goals of the public educators to be described and addressed in language acceptable to a broad range of persons. The crucial factor in utilizing each of these ambiguities to address the public educator goals is the role of problem posing pedagogy, empowering learners to develop broad skills, engage with social issues and become critically aware citizens.

In addition to those considered, a number of other educational terms are ambiguous, and can be exploited by a problem posing pedagogy. For example, each of the terms — economic awareness, enterprise, technological education, mathematical skills, discussion and project work — can be interpreted in a narrow way, or in a broader, emancipatory way, as part of problem posing pedagogy, challenging the utilitarian intentions of the National Curriculum.

Contradictions

The National Curriculum represents the most draconian move made by a modern British government to take control of the education system and to dictate its aims and outcomes. Although isolated aspects of the proposals may have merit, in view of the narrow range of aims and ideologies of the politicians directly wielding the power, the overall outcomes are at least potentially, profoundly anti-educational (Noss, 1989). However, the British education system, from politician to student, via the teacher, is not a monolith. At each level, there are contradictions and contradictory forces at work, which can be exploited to potentially subvert the anti-educational intentions of the National Curriculum.

First of all, there are the opposing ideologies of the different groups behind the National Curriculum. These include both the industrial trainers and technological pragmatists, with narrow and broad views of utility in education. This means that the specification of the National Curriculum includes a number of ambiguous terms, like those discussed above, which can be interpreted to serve an empowering pedagogy.

Secondly, some of the means adopted to further utilitarian ends, such as applications of mathematics and technology and computer programming, can equally serve emancipatory ends. For the meaning of applications of mathematics is broad enough to include problem solving and posing, and the examples given (in Attainment Targets 1 and 9) encourage rather than exclude this interpretation. Computer programming in the language Logo is used to exemplify the requirements in mathematics. But this language is promoted as most suitable for problem posing and investigation, for learners of all school ages (Papert, 1980). Fairly widespread experimental work has confirmed this potential, in practice (Noss, 1988, 1989a).

Finally, although the National Curriculum in mathematics provides a rigid specification of the content and means of assessing school mathematics, it nonetheless leaves the pedagogy unspecified. There is no statutory control over the teaching approach adopted. This leaves open a 'window of opportunity'. For a problem posing pedagogy, based on an appropriate philosophy of mathematics, can fulfil the statutory obligations in terms of content and assessment and yet still be emancipatory. Provided the curriculum content is understood in terms of social constructivist, or at the very least a progressive absolutist epistemology, this philosophy can be reflected in the pedagogical approach adopted. Thus, for example, work with young children in the USA (Cobb, 1987; Yackel, 1987) and in Britain (PrIME, 1987; Rowland, 1989) confirms that a conventional specification of mathematics content can be used as the basis for a problem posing pedagogy. In fact, the pedagogical suggestions that accompany the National Curriculum in mathematics (National Curriculum Council, 1989) promote a problem solving approach. Thus an invitation to use an inquiry orientated teaching approach, which allows the deployment of a problem posing pedagogy, is built into the system.

A number of authors have remarked how contradictions within an educational system can be used to subvert the inherent reproductive forces at work in general, and for mathematics in particular (Gintis and Bowles, 1980; Noss, 1989). A related argument is that these forces lead to resistance among learners (Giroux, 1983; Mellin-Olsen, 1987). What is argued above is that the contradictions allow teachers to challenge such forces through a problem posing pedagogy. However, for this to be emancipatory it must succeed in engaging the learners. As we have seen, it also needs to be based on a view of mathematics, and knowledge in general, which admits learners and hence also teachers as epistemological agents.

This raises the issue of the teacher as a 'reflective practitioner' (Schon, 1983) and a researcher (Stenhouse, 1975; Meighan, 1986). A pedagogy which empowers learners as epistemological agents depends on the teachers operating at this level too. Thus the teacher may create mathematics in his or her classroom, but as a professional, is also creating educational knowledge. Although this will not be elaborated further here, it can be said that a major aim of this book, by exploring the philosophical foundations of mathematics education, is to provide a thinking tool for the teacher as a reflective practitioner and researcher.

4. Conclusion

The theme of this chapter is that by reflecting on the nature of mathematics as a problem posing and solving activity, a problem posing pedagogy becomes emancipatory. As part of the public educator approach to teaching mathematics, it allows the highest social goals of mathematics education to be addressed. These goals include the fulfilment of individuals' potential as human beings, a move to greater awareness of social issues and the need for social change, and the fight against injustice, especially racism and sexism. However, these social goals are part of, and not opposed to the development of individuals' skills and creativity in mathematics. Of course, what lies at the heart of the public educator ideology is the social constructivist philosophy of mathematics. Each of the five ideologies of mathematics education is more or less driven by a philosophy of mathematics. Indeed, it is argued throughout that the philosophy of mathematics, or philosophies of mathematics underpin all mathematics curricula and teaching. This is the central claim and theme of this book. Namely that:

> whether one wishes it or not, all mathematical pedagogy, even if scarcely coherent, rests on a philosophy of mathematics.
>
> (Thom, 1971, p. 204)

And to take the argument back to the start of the book:

> The issue, then, is not, What is the best way to teach? but, What is mathematics really all about? . . . Controversies about . . . teaching cannot be resolved without confronting problems about the nature of mathematics.
>
> (Hersh, 1979, p. 34)

Notes

1 Adapted from Ernest (1984).
2 This account understresses the real difference between the progressive and public educators arising from their epistemologies. At root, the former group still believes in 'right answers' and absolute truth, an attitude which is likely to be manifested in the classroom. The latter group, with its social constructivist epistemology, is able to accept student solutions and constructions, and tolerate their lack of closure, without imposing such rigid preconceptions. (See below).
3 Some very successful implementations of problem posing and solving pedagogy have begun through supported changes in classroom practice, which seem to feed through into changes in teacher beliefs. For example, see the work of Cobb, Wood and Yackel in the USA (Keitel *et al.*, 1989) and the PrIME Project (1987) in the UK.
4 The view of industrial trainers and other members of the extreme right is that citizenship has no place in the school curriculum however it is interpreted (O'Hear, 1990).

References

ABBS, P. (1987) 'Training Spells the Death of Education', *The Guardian*, 5 January 1987.

ABIMBOLA, I. O. (1988) 'The Problem of Terminology in the Study of Student Conceptions in Science', *Science Education*, 72, 2, pp. 175–184.

ABRAHAM, J. and BIBBY, N. (1988) 'Mathematics and Society: Ethnomathematics and the Public Educator Curriculum', *For the Learning of Mathematics*, 8, 2, pp. 2–11.

ABRAHAM, J. and BIBBY, N. (1989) 'Human Agency: The Black Box of Mathematics in the Curriculum', *Zentralblatt fur Didaktik der Mathematik*, 21, 5, pp. 183–188.

ADORNO, R., FRENKEL-BRUNSWICK, E. LEVINSON, D. and SANFORD, R. (1950) *The Authoritarian Personality*, New York, Harper.

AGASSI, J. (1980) 'On Mathematics Education: the Lakatosian Revolution', *For the Learning of Mathematics*, 1, 1, pp. 27–31.

AGASSI, J. (1982) 'Mathematics Education as Training for Freedom', *For the Learning of Mathematics*, 2, 3, pp. 28–32.

AICHELE, D. B. and REYS, R. E. eds (1977) *Readings in Secondary School Mathematics*, Boston, Prindle, Weber & Schmidt.

ALEXANDER, R. (1984) *Primary Teaching*, Eastbourne, Holt, Rinehart and Winston.

ALEKSANDER, I. (1988) 'Putting down the professionals', *New Scientist*, 25 February 1988, pp. 65–63.

ALTHUSSER, L. (1971) 'Ideology and ideological state apparatuses', in *Lenin and Philosophy and Other Essays*, London, New Left Books.

ANDERSON, R. J., HUGHES, J. A. and SHARROCK, W. W. (1986) *Philosophy and the Human Sciences*, Beckenham, Croom Helm.

APPEL, K. and HAKEN, W. (1978) The Four-Color Problem, in Steen (1978), pp. 153–90.

APPLE, M. W. (1979) *Ideology and Curriculum*, London, Routledge and Kegan Paul.

APPLE, M. W. (1982) *Education and Power*, Boston, Routledge and Kegan Paul.

ASHLOCK, R. B. (1976) *Error Patterns in Computation*, Columbus, Ohio, Merrill.

ASSESSMENT OF PERFORMANCE UNIT (1985) *A Review of Monitoring in Mathematics 1978 to 1982* (2 vols), London, Department of Education and Science.

ASSOCIATION OF TEACHERS IN COLLEGES AND DEPARTMENT OF EDUCATION (1966) *Teaching Mathematics*, London, Association of Teachers in Colleges and Departments of Education.

ASSOCIATION OF TEACHERS OF MATHEMATICS (1966) *The Development of Mathematical Activity in Children and the Place of the Problem in this Development*, Nelson, Lancashire, Association of Teachers of Mathematics.

ASSOCIATION OF TEACHERS OF MATHEMATICS (1986) *Proposal for an all coursework GCSE examination in mathematics*, Derby, Association of Teachers of Mathematics.

AUSUBEL, D. P. (1968) *Educational Psychology, a cognitive view*, New York, Holt, Rinehart and Winston.

AYER, A. J. (1946) *Language, Truth and Logic*, London, Gollancz.

AYER, A. J. (1956) *The Problem of Knowledge*, Harmondsworth, Penguin Books.

BACON, F. (1960) *The New Organon*, Indianapolis, Bobbs-Merrill.

BALL, D., HIGGO, J., OLDKNOW, A., STRAKER, A. and WOOD, J. (1987) *Will Mathematics Count?*, Hatfield, England, AUCBE/Council for Educational Technology.

BANTOCK, G. H. (1975) 'Progressivism and the Content of Education', in Cox and Boyson, pp. 14–20.

BARKER, S. F. (1964) *Philosophy of Mathematics*, Englewood Cliffs, New Jersey, Prentice-Hall.

BARTON, L., MEIGHAN, S. and WALKER, S. (1980) *Schooling, Ideology and the Curriculum*, Lewes, Falmer Press.

BARWISE, J. (ed.) (1977) *Handbook of Mathematical Logic*, Amsterdam, North-Holland.

BARWISE, J. and PERRY, J. (1982) *Situations and Attitudes*, mimeographed manuscript (revised version published by Bradford Books).

BASH, L. and COULBY, D. (1989) *The Education Reform Act*, London, Cassell.

BAUERSFELD, H. (1979) *Hidden Dimensions in the So-called Reality of a Mathematics Classroom*, London, Centre for Science Education, Kings College, London University.

BECK, J. (1981) 'Education, Industry and the Needs of the Economy', *Education for Teaching*, 11, 2, pp. 87–106.

BECK, J., JENKS, C., KEDDIE, N. and YOUNG, M. F. D. (eds.) (1976) *Worlds Apart, Readings for a Sociology of Education*, London, Collier Macmillan.

BECKER, H. (1963) *Outsiders*, Oxford, Free Press.

BEGLE, E. G. (1979) *Critical Variables in Mathmatics Education*, Washington, DC, Mathematical Association of America & National Council of Teachers of Mathematics.

BELENKY, M. F., CLINCHY, B. M., GOLDBERGER, N. R. and TARULE, J. M. (1985) 'Epistemological Development and the Politics of Talk in Family Life', *Journal of Education*, 167, 3, pp. 9–27.

BELENKY, M. F., CLINCHY, B. M., GOLDBERGER, N. R. and TARULE, J. M. (1986) *Women's Ways of Knowing*, New York, Basic Books.

BELISLE, C. and SCHIELE, B. (eds.) (1984), *Les Savoirs Dans Les Practiques Quotidiennes*, Paris, Editions du Centre National De La Recherche Scientifique.

BELL, A. W., COSTELLO, J. and KUCHEMANN, D. (1983) *A Review of Research in Mathematical Education (Part A)*, Windsor, NFER-Nelson.

BELL, J., COLE, J., PRIEST, G. and SLOMSON, A. (eds) (1973) *Proceedings of the Bertrand Russell Memorial Logic Conference (Denmark 1971)*, Leeds, Bertrand Russell Memorial Logic Conference, Mathematics Department, The University.

BELL, J. L. (1981) 'Category Theory and the Foundations of Mathematics', *British Journal for the Philosophy of Science*, 12, 4, pp. 349–358.

BELL, J. L. and MACHOVER, M. (1977) *A Course in Mathematical Logic*, Amsterdam, North Holland.

BELL, J. L. and SLOMSON, A. B. (1971) *Models and Ultraproducts*, Amesterdam, North-Holland.

BENECERRAF, P. and PUTNAM, H. (eds.) (1964) *Philosophy of Mathematics: Selected Readings*, Englewood Cliffs, NJ, Prentice-Hall.

BENJAMIN, H. (1971) 'The Sabre Tooth Curriculum', in Hooper (1971), pp. 7–19.

BENTON, T. (1977) 'Education and Politics', in Macdonald *et al.*, (1977), pp. 124–131.

BERGERON, J. and HERSCOVICS, N. (eds.) (1983) *Proceedings of the 5th PME-NA, Montreal 29 September–1 October 1983* (2 Vols), Montreal, PME-NA.

BERGERON, J. C., HERSCOVICS, N. and KIERAN, C. (eds.) (1987) *Proceedings of PME 11 Conference* (3 volumes), Montreal, University of Montreal.

BERGERON, J. C., HERSCOVICS, N. and MOSER, J. (1986), 'Long Term Evolution of Student's Conceptions: An Example from Addition and Subtraction', in Carss, pp. 275–277.

BERNAYS, P. (1934) 'On Platonism in Mathematics', translated in Benecerraf and Putnam (1964), pp. 274–286.

BERNSTEIN, B. (1971) 'On the Classification and Framing of Educational Knowledge', in Young (1971), pp. 47–69.

BERRILL, R. (1982) 'Mathematics Counts: A Commentary on the Cockcroft Report', *The Durham and Newcastle Research Review*, 10, 49, pp. 23–26.

BERRY, J. S., BURGHES, D. N., HUNTLEY, I. D., JAMES, D. J. G. and MOSCARDINI, A. O. (eds.) (1984) *Teaching and Applying Mathematical Modelling*, Chichester, Ellis Horwood.

BIBBY, N. and ABRAHAM, J. (1989) 'Social History of Mathematics Controversies: Some Implications for the Curriculum', in Keitel *et al.* (1989), pp. 56–57.

BIGGS, E. E. (ed.) (1965) *Mathematics in Primary Schools* (Curriculum Bulletin No. 1), London, Her Majesty's Stationery Office.

BISHOP, A. J. (1985) 'The Social Construction of Meaning: A Significant Development for Mathematics Education?' *For the Learning of Mathematics*, 5, 1, pp. 24–28.

BISHOP, A. J. (1988) *Mathematical Enculturation*, Dordrecht, North Holland, Kluwer.

BISHOP, A. J. (1988a) 'Mathematics Education in its Cultural Context', *Educational Studies in Mathematics*, 19, pp. 179–191.

BISHOP, A. J. and NICKSON, M. (1983) *A Review of Research on Mathematical Education, Part B: Research on the Social Context of Mathematics Education*, Windsor, NFER-Nelson.

BISHOP, E. (1967) *Foundations of Constructive Analysis*, New York, McGraw-Hill.

BLAKE, R. M., DUCASSE, C. J. and MADDEN, E. H. (1960) *Theories of Scientific Method*, Seattle, University of Washington Press.

BLANCHE, R. (1966) *Axiomatics*, London, Routledge & Kegan Paul.

BLENKIN, G. M. and KELLY, A. V. (1981) *The Primary Curriculum*, London, Harper & Row.

BLOOM, B. S. (ed.) (1956) *Taxonomy of Educational Objectives 1, Cognitive Domain*, New York, David McKay.

BLOOR, D. (1973) 'Wittgenstein and Mannheim on the Sociology of Mathematics', *Studies in the History and Philosophy of Science*, 2, pp. 173–191.

BLOOR, D. (1976) *Knowledge and Social Imagery*, London, Routledge & Kegan Paul.

BLOOR, D. (1978) 'Polyhedra and the Abominations of Leviticus', *Studies in the History and Philosophy of Science*, 29, 4, pp. 245–272.

BLOOR, D. (1983) *Wittgenstein: A Social Theory of Knowledge*, London, Macmillan.

BLOOR, D. (1984) 'A Sociological Theory of Objectivity', in Brown (1984), pp. 229–245.

BORBAS, A. (ed.) (1988) *Proceedings of 12th International PME Conference* (2 volumes), 20–25 July 1988, Veszprem, Hungary, OOK.

BOURBAKI, N. (1948) 'L'Architecture des Mathématiques', translated in Lionnais (1971), 23–36.

BOURDIEU, P. and PASSERON, J. C. (1977) *Reproduction in Education, Society and Culture*, London, Sage.

BOWLES, S. and GINTIS, H. (1976) *Schooling in Capitalist America*, New York, Basic Books.

BRENT, A. (1978) *Philosophical Foundations for the Curriculum*, London, George Allen & Unwin.

BROUWER, L. E. J. (1913) 'Intuitionism and Formalism', *Bulletin of the American Mathematical Society*, 20, pp. 81–96, reprinted in Benecerraf and Putnam (1964), pp. 66–77.

BROUWER, L. E. J. (1927) 'Intuitionistic reflections on formalism', in Heijenoort (1967), pp. 490–492.

BROWN, A. (1985) *Trials of Honeyford: Problems in Multicultural Education*, London, Centre for Policy Studies.

BROWN, C. A. (1986) 'A Study of the Socialization to Teaching of a Beginning Secondary Mathematics Teacher', in Burton and Hoyles (1986), pp. 336–341.

BROWN, G. and DESFORGES, C. (1979) *Piaget's Theory: A Psychological Critique*, London, Routledge and Kegan Paul.

BROWN, J. S., COLLINS, A. and DUGUID, P. (1989) 'Situated Cognition and the Culture of Learning', *Educational Researcher*, 18, 1, pp. 32–42.

BROWN, M. (1988) 'Teachers as Workers and Teachers as Learners', paper presented at *Sixth International Congress of Mathematics Education*, Budapest, July 27–August 4, 1988.

BROWN, S. C. (ed.) (1984) *Objectivity and Cultural Divergence* (Royal Institute of Philosophy lecture series, 17), Cambridge: Cambridge University Press.

BROWN, S. I. (1984) 'The Logic of Problem Generation: from Morality and Solving to De-Posing and Rebellion', *For the Learning of Mathematics*, 4, 1, pp. 9–20.

BROWN, S., FAUVEL, J. and FINNEGAN, R. (1981) *Conceptions of Inquiry*, London, Methuen, in association with the Open University Press.

BROWNELL, W. A. (1942) 'Problem Solving', *The Psychology of Learning* (41st NSSE Yearbook, Part II), Chicago, National Society for the Study of Education.

BRUNER, J. (1960) *The Process of Education*, Cambridge, Massachusetts, Harvard University Press.

BRYANT, P. E. (1974) *Perception and Understanding in Young Children*, London, Methuen.

BUCHER, R. and STRAUSS, A. (1961) 'Professions in process', *American Journal of Sociology*, 66, pp. 325–334.
BUERK, D. (1982) 'An Experience with Some Able Women Who Avoid Mathematics', *For the Learning of Mathematics*, 3, 2, pp. 19–24.
BURGHES, D. N. (1984) 'Mathematical Investigations', *Teaching Mathematics and its Applications*, 3, 2, pp. 47–52.
BURGHES, D. N. (1989) 'Mathematics Education for the Twenty-first Century: It's Time for a Revolution!' in Ernest (1989), pp. 83–93.
BURHARDT, H. (1979) 'Learning to Use Mathematics', *Bulletin of the Institute of Mathematics and its Applications*, 15, 10, pp. 238–243.
BURKHARDT, H. (1981) *The Real World and Mathematics*, Glasgow, Blackie.
BURTON, L. (1984) *Thinking Things Through*, Oxford, Blackwell.
BURTON, L. (ed.) (1986) *Girls into maths can go*, London, Holt, Rinehart and Winston.
BURTON, L. and HOYLES, C. (eds.) (1986) *Proceedings of PME-10*, London, University of London Institute of Education.
BUXTON, L. (1981) *Do You Panic About Maths?*, London, Heinemann.
GALLAGHAN, J. (1976) What the PM said (Text of Ruskin College Speech), *Times Educational Supplement*, 22 October, 1976, pages 1 and 72.
CAMBRIDGE CONFERENCE ON SCHOOL MATHEMATICS (1963) *Goals for School Mathematics*, Boston, Houghton Mifflin Co.
CAMPAIGN FOR REAL EDUCATION (1987) *Campaign for Real Education Credo*, York, Campaign for Real Education.
CAMPAIGN FOR REAL EDUCATION (1989) *Campaign for Real Education Newsletter*, 3, 3.
CARIER, C. J. (1976) 'Business values and the educational state', in Dale *et al.* (1976), pp. 21–31.
CARNAP, R. (1931) The Logicist Foundations of Mathematics, *Erkenntnis* 1931, in Benecerraf and Putnam, (1964), pp. 31–41.
CARPENTER, T. P, CORBITT, M. K., KEPNER, H. S., LINDQUIST, M. M. and REYS, R. E. (1981) *Results from the Second Mathematics Assessment of the NAEP*, Reston, Virginia, National Council of Teachers of Mathematics.
CARR, W. and KEMMIS, W. (1986) *Becoming Critical*, Lewes: Falmer Press.
CARRAHER, T. N. (1988) 'Street Mathematics and School Mathematics', in Borbas (1988), 1, pp. 1–23.
CARRINGTON, B. and TROYNA, B. (1988) *Children and Controversial Issues*, Basingstoke, Falmer.
CARSS, M. (ed.) (1986) *Proceedings of the Fifth International Conference on Mathematical Education*, Boston, Birkhauser.
CENTRE FOR POLICY STUDIES (1987) *More To Do*, London, Centre for Policy Studies.
CENTRE FOR POLICY STUDIES (1988) *The Power of Ideas*, London, Centre for Policy Studies.
CHARDIN, T. de (1966) *The Phenomenon of Man*, London, Fontana Books.
CHICKERING, A. (ed.) (1981) *The Modern American College*, San Francisco, Jossey-Bass, pp. 76–116.
CHILDS, D. (ed.) (1977) *Readings in Psychology for the Teacher*, London, Holt Rinehart and Winston.
CHISHOLM, R. (1966) *Theory of Knowledge*, Englewood Cliffs, Prentice Hall.
CHITTY, C. (1987) 'The Commodification of Education', *Forum*, 24, 3, pp. 66–68.
CHRISTIANSEN, B., HOWSON, A. G. and OTTE, M. (eds.) (1986) *Perspectives on Mathematics Education*, Dordrecht, Reidel.
CHRISTIANSEN, B. and WALTHER, G. (1986) 'Task and Activity', in Christiansen *et al.* (1986), pp. 243–307.
CLARK, C. M. and PETERSON, P. L. (1986) 'Teachers' thought Processes', in Wittrock (1986), pp. 255–296.
CLEMENTS, K. and ELLERTON, N. (eds.) (1989) *School Mathematics: the challenge to change*, Victoria, Australia, Deakin University Press.
COBB, P. (1986) 'Contexts, goals, beliefs, and learning mathematics', *For the Learning of Mathematics*, 6, 2, pp. 2–9.
COBB, P. (1987) 'A Year in the Life of a Second Grade Class: Cognitive Perspective', in Bergeron *et al.* (1987), 3, pp. 201–207.

COBB, P. (1988) *Multiple Perspectives*, paper presented at American Educational Research Association Annual Conference, New Orleans.

COBB, P. and STEFFE, L. P. (1983) 'The Constructivist Researcher as Teacher and Model Builder' *Journal for Research in Mathematics Education*, 14, 2, pp. 83–94.

COCKCROFT, W. H. (chair) (1982) *Mathematics Counts*, London, HMSO.

COHEN, M. R. and NAGEL, E. (1963) *An Introduction to Logic*, London, Routledge & Kegan Paul.

COHEN, P. J. (1966) *Set Theory and the Continuum Hypothesis*, New York, Benjamin.

COHEN, R. S., STACHEL, J. and WARTOFSKY, M. W. (eds.) (1974) *For Dirk Struik* (Boston Studies in the Philosophy of Science 15), Dordrecht, Reidel.

COMMISSION FOR RACIAL EQUALITY (1985) *Swann: A Response from the Commission for Racial Equality*, London, Commission for Racial Equality.

COMMISSION FOR RACIAL EQUALITY (1990) *From Cradle to School, London*, Commission for Racial Equality.

CONFREY, J. (1981) 'Conceptual Change Analysis: Implications for Mathematics and Curriculum', *Curriculum Inquiry*, 11, 5, pp. 243–257.

COONEY, T. J. (1983) 'Espoused Beliefs and Beliefs in Practice: the Cases of Fred and Janice', in Bergeron and Herscovics (1983), 2, pp. 162–169.

COONEY, T. J. (1985) 'A Beginning Teacher's View of Problem Solving', *Journal for Research in Mathematics Education*, 16, 5, pp. 324–336.

COONEY, T. J. (1988) 'The Issue of Reform', *Mathematics Teacher*, 80, pp. 352–363.

COONEY, T. J. and JONES, D. A. (1988) 'The Relevance of Teachers' and Students' Beliefs for Research in Mathematics Teacher Education', presented at PME 12 Conference, 20–25 July 1988, Veszprem, Hungary.

COOPER, B. (1985) *Renegotiating Secondary School Mathematics*, Lewes, Falmer Press.

COOPER, T. (1989) 'Negative Power, Hegemony and the Primary Mathematics Classroom: A Summary', in Keitel *et al.* (1989), pp. 150–154.

COPES, L. (1982) 'The Perry Development Scheme: a Metaphor for Learning and Teaching Mathematics', *For the Learning of Mathematics*, 3, 1, pp. 38–44.

COPES, L. (1988) 'The Perry Development Scheme and its Implications for Teaching Mathematics', in Pereira-Mendoza (1988), pp. 131–141.

COSIN, B. (1972) *Ideology*, Milton Keynes, Open University Press.

COTTON, A. (ed.) (1989) 'Anti-Racist Education in the National Curriculum', unpublished paper.

COTTON, J. W. and KLATSKY, R. L. (eds.) (1978) *Semantic Factors in Cognition*, Hillsdale, New Jersey, Erlbaum.

COX, C. and MARKS, J. (1982) *The Right to Learn*, London, Centre for Policy Studies.

COX, C., DOUGLAS-HOME, J., MARKS, J., NORCROSS, L. and SCRUTON, R. (1987) *Whose Schools? A Radical Manifesto*, London, Hillgate Group.

COX, C. B. and BOYSON, R. (eds.) (1975) *Black Paper 1975: The Fight for Education*, London, Dent and Sons.

COX, C. B. and DYSON, A. E. (eds.) (1969) *Fight for Education*, London, Critical Quarterly Society.

COX, C. B. and DYSON, A. E. (eds.) (1969a) *Black Paper Two*, London, Critical Quarterly Society.

COX, C. B. and DYSON, A. E. (eds.) (1970) *Black Paper Three*, London, Critical Quarterly Society.

CRANE, D. (1972) *Invisible Colleges*, Chicago, University of Chicago Press.

CRAWFORD, K. (1989) 'Knowing what versus Knowing How', in Keitel *et al.* (1989), pp. 22–24.

CROWE, M. J. (1975) 'Ten "Laws" Concerning Patterns of Change in the History of Mathematics', *Historia Mathematica*, 2, pp. 161–166.

CUMMINGS, A. L. and MURRAY, H. G. (1989) 'Ego Development and its Relation to Teacher Education', *Teaching and Teacher Education*, 5, 1, pp. 21–32.

CURRY, H. B. (1951) *Outlines of a Formalist Philosophy of Mathematics*, Amsterdam, North Holland.

DALE, R. (1985) *Education, Training and Employment: Towards a new vocationalism?*, Oxford, Pergamon.

DALE, R., ESLAND, G. and MACDONALD, M. (1976) *Schooling and Capitalism*, London, Routledge and Kegan Paul.

D'AMBROSIO, U. (1985) *Socio-cultural bases for Mathematics Education*, Campinas, Brazil, UNICAMP.

D'AMBROSIO, U. (1985a) 'Ethnomathematics and its place in the history and pedagogy of mathematics', *for the Learning of Mathematics*, 5, 1, pp. 44–48.

DAMEROW, P., DUNKLEY, M. E., NEBRES, B. F. and WERRY, B., (eds.) (1986) *Mathematics for All*, Paris, UNESCO.

DAVIES, B. (1976) *Social Control and Education*, London, Methuen.

DAVIS, C. (1974) Materialist philosophy of mathematics, in Cohen *et al.* (1974).

DAVIS, P. J. (1972) 'Fidelity in Mathematical Discourse: Is One and One Really Two' *American Mathematical Monthly*, 79, 3, pp. 252–263.

DAVIS, P. J. and HERSH, R. (1980) *The Mathematical Experience*, Boston, Birkhauser.

DAVIS, P. J. and HERSH, R. (1988) *Descartes' Dream*, London, Penguin Books.

DAVIS, R. B. (1967) *Mathematics Teaching — With Special Reference to Epistemological Problems*, Athens, Georgia, University of Georgia.

DAWSON, K. and WALL, P. (1969) *Society and Industry in the 19th Century: 4. Education*, Oxford, Oxford University Press.

DEARDEN, R. F. (1968) *The Philosophy of Primary Education*, London, Routledge & Kegan Paul.

DENVIR, B. and BROWN, M. (1986) 'Understanding of Number Concepts in Low Attaining 7–9 Year Olds: I and II', *Educational Studies in Mathematics*, 17, pp. 15–36 and pp. 143–164.

DEPARTMENT OF EDUCATION AND SCIENCE (1983) *Blueprint for Numeracy: An employers' guide to the Cockcroft Report*, London, Department of Education and Science.

DEPARTMENT OF EDUCATION AND SCIENCE (1987) *Interim Report of the Mathematics Working Group*, London, Department of Education and Science.

DEPARTMENT OF EDUCATION AND SCIENCE (1987a) *The National Curriculum: A Consultation Document*, Department of Education and Science.

DEPARTMENT OF EDUCATION AND SCIENCE (1988) *Final Report of the Mathematics Working Group*, London, Department of Education and Science.

DEPARTMENT OF EDUCATION AND SCIENCE (1988a) *National Curriculum: Task Group on Assessment and Testing: A Report*, London, Department of Education and Science.

DEPARTMENT OF EDUCATION AND SCIENCE (1989) *The Parliamentary Orders for Mathematics*, London, Department of Education and Science.

DEPARTMENT OF EDUCATION AND SCIENCE (1989a) *National Curriculum: From Policy to Practice*, London, Department of Education and Science.

DESCARTES, R. (1931) *Philosophical Works*, Volume 1, Cambridge, Cambridge University Press (reprinted by Dover Press, New York, 1955).

DESFORGES, C. W. and COCKBURN, A. (1987) *Understanding the Mathematics Teacher*, Basingstoke, Falmer Press.

DESSART, D. J. (1981) Curriculum, in Fennema (1981), pp. 1–21.

DEWEY, J. (1916) *Democracy and Education*, New York: Collier-Macmillan.

DEWEY, J. (1938) *Logic: The Theory of Inquiry*, New York, Holt, Rinehart & Winston.

DEWEY, J. (1950) *Reconstruction in Philosophy*, New York, Mentor.

DEWEY, J. (1966) *Selected Educational Writings* (F. W. Garforth ed.), London, Heinemann.

DICKENS, C. (1854) *Hard Times*, (reprinted by Everyman Books, London, Dent and Sons).

DICKSON, D. (1974) *Alternative Technology and the Politics of Technical Change*, Glasgow, Fontana/Collins.

DIENES, Z. P. (1960) *Building up Mathematics*, London, Hutchinson.

DONALDSON, M. (1978) *Children's Minds*, Glasgow: Fontana/Collins.

DORE, R. P. (1976) *The Diploma Disease*, London, Allen and Unwin.

DOUGLAS, M. (1966) *Purity and Danger*, London, Routledge & Kegan Paul.

DOWLING, P. (1988) 'The contextualising of Mathematics: Towards a Theoretical Map', unpublished paper, University of London Institute of Education, (to appear in Harris, M. ed., forthcoming, *Mathematics in Work*, Falmer).

DRIVER, R. (1983) *The Pupil as Scientist?*, Milton Keynes, Open University Press.

DUBINSKY, E. (1988) 'On helping students construct the concept of quantification, in Borbas (1988), 1, pp. 255–262.

DUBINSKY, E. (1989) 'Development of the Process Conception of Function by Pre-service Teachers in a Discrete Mathematics Course', presented at *Psychology of Mathematics Education, 13th Conference*, Paris, 9–13 July 1989.

DUGAS, L. (1989) 'The Problematics of Political Polls: Mathematics Curriculum for Social Understanding', presented at the annual meeting of the National Association for Science, Technology and Society, Washington DC, February 1989.

DUMMETT, M. (1959) Wittgenstein's Philosophy of mathematics, *Philosophical Review*, 68, pp. 324–348. (Reprinted in Benecerraf and Putnam, 1964).

DUMMETT, M. (1973) 'The Philosophical Basis of Intuitionistic Logic', in Gandy (1973).

DUMMETT, M. (1977) *The Elements of Intuitionism*, Oxford, Oxford University Press.

EASLEA, B. (1983) *Fathering the Unthinkable: Masculinity, Scientists and the Nuclear Arms Race*, London, Pluto Press.

EASLEY, J. (1984) 'Is there Educative Power in Students' Alternative Frameworks?' *Problem Solving*, 6, 2, pp. 1–4.

ELIOT, T. S. (1948) *Notes Towards the Definition of Culture*, London, Faber.

ELLUL, J. (1980) *The Technological system*, New York, Continuum.

ERLWANGER, S. H. (1973) 'Benny's Conception of Rules and Answers in IPI Mathematics', *Journal of Children's Mathematical Behaviour*, 1, 2, pp. 7–26.

ERNEST, P. (1984) 'Investigations', *Teaching Mathematics and its Applications*, 3, 3, pp. 81–86.

ERNEST, P. (1985) *Meaning and Intension in Mathematics*, unpublished doctoral dissertation, London, Chelsea College, University of London.

ERNEST, P. (1985a) 'The Philosophy of Mathematics and Mathematics Education', *International Journal for Mathematical Education in Science and Technology*, 16, 5, pp. 603–612.

ERNEST, P. (1986) 'Social and Political Values', *Mathematics Teaching*, 116, pp. 16–18 (reprinted in Ernest, 1989, pp. 197–202).

ERNEST, P. (1987) 'Mathematics, Education and Society', paper presented at RSPME Conference, University of London Institute of Education, London, December 1987.

ERNEST, P. (1988a) 'The Problem-solving Approach to Mathematics Teaching', *Teaching Mathematics and its Applications*, 7, 2, pp. 82–92.

ERNEST, P. (1988b) 'Mathematical Values: A Personal Experience', *Mathematical Education for Teaching*, 5, 1, pp. 40–43.

ERNEST, P. (ed.) (1988c) *The Social Context of Mathematics Teaching* (Perspectives 37) Exeter, University of Exeter School of Education.

ERNEST, P. (ed.) (1989) *Mathematics Teaching: The State of the Art*, Basingstoke, Falmer Press.

ERNEST, P. (1989a) 'Mathematics-Related Belief Systems', paper presented at *Psychology of Mathematics Education Conference 13*, 9–13 July 1989, Paris.

ERNEST, P. (1989b) 'The Impact of Beliefs on the Teaching of Mathematics', in Ernest (1989), pp. 249–254; Keitel *et al.* (1989), pp. 99–101.

ERNEST, P. (1989c) 'The Knowledge, Beliefs and Attitudes of the Mathematics Teacher: A Model', *Journal of Education for Teaching*, 15, 1, pp. 13–33.

ERNEST, P. (1989d) 'The Psychology of Teaching Mathematics: A Conceptual Framework', *Mathematical Education for Teaching*, 6, pp. 21–55.

ERNEST, P. (1989e) 'The Philosophy of the National Mathematics Curriculum', paper presented at RSPME conference, University of London Institute of Education, June 1989.

ERNEST, P. (1990) 'The Meaning of Mathematical Expressions: Does Philosophy Shed Any Light on Psychology?' *British Journal for the Philosophy of Science*, in press.

ESLAND, G. M. (1989) 'Teaching and Learning as the Organisation of Knowledge', in Young (1971), pp. 70–115.

EVANS, J. (1988) 'The Politics of Numeracy', in Ernest, P. (1988c), pp. 34–57 (reprinted in Ernest, 1989).

EVANS, J. (1988a) 'Contexts and Performance in Numerical Activity among Adults', in Borbas (1988), 1, pp. 296–303.

EVES, H. (1953) *An Introduction to the History of Mathematics*, New York, Holt Rinehart and Winston.

FEIGL, H. and SELLARS, W. (eds.) (1949) *Readings in Philosophical Analysis*, New York, Appleton-Century-Crofts.

FENNEMA, E. (ed.) (1981) *Mathematics Education Research*, Reston VA, National Council of Teachers of Mathematics.

FISHER, E. (1979) *Woman's Creation*, London, Wildwood House.

FLEW, A. (1976) *Sociology, Equality and Education*, London, Macmillan.

FOUCAULT, M. (1972) *The Archaeology of Knowledge*, London, Tavistock.

FOUCAULT, M. (1981) *The History of Sexuality (Part 1)*, Harmondsworth, Penguin Books.

FOX, L. H., FENNEMA, E. and SHERMAN, J. (1977) *Women and Mathematics: Research Perspectives for Change* (NIE Papers in Education and Work No. 8), Washington DC, National Institute of Education.

FRANKENSTEIN, M. (1983) 'Critical Mathematics Education: An Application of Paulo Freire's Epistemology', *Journal of Education*, 165, 4, pp. 315–339.

FRANKENSTEIN, M. (1989) *Relearning Mathematics: A Different Third R — Radical Maths*, London, Free Association Books.

FRANKENSTEIN, M. (1989a) 'Incorporating Race and Gender Issues into a Basic Mathematics Curriculum', unpublished paper.

FRANKENSTEIN, M. (1990) 'Critical Mathematical Literacy', in Noss, R., *et al.* (1990).

FRANKENSTEIN, M. and POWELL, A. B. (1988) 'Mathematics Education and Society: Empowering Non-Traditional Students', paper presented at ICME6 Conference, Budapest 27 July–13 August 1988 (short version in Keitel (1989), pp. 157–159).

FREGE, G. (1879) *Begriffsschrift*, translated, in Heijenoort (1967), pp. 1–82.

FREGE, G. (1893) *Grundgesetze der Arithmetik*, translated selections in Furth (1964).

FREIRE, P. (1972a) *Pedagogy of the Oppressed*, Harmondsworth, Penguin Books.

FREIRE, P. (1972b) *Cultural Action for Freedom*, Harmondsworth, Penguin Books.

FREUDENTHAL, H. (1973) *Mathematics as an Educational Task*, Dordrecht, Reidel.

FROOME, S. (1970) *Why Tommy Isn't Learning*, London, Tom Stacey Books.

FURTH, M. (ed.) (1964) *Gottlob Frege: The Basic Laws of Arithmetic*, Berkeley, University of California Press.

FURTHER EDUCATION UNIT (1982) *Basic Skills*, London, Further Education Unit, Department of Education and Science.

FURTHER EDUCATION UNIT (1984) *FEU Focus on Multicultural Society* (Focus No. 3, June 1984), London, Further Education Unit, Department of Education and Science.

FURTHER EDUCATION UNIT (1985) *Curriculum Development for a Multicultural Society*, London, Further Education Unit, Department of Education and Science.

GAGNE, R. M. (1977) Learning and Proficiency in Mathematics, in Aichele, and Reys (1977), 161–169.

GAINE, C. (1987) *No Problem Here*, London, Hutchinson.

GAMMAGE, P. (1976) *Human Development and the Curriculum* (Unit 6, Curriculum Design and Development, E203), Milton Keynes, The Open University.

GANDY, R. (ed.) (1973) *Proceedings of the Bristol Logic Colloquium*, Amsterdam, North-Holland.

GARLAND, R. (ed.) (1982) *Microcomputers and Children in the Primary School*, Lewes, Falmer Press.

GENTZEN, G. (1936) Die Widerspruchfreiheit der reinen Zahlentheorie, *Mathematische Annalen*, 112, pp. 493–565.

GERDES, P. (1981) 'Changing Mathematics Education in Mozambique', *Educational Studies in Mathematics*, 12, pp. 455–477.

GERDES, P. (1985) 'Conditions and strategies for emancipatory mathematics education in underdeveloped countries', *For the Learning of Mathematics*, 5, 1, pp. 15–20.

GIDDENS, A. (1983) 'Four Theses on Ideology', *Canadian Journal of Political and Social Theory*, 7, 1–2, pp. 18–21.

GILBERT, J. K. and WATTS, D. M. (1983) 'Concepts, Misconceptions and Alternative Conceptions: Changing Perspectives in Science Education', *Studies in Science Education*, 10, pp. 61–98.

GILL, D. and LEVIDOW, L. (eds.) (1987) *Anti-Racist Science Teaching*, London, Free Association Books.

GILLIGAN, C. (1982) *In a Different Voice*, Cambridge, Massachusetts, Harvard University Press.

GINSBURG, H. (1977) *Children's Arithmetic: The Learning Process*, New York, Van Nostrand.
GINTIS, H. and BOWLES, S. (1980) 'Contradiction and Reproduction in Educational Theory', in Barton, *et al.* (1980), pp. 51–65.
GIRLS AND MATHEMATICS UNIT (1988) *Girls and Mathematics: Some Lessons for the Classroom*, London, Economic and Social Research Council.
GIROUX, H. A. (1983) *Theory and Resistance in Education*, London, Heinemann.
GLASERSFELD, E. von (1983) Learning as a Constructive Activity, in Bergeron and Herscovics, 1, pp. 41–69.
GLASERSFELD, E. von (1984) 'An Introduction to Radical Constructivism', in Watzlawick (1984), pp. 17–40.
GLASERSFELD, E. von (1987) 'Learning as a Constructive Activity', in Janvier (1987).
GLASERSFELD, E. von (1989) 'Constructivism in Education', in Husen and Postlethwaite (1989), pp. 162–163.
GLASS, D. V. (1971) 'Education and Social Change in Modern England', in Hooper (1971), pp. 20–41.
GODEL, K. (1931) 'Uber formal unentscheidbare Satze der Principia Mathematica und verwandter Systeme I'. *Monatshefte fur Mathematik und Physik*, 38, pp. 173–198, translated in Heijenoort, J. von (1967), pp. 592–617.
GODEL, K. (1940) *The Consistency of the Continuum Hypothesis* (Annals of Mathematics Study 3), Princeton, New Jersey, Princeton University Press.
GODEL, K. (1964) 'What is Cantor's Continuum Problem?' in Benecerraf and Putnam (1964), pp. 258–71.
GOFFREE, F. (1985) 'The Teacher and Curriculum Development', *For the Learning of Mathematics*, 5, 2, pp. 26–27.
GOLBY, M. (1982) 'Computers in the Primary Curriculum', in Garland (1982).
GOLBY, M. (ed.) (1987) *Perspectives on the National Curriculum* (Perspectives 32), Exeter, University of Exeter School of Education.
GOLBY, M., GREENWALD, J. and WEST, R. (eds.) (1975) *Curriculum Design*, London, Croom Helm in association with Open University Press.
GOLD, K. (1990) 'Must we buy schooling from Marks & Spencer?' *The Observer*, 21 January 1990, 74.
GOLDSTEIN, H. and WOLF, A. (1983) 'The report whose sum total doesn't add up to much', *The Guardian*, 5 July 1983.
GOODMAN, P. (1962) *Compulsory Miseducation*, Harmondsworth, Penguin Books, 1971.
GORDON, P. (1978) 'Tradition and Change in the Curriculum', in Lawton *et al* (1978), pp. 121–127.
GORDON, P. (1989) 'The New Educational Right', *Multicultural Teaching*, 8, 1, pp. 13–15.
GORZ, A. (1989) *Critique of Economic Reason*, London, Verso.
GOW, D. (1988) 'Adding up to numeracy', *The Guardian*, 23 February, 1988.
GOW, D. and TRAVIS, A. (1988) 'Leak exposes Thatcher rift with Baker', *The Guardian*, 10 March 1988, 1.
GRABINER, J. (1986) 'The Centrality of Mathematics in the History of Western Thought', International Congress of Mathematicians, Berkeley, California, 1986 (*Mathematics Magazine*, 1988, pp. 220–230).
GRAMSCI, A. (1971) *Selections from Prison Notebooks* (edited and translated by Q. Hoare and G. Smith), New York, International Publishers.
GRIFFITHS, H. B. and HOWSON, A. G. (1974) *Mathematics: Society and Curricula*, Cambridge, Cambridge University Press.
GROUWS, D. A., COONEY, T. J. and JONES, D. (eds.) (1988) *Effective Mathematics Teaching*, Reston, Virginia, National Council of Teachers of Mathematics.
HAACK, S. (1979) 'Epistemology with a Knowing Subject', *Review of Metaphysics*, 33, pp. 309–335.
HADAMARD, J. (1945) *The Psychology of Invention in the Mathematical Field*, Princeton, New Jersey, Princeton University Press (reprinted, New York, Dover Press).
HADOW REPORT (1931) *Report of the Consultative Committee on the Primary School*, London, His Majesty's Stationery Office.
HALL, S. (1974) 'Education and the crisis of the urban school', in Raynor (1974), pp. 49–55.

HALLETT, M. (1979) 'Towards a Theory of Mathematical Research Programmes I and II', *British Journal for the Philosophy of Science*, 30, 1, pp. 1–25; 2, pp. 135–159.
HALLIDAY, M. A. K. (1978) *Language as a Social Semiotic*, London, Edward Arnold.
HALMOS, P. R. (1981) 'Does Mathematics Have Elements?' *Mathematical Intelligencer*, 3, 4, pp. 147–153.
HALMOS, P. R. (1985) 'Applied Mathematics is Bad Mathematics', in Steen (1985), pp. 9–20.
HAMLYN, D. W. (1978) *Experience and the Growth of Understanding*, London, Routledge & Kegan Paul.
HAMMERSLEY, M. (1977) *Teacher Perspectives* (E202 Schooling and Society: Block II, The Process of Schooling, Units 9 and 10), Milton Keynes, Open University Press.
HANSEN, S. and JENSEN, J. (1971) *The Little Red School Book*, London, Stage 1.
HARDY, G. H. (1967) *A Mathematician's Apology*, Cambridge, Cambridge University Press.
HARRE, R. (1989) 'Social construction of selves as a discursive practice', unpublished paper presented to the LMMG, London, 23 May 1989.
HART, K. (ed.) (1981) *Children's Understanding of Mathematics: 11–16*, London, John Murray.
HART, K. (1989) 'Why the National Curriculum doesn't quite add up', Letters, *The Guardian*, 22 November 1989.
HEIJENOORT, J. van (ed.) (1967) *From Frege to Godel: A Source Book in Mathematical Logic*, Cambridge, Massachusetts, Harvard University Press.
HEMPEL, C. G. (1945) 'On the Nature of Mathematical Truth', *American Mathematical Monthly*, 52, reprinted in Feigl and Sellars (1949).
HEMPEL, C. G. (1952) *Fundamentals of Concept Formation in Empirical Science*, Chicago, University of Chicago Press.
HEMPEL, C. G. (1966) *Philosophy of Natural Science*, Englewood Cliffs, New Jersey, Prentice-Hall.
HENRY, J. (1971) *Essays on Education*, Harmondsworth, Penguin.
HER MAJESTY'S INSPECTORATE (1977) *Curriculum 11–16*, London, Her Majesty's Stationery Office.
HER MAJESTY'S INSPECTORATE (1979) *Mathematics 5–11: A Handbook of Suggestions*, London, Her Majesty's Stationery Office.
HER MAJESTY'S INSPECTORATE (1985) *Mathematics from 5 to 16*, London, Her Majesty's Stationery Office.
HER MAJESTY'S INSPECTORATE (1987) *Responses to Mathematics from 5 to 16*, London, Her Majesty's Stationery Office.
HER MAJESTY'S INSPECTORATE (1989) *Girls Leaning Mathematics* (Education Observed 14), London, Department of Education and Science.
HERSH, R. (1979) 'Some Proposals for Reviving the Philosophy of Mathematics', *Advances in Mathematics*, 31, pp. 31–50.
HERSH, R. (1988) 'Mathematics has a Front and a Back', paper presented at Sixth International Congress of Mathematics Education, Budapest, July 27-August 4, 1988.
HEYTING, A. (1931) 'The Intuitionist Foundations of Mathematics', *Erkenntnis*, pp. 91–121, translated in Benecerraf and Putnam (1964), pp. 42–49.
HEYTING, A. (1956) *Intuitionism: An Introduction*, Amsterdam, North-Holland.
HIGGINSON, W. (1980) 'On the foundations of Mathematics Education', *For the Learning of Mathematics*, 1, 2, pp. 3–7.
HILBERT, D. (1925) 'On the Infinite', *Mathematische Annalen*, 95, pp. 161–90, translated in Benecerraf and Putnam (1964), pp. 134–51.
HIMMELFARB, G. (1987) *Victorian Values*, London, Centre for Policy Studies.
HIRST, P. H. (1974) *Knowledge and the Curriculum*, London, Routledge & Kegan Paul.
HIRST, P. H. and PETERS, R. S. (1970) *The Logic of Education*, London, Routledge & Kegan Paul.
HODKINSON, P. (1989) 'Crossing the Academic/Vocational Divide: Personal Effectiveness and Autonomy as an Integrating Theme in Post-16 Education', *British Journal of Educational Studies*, 37, 4, pp. 369–384.
HOLT, J. (1964) *How Children Fail*, Harmondsworth, Penguin Books, 1965.
HOLT, J. (1972) *Freedom and Beyond*, Harmondsworth, Penguin Books.
HOLT, M. (ed.) (1987) *Skills and Vocationalism: the easy answer*, Milton Keynes, Open University.
HOME OFFICE (1977) *Racial Discrimination: A Guide to the Race Relations Act 1976*, London, Her Majesty's Stationery Office for the Home Office.

HOOK, S. (ed.) (1950) *John Dewey: Philosopher of Science and Freedom*, New York, Dial Press.

HOOPER, R. (ed.) (1971) *The Curriculum: Context, Design and Development*, Edinburgh, Oliver and Boyd.

HORON, P.F. and LYNN, D.D. (1980), 'Learning Hierarchies Research', *Evaluation in Education*, 4, pp. 82–83.

HOWE, M.J.A. (1989) *Fragments of Genius*, London, Routledge.

HOWSON, A.G. (ed.) (1973) *Developments in Mathematical Education*, Cambridge, Cambridge University Press.

HOWSON, A.G. (1980) 'Socialist Mathematics Education: Does it Exist?' *Educational Studies in Mathematics*, 11.

HOWSON, A.G. (1982) *A History of Mathematics Education in England*, Cambridge, Cambridge University Press.

HOWSON, A.G. (1983) *A Review of Research in Mathematical Education, Part C: Curriculum Development and Curriculum Research*, Windsor, NFER-Nelson.

HOWSON, A.G. (ed.) (1987) *Challenges and Responses in Mathematics*, Cambridge, Cambridge University Press.

HOWSON, A.G., KAHANE, J.P., LAUGINIE, P. and TURCKHEIM, E. de (eds.) (1988) *Mathematics as a Service Subject*, Cambridge, Cambridge University Press.

HOWSON, A.G., KEITEL, C. and KILPATRICK, J. (1981) *Curriculum Development in Mathematics*, Cambridge, Cambridge University Press.

HOWSON, A.G. and MELLIN-OLSEN, S. (1986) 'Social Norms and External Evaluation', in Christiansen *et al.* (1986), pp. 1–48.

HOWSON, A.G. and WILSON, B. (eds.) (1986) *School Mathematics in the 1990s*, Cambridge, Cambridge University Press.

HUGHILL, B. (1990) 'Government lets parents choose schools by race', *The Observer*, 22 April, 1990, p. 1.

HUSEN, T. and POSTLETHWAITE, T.N. (eds.) (1989) *The International Encyclopedia of Education, Supplementary Volume*, Oxford, Pergamon Press.

ILLICH, I. (1970) *Deschooling Society*, New York, Harrow Books (Harper and Row).

INGLIS, F. (1975) Ideology and the Curriculum: the Value Assumptions of the Systems Builders, in Golby *et al*,. pp. 36–47.

INNER LONDON EDUCATION AUTHORITY (1985) *Everyone Counts*, London, Inner London Education Authority, Learning Resources Branch.

INSTITUTE OF MATHEMATICS AND ITS APPLICATIONS (1984) *Mathematical Needs of School Leavers Entering Employment: 1*, Southend, Essex, Institute of Mathematics and its Applications.

IRVINE, J., MILES, I. and EVANS, J. (eds.) (1979) *Demystifying Social Statistics*, London, Pluto Press.

ISAACSON, Z. (1989) Taking Sides, *Times Educational Supplement*, 27 October, 1989, p. 54.

ISAACSON, Z. (1990) 'Is there More than One Math?' in Noss *et al.* (1990).

JAKOBSON, R. (1956) 'Two aspects of language and two types of aphasic disturbances', in *Fundamentals of Language* (Janua Linguarum, 1), the Hague, pp. 55–82.

JANVIER, C. (ed.) (1987) *Problems of Representation in the Teaching and Learning of Mathematics*, Hillsdale NJ, Erlbaum.

JECH, T.J. (1971) *Lectures in Set Theory* (Lecture Notes in Mathematics 217), Berlin; Springer-Verlag.

JENKINS, D. (1975) 'Classic and Romantic in the Curriculum Landscape', in Golby, *et al.* (1975), pp. 15–25.

JOHNSON, D.C. (ed.) (1989) *Children's Mathematical Frameworks*, Windsor; NFER-Nelson.

JONES, K. (1989) *Right Turn: The Conservative Revolution in Education*, London, Hutchinson.

JOSEPH, G.G. (1987) 'Foundations of Eurocentrism in Mathematics', *Race and Class*, 28, 3, pp. 13–28.

JOSEPH, G.G. (1990) *The Crest of the Peacock*, London, Penguin Books.

JUDD, J. (1989) 'Thatcher changes course of history', *The Observer*, 20 August 1989.

KALMAR, L. (1967) 'Foundations of Mathematics — Whither Now?' in Lakatos (1967a), pp. 187–194.

KEITEL, C. (1987) 'A glance at SMP 11–16 from a distance', in Howson (1987).

KEITEL, C. (1975) 'Konzeptionen der Curriculumentwicklung im Bereich des Mathematikunterrichts', unpublished Ph.D. thesis, IDM, University of Bielefeld.

KEITEL, C., DAMEROW, P., BISHOP, A. and GERDES, P. (eds.) (1989) *Mathematics, Education and Society* (Science and Technology Education Document Series No. 35) Paris, UNESCO.
KELLY, G. A. (1955) *The Psychology of Personal Constructs*, Norton, New York.
KENNY, A. (1973) *Wittgenstein*, Harmondsworth, Penguin Books.
KEOHANE, K. W. and HART, K. (1989) Unpublished letter to Department of Education and Science on the National Curriculum and Nuffield Secondary Mathematics, dated 7 February 1989.
KEYS, W. and FOXMAN, D. (1989) *A World of Differences*, Windsor, NFER.
KILPATRICK, J. and WIRSZUP, I. (1976) Introduction, in Krutetskii (1976), pp. xi–xvi.
KIRBY, N. (1981) *Personal Values in Education*, London, Harper and Row.
KITCHENER, K. S. and KING, P. M. (1981) 'Reflective judgement: Concepts of justification and their relationship to age and education', *Journal of Applied Developmental Psychology*, 2, pp. 89–116.
KITCHER, P. (1984) *The Nature of Mathematical Knowledge*, New York, Oxford University Press.
KLINE, M. (1980) *Mathematics the Loss of Certainty*, Oxford, Oxford University Press.
KNEEBONE, G. T. (1963) *Mathematical Logic and the Foundations of Mathematics*, London, Van Nostrand.
KOESTLER, A. (1964) *The Sleepwalkers*, Harmondsworth, Penguin Books.
KOHLBERG, L. (1969) *Stages in the Development of Moral Thought and Action*, New York, Holt, Rinehart and Winston.
KOHLBERG, L. (1981) *The Philosophy of Moral Development*, San Francisco, Harper and Row.
KORNER, S. (1960) *The Philosophy of Mathematics*, London Hutchinson.
KRULIK, S. and REYS, R. E. (eds.) (1980) *Problem Solving in School Mathematics* (1980 Yearbook), Reston, VA, National Council of Teachers of Mathematics.
KRUTETSKII, V. A. (1976) *The Psychology of Mathematical Abilities in Schoolchildren* (Translated by J. Teller) Chicago, University of Chicago Press.
KUHN, T. S. (1970) *The Structure of Scientific Revolutions*, Chicago, Chicago University Press (2nd ed.).
LAKATOS, I. (1962) 'Infinite Regress and the Foundations of Mathematics', *Aristotelian Society Proceedings, Supplementary Volume No. 36*, pp. 155–184 (revised version in Lakatos, 1978).
LAKATOS, I. (1967) 'A renaissance of empiricism in the recent philosophy of mathemematics?' in Lakatos (1967a), pp. 199–202 (extended version: Lakatos, 1978a).
LAKATOS, I. (ed.) (1967a) *Problems in the Philosophy of Mathematics*, Amsterdam, North Holland Publishing Company.
LAKATOS, I. (1970) 'Falsification and the Methodology of Scientific Research Programmes', in Lakatos and Musgrave (1970), pp. 91–196.
LAKATOS, I. (1976) *Proofs and Refutations*, Cambridge, Cambridge University Press.
LAKATOS, I. (1978) *Mathematics, Science and Epistemology (Philosphical Papers Vol. 2)*, Cambridge, Cambridge University Press.
LAKATOS, I. (1978a) 'A renaissance of empiricism in the recent philosophy of mathematics?' in Lakatos (1978), pp. 24–42.
LAKATOS, I. (1978b) *The Methodology of Scientific Research Programmes (Philosophical Papers Volume 1)*, Cambridge, Cambridge University Press.
LAKATOS, I. and MUSGRAVE, A. (1970) *Criticism and the Growth of Knowledge*, Cambridge, Cambridge University Press.
LANGFORD, G. (1987) 'Appeals to Relevance and the Utilitarian Attitude in Education', in Preece (1987), pp. 14–24.
LAPOINTE, A. E., MEAD, N. A. and PHILLIPS, G. W. (1989) *A World of Differences*, Princeton, New Jersey, Educational Testing Service.
LAUDAN, L. (1977) *Progress and Its Problems*, Berkeley, University of California Press.
LAVE, J. (1988) *Cognition in Practice*, Cambridge, Cambridge University Press.
LAWLOR, S. (1988) *Correct Core*, London, Centre for Policy Studies.
LAWTON, D. (1984) *The Tightening Grip* (Bedford Way Papers 21), London, University of London Institute of Education.
LAWTON, D. (1988) 'Ideologies of Education', in Lawton and Chitty (1988), 10–20.
LAWTON, D. and CHITTY, C. (eds.) (1988) *The National Curriculum* (Bedford Way Papers 33), London, University of London Institute of Education.

LAWTON, D., GORDON, P., ING, M., GIBBY, B., PRING, R. and MOORE, T. (1978) *Theory and Practice of Curriculum Studies*, London, Routledge & Kegan Paul.

LAWTON, D. and PRESCOTT, W. (1976) Curriculum Change and social Change, in Prescott (1976).

LAWVERE, F. W. (1966) 'The Category of Categories as a Foundation for Mathematics', *Proceedings of La Jolla Conference in Categorical Algebra*, Heidelberg, Springer-Verlag, pp. 1–20.

LAYTON, D. (1973) *Science for the People*, London, Allen and Unwin.

LEFEBVRE, H. (1972) *The Sociology of Marx*, Harmondsworth, Penguin Books.

LEONT'EV, A. N. (1978) *Activity, Consciousness and Personality*, Englewood Cliffs, New Jersey, Prentice-Hall.

LERMAN, S. (1983) 'Problem-solving or knowledge-centred: the influence of philosophy on mathematics teaching', *International Journal for Mathematical Education in Science and Technology*, 14, 1, pp. 59–66.

LERMAN, S. (1986) 'Alternative Views of the Nature of Mathematics and their Possible Influence on the Teaching of Mathematics', unpublished Ph.D. Thesis, King's College University of London.

LERMAN, S. (1988) 'Learning Mathematics as a Revolutionary Activity', paper presented at RSPME Conference, Polytechnic of the South Bank, London, April 1988.

LERMAN, S. (1989) 'Constructivism, Mathematics and Mathematics Education', *Educational Studies in Mathematics*, 20, pp. 211–223.

LERMAN, S. (1989a) 'Investigations: Where to Now?' in Ernest (1989), pp. 73–80.

LESTER, F. K. (1980) 'Research on Mathematical Problem Solving', in Shumway, R. J. (ed.) (1980) *Research in Mathematics Education*, Reston, Virginia, National Council of Teachers of Mathematics, pp. 286–323.

LETWIN, O. (1988) *Aims of Schooling*, London, Centre for Policy Studies.

LICKONA, T. (ed.) (1976) *Moral Development and Behaviour*, New York, Holt, Rinehart and Winston.

LIGHTHILL, J. (1973) Presidential Address, in Howson (1973), 88–100.

LILLARD, P. P. (1973) *Montessori A Modern Approach*, New York, Schocken Books.

LINDQUIST, M. M. (ed.) (1989) *Results from the Fourth Mathematics Assessment of the NAEP*, Reston, Virginia, National Council of Teachers of Mathematics.

LINGARD, D. (1984) 'Myths', *Mathematics Teaching*, 107, pp. 14–17.

LIONNAIS, F. Le (ed.) (1971) *Great Currents of Mathematical Thought* (Vol 1), New York, Dover.

LISTER, I. (ed.) (1974) *Deschooling*, Cambridge, Cambridge University Press.

LOEVINGER, J. (1976) *Ego Development*, San Francisco, Jossey Bass.

LORENZ, K. (1977) *Behind the Mirror: A Search for a Natural History of Human Knowledge* (translated by R. Taylor), New York, Harcourt Brace Jovanovich.

LOSEE, J. (1987) *Philosophy of Science and Historical Enquiry*, Oxford, Clarendon Press.

LOVEJOY, A. O. (1953) *The Great Chain of Being*, Cambridge, Massachusetts, Harvard University Press.

LYOTARD, J. F. (1984) *The Postmodern Condition: A Report on Knowledge*, Manchester, Manchester University Press.

MCCLEARY, J. and MCKINNEY, A. (1986) 'What Philosophy of Mathematics Isn't', *The Mathematical Intelligencer*, 8, 3, pp. 51–53 and 77.

MACDONALD, M. (1977) *Culture, Class and the Curriculum (E202 Schooling and Society: Unit 16)*, Milton Keynes, Open University Press.

MACDONALD, M., DALE, R. and WHITTY, G. (1977) *Revision II (E202 Schooling and Society: Unit 20)*, Milton Keynes, Open University Press.

MACHOVER, M. (1983) 'Towards a New Philosophy of Mathematics', *British Journal for the Philosophy of Science*, 34, pp. 1–11.

MACKENZIE, D. (1981) *Statistics in Britain, 1865–1930*, Edinburgh, University of Edinburgh Press.

MACLANE, S. (1981) 'Mathematical Models: A Sketch for the philosophy of mathematics', *American Mathematical Monthly*, 88, pp. 462–472.

MADDY, P. (1984) 'New Directions in the Philosophy of Mathematics', *Philosophy of Science Proceedings 1984* (Volume 2), East Lansing, Michigan, Philosophy of Science Association, pp. 427–448.

MARCUSE, H. (1964) *One Dimensional Man*, London, Routledge & Kegan Paul.

MARX, K. (1967) *Capital*, Volume I, New York, International Publishers.

MASON, J. with BURTON, L. and STACEY, K. (1982) *Thinking Mathematically*, London, Addison-Wesley.

MATHEMATICAL ASSOCIATION (1956) *The Teaching of Mathematics in Primary Schools*, London, G. Bell and Sons.

MATHEMATICAL ASSOCIATION (1970) *Primary Mathematics: A Further Report*, Leicester, Mathematical Association.

MATHEMATICAL ASSOCIATION (1976) *Why, What and How?*, Leicester, Mathematical Association.

MATHEMATICAL ASSOCIATION (1988) *Mathematics in a Multicultural Society*, Leicester, The Mathematical Association.

MATHEMATICAL ASSOCIATION (1988a) *Response to the Final Report of the Mathematics Working Group of the National Curriculum*, Leicester, Mathematical Association.

MAW, J. (1988) 'National Curriculum Policy: coherence and progression?', in Lawton and Chitty (1988), 49–61.

MAXWELL, J. (1984) *Is Mathematics Education Politically Neutral?*, Unpublished M. Ed. dissertation, University of Birmingham.

MAXWELL, J. (1989) 'Mathephobia', in Ernest (1989), pp. 221–226.

MAYHEW, J. (1987) 'Statements about mathematics and mathematics education, Some possible consequences for the National Mathematics Curriculum (NMC)', unpublished paper dated 7 September 1978.

MEAD, G. H. (1934) *Mind, Self and Society*, Chicago, University of Chicago Press.

MEHRTENS, H. (1976) 'T. S. Kuhn's Theories and Mathematics: A Discussion Paper on the "New Historiography" of Mathematics', *Historia Mathematica*, 3, pp. 297–320.

MEIGHAN, R. (1986) *A Sociology of Education*, Eastbourne, Holt, Rinehart and Winston.

MELLIN-OLSEN, S. (1981) 'Instrumentalism as an Educational Concept', *Educational Studies in Mathematics*, 12, pp. 351–367.

MELLIN-OLSEN, S. (1987) *The Politics of Mathematics Education*, Dordrecht, Reidel.

MILL, J. S. (1961) *A System of Logic (Eighth Edition)*, London, Longman, Green and Company.

MILLER, J. (1983) *States of Mind*, London, BBC Publications.

MILLS, C. W. (1970) *The Sociological Imagination*, Harmondsworth, Penguin Books.

MINISTRY OF EDUCATION (1958) *Teaching Mathematics in Secondary Schools*, London, Her Majesty's Stationery Office.

MOON, B. (1986) *The 'New Maths' Curriculum Controversy*, Lewes, Falmer Press.

MORRELL, F. (1990) 'Rights of Passage', *The Times Educational Supplement*, 2 February 1990, p. 38.

MORRIS, C. (1945) *Foundations of the Theory of Signs*, Chicago, University of Chicago Press.

MORRIS, R. (ed.) (1981) *Studies in Mathematics Education*, Vol. 2, Paris, UNESCO.

NASH, I. (1988) 'New maths chairman obey's Baker's Call', *Times Educational Supplement*, 11 March 1988.

NATIONAL CONSUMER COUNCIL (1986) *Classroom Advertisements*, London, National Consumer Council.

NATIONAL COUNCIL OF TEACHERS OF MATHEMATICS (1970) *A History of Mathematics Education* (32nd Yearbook), Washington DC, National Council of Teachers of Mathematics.

NATIONAL COUNCIL OF TEACHERS OF MATHEMATICS (1980) *An Agenda for Action*, Reston, Virginia, National Council of Teachers of Mathematics.

NATIONAL COUNCIL OF TEACHERS OF MATHEMATICS (1981) *Priorities in School Mathematics*, Reston, Virginia, National Council of Teachers of Mathematics.

NATIONAL COUNCIL OF TEACHERS OF MATHEMATICS (1989) *Curriculum and Evaluation Standards for School Mathematics*, Reston, Virginia, National Council of Teachers of Mathematics.

NATIONAL CURRICULUM COUNCIL (1988) *Consultation Report: Mathematics*, York, National Curriculum Council.

NATIONAL CURRICULUM COUNCIL (1989) *Mathematics, Non-Statutory Guidance*, York, National Curriculum Council.

NEILL, A. S. (1968) *Summerhill*, Harmondsworth, Penguin.

NEUMANN, J. von (1931) 'The Formalist Foundations of Mathematics', *Erkenntnis*, 1931, pp. 91–121, translated in Benecerraf and Putnam (1964), 50–54.

NEWMAN, J. R. (ed.) (1956) *The World of Mathematics* (4 volumes), New York, Simon and Schuster.

NICKSON, M. (1981) *Social Foundations of the Mathematics Curriculum: A Rationale for Change*, unpublished doctoral dissertation, London, University of London Institute of Education.

NILSSON, N. (1971) *Problem Solving Methods in Artificial Intelligence*, New York, McGraw-Hill.

NISS, M. (1983) 'Considerations and Experiences Concerning Integrated Courses in Mathematics and Other Subjects', in Zweng (1983), pp. 247–249.

NODDINGS, N. (1987) Politicizing the mathematics classroom, *Zentralblatt fur Didaktik der Mathematik*, 21, 6, pp. 221–224.

NOSS, R. (1988) 'The computer as a Cultural Influence in Mathematical Learning', *Educational Studies in Mathematics*, 19, 2, pp. 251–268.

NOSS, R. (1989) 'The National Curriculum and Mathematics: a case of divide and rule? Paper presented at RSPME Conference, June 1989, Institute of Education, University of London.

NOSS, R. (1989a) 'Just Testing: A Critical Review of Recent Change in the United Kingdom School Mathematics Curriculum', in Clements and Ellerton (1989), pp. 155–169.

NOSS, R., BROWN, A., DRAKE, P., DOWLING, P., HARRIS, M., HOYLES, C. and MELLIN-OLSEN, S. (eds.) (1990) *Political Dimensions of Mathematics Education: Action and Critique (Proceedings of the First International Conference, April 1–4, 1990)*, London, University of London Institute of Education.

NOSS, R., GOLDSTEIN, H. and HOYLES, C. (1989) 'Graded Assessment and Learning Hierarchies in Mathematics', *British Educational Research Journal*, 15, 3.

NOVAK, J. (ed.) (1987) *Proceedings of the Second International Seminar on Misconceptions and Educational Strategies in Science and Mathematics* (July 1987), 3 Vols, Ithaca, New York, Cornell University.

NUFFIELD MATHEMATICS TEACHING PROJECT (1965) *I do and I understand (Draft edition)*, London, Nuffield Foundation.

OAKSHOTT, M. (1944) 'Learning and Teaching', in Peters (1967), pp. 156–176.

O'HEAR, A. (1990) '2nd Opinion', *The Times Educational Supplement*, 9 March 1990.

ONIONS, C. T. (ed.) (1967) *The Shorter Oxford Dictionary*, Oxford, Clarendon Press.

OPEN UNIVERSITY (1986) *Girls into Mathematics*, Cambridge, Cambridge University Press.

OPREA, J. M. and STONEWATER, J. (1987) 'Mathematics Teachers' Belief Systems and Teaching Styles: Influences on Curriculum Reform', in Bergeron *et al.* (eds.) (1987), 1, pp. 156–162.

ORMELL, C. (1975) 'Towards a naturalistic mathematics in the sixth form', *Physics Education*, July 1985, pp. 349–354.

ORMELL, C. (1980) 'Mathematics', in Straughan and Wrigley (1980), pp. 213–229.

ORMELL, C. (1985) *The Applicability of Mathematics*, Norwich, Mathematics Applicable Group, School of Education, University of East Anglia.

OXENHAM, J. (ed.) (1984) *Education Versus Qualifications?*, London, Allen and Unwin.

PALMER, F. (ed.) (1986) *Anti-Racism — An Assault on Education and Value*, London, Sherwood Press.

PAPERT, S. (1980) *Mindstorms: Children Computers and Powerful Ideas*, Brighton, Harvester.

PAPERT, S. (1988) 'A Critique of Technocentrism in Thinking About the School of the Future', in Sendov and Stanchev (1988), pp. 3–18.

PEARCE, G. and MAYNARD, P. (eds.) (1973) *Conceptual Change*, Dordrecht, Reidel.

PEREIRA-MENDOZA, L. (ed.) (1988), *Proceedings Annual Meeting of Canadian Mathematics Education Study Group*, June 2–6, 1988, Manitoba, Canada, University of Manitoba.

PERRY, W. G. (1970) *Forms of Intellectual and Ethical Development in the College Years: A Scheme*, New York, Holt, Rinehart and Winston.

PERRY, W. G. (1981) 'Cognitive and ethical growth: the making of meaning', in Chickering (1981), pp. 76–116.

PETERS, R. S. (ed.) (1967) *The Concept of Education*, London, Routledge & Kegan Paul.

PETERS, R. S. (ed.) (1969) *Perspectives on Plowden*, London, Routledge & Kegan Paul.

PETERS, R. S. (ed.) (1975) *The Philosophy of Education*, Oxford, Oxford University Press.

PFUNDT, H. and DUIT, R. (1988) *Bibliography: Students' Alternative Frameworks and Science Education* (IPN Reports-in-Brief 34) Kiel, Federal Republic of Germany, Institute for Science Education, University of Kiel.

PHENIX, P. H. (1964) *Realms of Meaning*, New York, McGraw-Hill.

PHILLIPS, E. R. and KANE, R. B. (1973) 'Validating Learning Hierarchies for Sequencing Mathematical Tasks', in 'Elementary School Mathematics', *Journal for Research in Mathematics Education*, 4, 3, 1973, pp. 141–151.

PIAGET, J. (1972) *Psychology and Epistemology: Towards a Theory of Knowledge*, Harmondsworth, Penguin Books.

PIAGET, J. (1977) *The Principles of Genetic Epistemology*, London: Routledge & Kegan Paul.

PIGEON, D. (1977) Intelligence: a changed view, in Childs (1977), pp. 240–242.

PIMM, D. (1987) *Speaking Mathematically*, London: Routledge & Kegan Paul.

PIRIE, S. (1987) *Mathematical Investigations in Your Classroom*, Basingstoke, Macmillan.

PLATO (1941) *The Republic of Plato* (Translated and annotated by F. M. Cornford), Oxford, The Clarendon Press.

PLOWDEN COMMITTEE (1967) *Children and Their Primary Schools (Report of Central Advisory Council for Education, England)*, London, Her Majesty's Stationery Office.

PLUMB, J. H. (1950) *England in the Eighteenth Century*, Harmondsworth, Penguin Books.

POINCARÉ, H. (1956) 'Mathematical Creation', in Newman (1956), pp. 2041–2050.

POLANYI, M. (1958) *Personal Knowledge*, London, Routledge & Kegan Paul.

POLLAK, H. (1988) 'Mathematics as a Service Subject — Why?' in Howson *et al.* (1988).

POLLARD, A. (ed.) (1987) *Children and their Primary Schools*, Lewes, East Sussex, Falmer Press.

POLLARD, A., PURVIS, J. and WALFORD, G. (eds.) (1988) *Education, Training and the New Vocationalism: Experience and Policy*, Milton Keynes, Open University Press.

POLYA, G. (1945) *How to Solve it*, Princeton, New Jersey, Princeton University Press.

POPPER, K. (1959) *The Logic of Scientific Discovery*, Hutchinson, London.

POPPER, K. R. (1979) *Objective Knowledge (Revised Edition)*, Oxford, Oxford University Press.

POSTMAN, N. and WEINGARTNER, C. (1969) *Teaching as a Subversive Activity*, Harmondsworth, Penguin Books, 1971.

PRAIS, S. J. (1987) *National Curriculum Mathematics Working Group: Interim Report, Note of Dissent* (dated 30 November 1987), London, National Institute of Economic and Social Research.

PRAIS, S. J. (1987a) 'Further Response to National Curriculum Mathematics Working Group' (unpublished paper dated 10 December 1987).

PREECE, P. (ed.) (1987) *Philosophy and Education* (Perspectives 28), Exeter, University of Exeter School of Education.

PRESCOTT, W. (ed.) (1976) *The Child, the School and Society (E203: Curriculum Design and Development, Units 5, 6, 7, & 8)*, Milton Keynes, Open University Press.

PRESTON, M. (1975) *The Measurement of Affective Behaviour in CSE Mathematics* (Psychology of Mathematics Education Series), London, Centre for Science Education, Chelsea College.

PRIEST, G. (1973) 'A bedside reader's guide to the conventionalist philosophy of mathematics', in Bell *et al.* (1973), pp. 115–132.

PRIME PROJECT (1987) *One Year of CAN*, Cambridge, PrIME Project, Homerton College.

PRING, R, (1984) *Personal and Social Education in the Curriculum*, London, Hodder and Stoughton.

PUTNAM, H. (1972) *Philosophy of Logic*, London, George Allen & Unwin.

PUTNAM, H. (1975) *Mathematics, Matter and Method* (Philosophical Papers Vol. 1), Cambridge, Cambridge University Press.

QUINE, W. V. O. (1936) 'Truth by Convention', reprinted in Feigl and Sellars (1949), pp. 250–273.

QUINE, W. V. O. (1948) 'On What There Is', reprinted in Quine (1953), pp.1–19.

QUINE, W. V. O. (1951) 'Two dogmas of empiricism', reprinted in Quine (1953), pp. 20–46.

QUINE, W. V. O. (1953) *From a Logical Point of View*, New York, Harper Torchbooks.

QUINE, W. V. O. (1960) *Word and Object*, Cambridge, Massachusetts, Massachusetts Institute of Technology Press.

QUINE, W. V. O. (1966) *The Ways of Paradox and other essays*, New York, Random House.

QUINTON, A. (1963) The A Priori and the Analytic, *Proceedings of the Aristotelian Society*, 64, pp. 31–54.

RABAN, J. (1989) *God, Man and Mrs Thatcher* (Counterblasts 1), London, Chatto.

RAFFE, D. (1985) 'Education and training initiatives for 14–18s', in Watts (1985), pp. 19–23.

RAMSDEN, M. (1986) *Competing Ideologies of the Child*, (Unpublished course notes), Exeter, University of Exeter, School of Education.
RAYNOR, J. (1972) *The Curriculum in England*, The Curriculum: context, Design and Development, Unit 2 (E283), Milton Keynes, Open University Press.
RAYNOR, J. (1974) *Why Urban Education?* (E351 Urban Education Block 1), Milton Keynes: The Open University.
RAYNOR, J. and HARDEN, J. (eds.) (1973) *Equality and City Schools, Readings in Urban Education* (Vol. 2), London, Routledge & Kegan Paul.
REED, C. (1988) 'Reform and Rage', *The Guardian*, 27 September 1988.
RESNICK, L. B. and FORD, W. W. (1981) *The Psychology of Mathematics for Instruction*, Hillsdale, New Jersey, Lawrence Erlbaum.
RESTIVO, S. (1984) 'Representations and the Sociology of Mathematical Knowledge', in Belisle and Schiele (1984), pp. 66–93.
RESTIVO, S. (1985) *The Social Relations of Physics, Mysticism and Mathematics*, Dordrecht, Pallas Paperbacks, Reidel Publishing Company.
RESTIVO, S. (1988) 'The Social Construction of Mathematics', *Zentralblatt fur Didaktik der Mathematik*, 20, 1, pp. 15–19.
RESTIVO, S. (1988a) 'Modern Science as a Social Problem', *Social Problems*, 35, 3, pp. 206–225.
REYNOLDS, J. and SKILBECK, M. (1976) *Culture and the Classroom*, London, Open Books.
RICHARDS, C. (1984) *The Study of Primary Education: A Source Book*, (Vol. 1), Lewes, Falmer Press.
RICHARDS, J. L. (1980) 'The Art and Science of British Algebra: A Study in the Perception of Mathematical Truth', *Historia Mathematica*, 7, 3, pp. 343–365.
RICHARDS, J. L. (1989) *Mathematical Visions*, London, Academic Press.
RIDGEWAY, A. (ed.) (1949) *Everyman's Encyclopedia* (Vol. 4), London, J. M. Dent and Sons.
ROBITAILLE, D. and DIRKS, M. (1982) 'Models for the Mathematics Curriculum', *For the Learning of Mathematics*, 2, 3, pp. 3–21.
ROBITAILLE, D. F. and GARDEN, R. A. (eds.) (1989) *The IEA Study of Mathematics II: Contexts and Outcomes of School Mathematics*, Oxford, Pergamon.
ROGERS, C. (1961) *On Becoming a Person*, Boston, Houghton Mifflin.
ROGERSON, A. (1986) 'The Mathematics in Society Project: a new conception of mathematics', *International Journal for Mathematical Education in Science and Technology*, 17, 5, pp. 611–616.
ROGERSON, A. (ed.) (undated) *Mathematics in Society, The Real Way to Apply Mathematics?*, Victoria, Australia, The Mathematics in Society Project.
RORTY, R. (1979) *Philosophy and the Mirror of Nature*, Princeton, New Jersey, Princeton University Press.
ROUSSEAU, J. J. (1762) *Emile or Education* (translated in Everyman Books, London, Dent and Sons, 1918).
ROWLAND, T. (1989) *CAN in Suffolk*, Ipswich, Suffolk Education Committee.
RUMELHART, D. E. and NORMAN, D. A. (1978) 'Accretion, tuning and restructuring: Three modes of learning', in Cotton and Klatsky (1978).
RUSSELL, B. (1902) 'Letter to Frege', in Heijenoort (1967), pp. 124–125.
RUSSELL, B. (1919) *Introduction to Mathematical Philosophy*, London, George Allen and Unwin.
RUTHVEN, K. (1986) 'Differentiation in mathematics: A critique of Mathematics Counts and Better Schools', *Cambridge Journal of Education*, 16, 1, pp. 41–45.
RUTHVEN, K. (1987) 'Ability Stereotyping in Mathematics', *Educational Studies in Mathematics*, 18, pp. 243–253.
RYLE, G. (1949) *The Concept of Mind*, London, Hutchinson.
SAATY, T. L. and WEYL F. J. (eds.) (1969) *The Spirit and Uses of the Mathematical Sciences*, New York, McGraw-Hill.
SALNER, M. (1986) 'Adult Cognitive and Epistemological Development in Systems Education', *Systems Research*, 3, 4, pp. 225–232.
SAPIR, E. (1949) *Language*, New York, Harcourt, Brace and Company.
SCHOENFELD, A. H. (1985) *Mathematical Problem Solving*, London, Academic Press.
SCHON, D. (1983) *The Reflective Practitioner*, London, Temple Smith.

SCHOOL EXAMINATIONS COUNCIL (1985) *National Criteria in Mathematics for the General Certificate of Secondary Education*, London, Her Majesty's Stationery Office.
SCHOOL OF BARBIANA (1970) *Letter to a Teacher*, Harmondsworth, Penguin Books.
SCHWAB, J.J. (1977) 'Structure of the Disciplines: Meanings and Significances', in Golby *et al.* (1975), pp. 249–267.
SCOTT-HODGETTS, R. (1988) 'The National Curriculum: Implications for the Sociology of Mathematics Classrooms', paper presented at RSPME Conference, Polytechnic of the South Bank, London, December 1988.
SELLS, L. (1973) 'High school mathematics as the critical filter in the job market', *Proceedings of the Conference on Minority Graduate Education*, Berkeley, University of California, pp. 37–49.
SELLS, L. (1976) 'The mathematics filter and the education of women and minorities', presented at annual meeting of American Association for the Advancement of Science, Boston, February 1976.
SELMAN, R.L. (1976) 'Social Cognitive Understanding', in Lickona (1976).
SENDOV, B. and STANCHEV, I. (eds.) (1988) *Children in the Information Age*, Oxford, Pergamon.
SFARD, A. (1987) Two Conceptions of Mathematical Notions: Operational and Structural, in Bergeron *et al.* (1987), 3, pp. 162–169.
SFARD, A. (1989) 'Translation from Operational to Structural Conception: The Notion of Function Revisited', in Vergnaud *et al.* (1989), 3, pp. 151–158.
SHARMA, S. and MEIGHAN, R. (1980) 'Schooling and Sex Roles: the case of GCE 'O' level mathematics', *British Journal of Sociology of Education*, 1, 2, pp. 193–205.
SHEFFLER, I. (1965) *Conditions of Knowledge: An Introduction to Epistemology and Education*, Chicago, Scott Foresman.
SHIRLEY, L. (1986) 'Editorial', *International Study Group on the Relations Between the History and Pedagogy of Mathematics Newsletter*, No. 13, 2–3.
SHUKLA, K. (ed.) (1989) *Mainstream Curricula in a Multicultural Society*, London, Further Education Unit, Department of Education and Science.
SILBERMAN, C.E. (ed.) (1973) *The Open Classroom Reader*, New York, Vintage Books (Random House).
SIMON, B. (1976) *Social Control and Education*, London, Methuen.
SKEMP, R.R. (1971) *The Psychology of Learning Mathematics*, Harmondsworth, Penguin Books.
SKEMP, R.R. (1976) 'Relational understanding and instrumental understanding', *Mathematics Teaching*, 77, pp. 20–26.
SKEMP, R.R. (1979) *Intelligence, Learning and Action*, New York, Wiley.
SKILBECK, M. (1976) *Ideology, Knowledge and the Curriculum* (Unit 3, E203 Curriculum Design and Development), Milton Keynes, The Open University.
SKOVSMOSE, O. (1985) 'Mathematical Education versus Critical Education', *Educational Studies in Mathematics*, 16, pp. 337–354.
SKOVSMOSE, O. (1988) 'Mathematics as a Part of Technology', *Educational Studies in Mathematics*, 19, 1, pp. 23–41.
SMITH, D. and TOMLINSON, S. (1989) *The School Effect: The Study of Multi-Racial Comprehensives*, Lancaster, Policy Studies Institute, University of Lancaster.
SNEED, J. (1971) *The Logical Structure of Mathematical Physics*, Dordrecht, Reidel.
SOCKETT, H. (1975) 'Curriculum Planning: Taking a Means to an End', in Peters (1975), pp. 150–160.
SOLOMON, Y. (1989) *The Practice of Mathematics*, London, Routledge.
SPARROW, J. (1970) 'Egalitarianism and an Academic Elite', in Cox and Dyson (1970).
STABLER, E.R. (1953) *Introduction to Mathematical Thought*, Reading, Massachusetts, Addison-Wesley.
STANFIELD-POTWOROWSKI, J. (1988) 'Socialising Mathematics', *Mathematics Teaching*, 125, pp. 3–8.
STAPLEDON, O. (1937) *Star Maker*, Harmondsworth, Penguin Books.
STEEN, L.A. (ed.) (1978) *Mathematics Today*, New York, Springer Verlag.
STEEN, L.A. (ed.) (1985) *Mathematics Tomorrow*, New York, Springer Verlag.
STEEN, L.A. (1988) 'The Science of Patterns', *Science*, 240, 4852, pp. 611–616.
STEFFE, L.P., GLASERSFELD, E. von, RICHARDS, J., and COBB, P. (1983) *Children's Counting Types: Philosophy, Theory, and Application*, New York, Praeger.

STEINER, H. G. (1987) 'Philosophical and Epistemological Aspects of Mathematics and their Interaction with Theory and Practice in Mathematics Education', *For the Learning of Mathematics*, 7, 1, pp. 7–13.

STENHOUSE, L. (1975) *An Introduction to Curriculum Research and Development*, London, Heinemann.

STOLZENBERG, G. (1984) 'Can an Inquiry into the Foundations of Mathematics Tell Us Anything Interesting About Mind?' in Watzlawick (1984), pp. 257–308.

STONEWATER, J. K., STONEWATER, B. B. and PERRY, B. E. (1988) 'Using Developmental Clues to Teach Problem Solving', *School Science and Mathematics*, 88, 4, pp. 272–283.

STRAUGHAN, R. and WRIGLEY, J. (eds) (1980) *Values and Evaluation in Education*, London, Harper and Row.

STRAW, J. (1988) 'Lead story', *The Guardian*, 10 March 1988, p. 1.

SWANN, Lord, (chair) (1985) *Education for All*, London, Her Majesty's Stationery Office.

SWETZ, F. (1978) *Socialist Mathematics Education*, Southampton, Pennsylvania, Burgundy Press.

SZABO, A. (1967) 'Greek Dialectic and Euclid's Axiomatics', in Lakatos (1967a), pp. 1–27.

TABA, H. (1962) *Curriculum Development: Theory and Practice*, New York, Harcourt Brace and World.

TANSLEY, A. E. and GULLIFORD, R. (1960) *The Education of Slow Learning Children*, London, Routledge and Kegan Paul.

TARSKI, A. (1936) 'Der Wahrheitsbegriff in den formaliesierten Sprachen', *Studia Philosphica*, 1, pp. 261–405, translated in Tarski (1954).

TARSKI, A. (1954) *Logic, Semantics and Metamathematics*, Oxford, Oxford University Press.

THATCHER, M. (1987) Address to Conservative Party Conference, 1987.

THOM, R. (1973) 'Modern mathematics: does it exist?' in Howson (1973), pp. 194–209.

THOMPSON, A. G. (1984) 'The Relationship Between Teachers Conceptions of Mathematics and Mathematics Teaching to Instructional Practice', *Educational Studies in Mathematics*, 15, pp. 105–127.

THWAITES, B. (1979) 'The Development of School Mathematics: Some General Principles', in Tom (1979), pp. 47–51.

TOBIAS, S. (1978) *Overcoming Math Anxiety*, Boston, Houghton Mifflin.

TOLKIEN, J. R. R. (1954) *The Lord of the Rings* (3 Vols), London, George Allen & Unwin.

TOM, M. E. A., EL (ed.) (1979) *Developing Mathematics in Third World Countries*, Amsterdam, North Holland.

TOULMIN, S. (1972) *Human Understanding, I*, Oxford, Clarendon Press.

TRAVERS, K. and WESTBURY, I. (eds.) (1989) *The IEA Study of Mathematics I: International Analysis of Mathematics Curricula*, Oxford, Pergamon.

TREVELYAN, G. M. (1944) *English Social History*, London, Longman, Green and Company.

TYMOCZKO, T. (1979) 'The Four-Color Problem and its Philosophical Significance', *The Journal of Philosophy*, 76, 2, pp. 57–83.

TYMOCZKO, T. (1985) 'Godel, Wittgenstein and the Nature of Mathematical Knowledge', *Philosophy of Science Association; Proceedings 1985*, 2, pp. 449–468.

TYMOCZKO, T. (ed.) (1986) *New Directions in the Philosophy of Mathematics*, Boston, Birkhauser.

TYMOCZKO, T. (1986a), 'Making Room for Mathematicians in the Philosophy of Mathematics', *The Mathematical Intelligencer*, 8, 3, pp. 44–50.

VERGNAUD, G. (1983) 'Why is an Epistemological Perspective a Necessity for Research in Mathematics Education?' in Bergeron and Herscovics (1983), 1, pp. 2–20.

VERGNAUD, G., ROGALSKI, J. and ARTIGUE, M. (eds.) (1989) *Proceedings of PME-13 (Paris 9–13 July 1989)*, Paris, Centre National de la Recherche Scientifique, Université René Descartes.

VOLMINK, J. (1990) 'The Constructivist Foundation of Ethnomathematics', paper presented at Political Dimensions of Mathematics Education: Action and Critique, First International Conference, April 1–4, 1990, University of London Institute of Education.

VYGOTSKY, L. S. (1962) *Thought and Language*, Cambridge, Massachusetts, Massachusetts Institute of Technology Press.

WAISMANN, F. (1951) *Introduction to Mathematical Thinking*, London, Hafner Publishing Company.

WALDEN, R. and WALKERDINE, V. (1982) *Girls and Mathematics: The Early Years* (Bedford Way Papers 8), London, Institute of Education, University of London.

WALKERDINE, V. (1988) *The Mastery of Reason*, London, Routledge.
WALKERDINE, V. (1989) 'Femininity as Performance', *Oxford Review of Education*, 15, 3, pp. 267–279.
WALKERDINE, V. AND THE GIRLS AND MATHEMATICS UNIT (1989) *Counting Girls Out*, London, Virago.
WARNOCK, M. (1989) *Universities: Knowing Our Minds*, London, Chatto and Windus.
WATTS, A. G. (ed.) (1985) *Education and Training 14–18: Policy and Practice*, Cambridge, CRAC.
WATZLAWICK, P. (ed.) (1984) *The Invented Reality*, New York, Norton & Co.
WEBER, M. (1964) *The Theory of Social and Economic Organization*, New York, Free Press.
WERTSCH, J. V. (1985) *Vygotsky and the Social Formation of Mind*, Cambridge, Harvard University Press.
WHEELER, D. H. (ed.) (1967) *Notes on Mathematics in Primary Schools*, Cambridge, Cambridge University Press.
WHITE, F. C. (1982) 'Knowledge and Relativism I', *Educational Philosophy and Theory*, 14, 2, pp. 1–20.
WHITE, L. A. (1975) *The Concept of Cultural Systems*, New York, Columbia University Press.
WHITE, M. (1950) 'The analytic and the synthetic: an untenable dualism', in Hook (1950), pp. 316–330.
WHITEHEAD, A. N. (1932) *The Aims of Education*, reprinted, London, Ernest Benn, 1959.
WHITEHEAD, A. N. and RUSSELL, B. (1910–13) *Principia Mathematica* (3 Vols), Cambridge, Cambridge University Press.
WHITTY, G. (1977) *School Knowledge and Social Control* (Units 14–15 of E202, Schooling and Society), Milton Keynes, Open University.
WHYLD, J. (ed.) (1983) *Sexism in the Secondary Curriculum*, London, Harper & Row.
WIGNER, E. P. (1960) 'The unreasonable effectiveness of mathematics in the physical sciences', reprinted in Saaty and Weyl (1969), pp. 123–140.
WILDER, R. L. (1965) *Introduction to the Foundations of Mathematics*, New York, John Wiley & Sons.
WILDER, R. L. (1974) *Evolution of Mathematical Concepts*, London, Transworld Books.
WILDER, R. L. (1981) *Mathematics as a Cultural System*, Oxford, Pergamon Press.
WILLIAMS, J. D. (1971) *Teaching Technique in Primary Mathematics*, Windsor, National Foundation for Educational Research.
WILLIAMS, M. (1989) Vygotsky's Social Theory of Mind, *Harvard Educational Review*, 59, 1, pp. 108–126.
WILLIAMS, R. (1961) *The Long Revolution*, Harmondsworth, Penguin Books.
WILLIAMS, R. (1976) 'Base and superstructure in Marxist cultural theory', in Dale *et al.* (1976), pp. 202–210.
WILLIAMS, R. (1977) 'Ideology', in Open University (1977) *The Curriculum and Cultural Reproduction* (E202 Schooling and Society Units 18, 19 & 20), Milton Keynes, Open University Press, pp. 122–123.
WILTSHIRE EDUCATION AUTHORITY (1987) *Mathematics for All*, Trowbridge, Wiltshire Education Authority.
WITTGENSTEIN, L. (1922) *Tractatus Logico-Philosophicus*, London, Routledge & Kegan Paul.
WITTGENSTEIN, L. (1953) *Philosophical Investigations* (translated by G. E. M. Anscombe), Oxford, Basil Blackwell.
WITTGENSTEIN, L. (1978) *Remarks on the Foundations of Mathematics* (Revised Edition), Cambridge, Massachusetts, Massachusetts Institute of Technology Press.
WITTROCK, M. C. (ed) (1986) *Handbook of Research on Teaching* (3rd edition), New York, Macmillan.
WOMEN'S NATIONAL COMMITTEE (1985) *Do Girls Give Themselves A Fair Chance?*, London, Women's National Committee.
WOOZLEY, A. D. (1949) *Theory of Knowledge*, London, Hutchinson.
WRIGLEY, J. (1958) 'The factorial nature of ability in elementary mathematics', *British Journal of Educational Psychology*, 28, pp. 61–78.
YACKEL, E. (1987) 'A Year in the Life of a Second Grade Class: A Small Group Perspective', in Bergeron *et al.* (1987), 3, pp. 208–214.
YOUNG, H. (1989) *One of Us*, London, Macmillan.
YOUNG, M. F. D. (ed.) (1971) *Knowledge and Control*, London, Collier-Macmillan.
YOUNG, M. F. D. (1971a) 'An Approach to the Study of Curricula as Socially Organized Knowledge', in Young (1971), pp. 19–46.

YOUNG, M. F. D. and WHITTY, G. (eds) (1977) *Society, State and Schooling*, Lewes, Sussex, Falmer Press.
ZASLAVSKY, C. (1973) *Africa Counts*, Boston, Prindle, Weber and Schmidt.
ZELDIN, D. (1974) *Community Participation*, (E351 Urban Education, Block 6 Whose Schools? Part 3), Milton Keynes: The Open University.
ZWENG, M., GREEN, T., KILPATRICK, J., POLLAK, H. and SUYDAM, M. (eds) (1983) *Proceedings of the Fourth International Congress on Mathematical Education*, Boston, Birkhauser.

Index